Logic-Driven Traffic Big Data Analytics

Shaopeng Zhong · Daniel (Jian) Sun

Logic-Driven Traffic Big Data Analytics

Methodology and Applications for Planning

Springer

Shaopeng Zhong
School of Transportation and Logistics
Dalian University of Technology
Dalian, China

Institute of Smart City and Intelligent Transportation
Southwest Jiaotong University
Chengdu, China

Daniel (Jian) Sun
College of Future Transportation
Chang'an University
Xi'an, China

Institute of National Security
Shanghai Jiao Tong University
Shanghai, China

National Natural Science Foundation of China 71971038 71701030
Fundamental Research Funds for the Central Universities of China DUT20GJ210
National Natural Science Foundation of China 72150410445 52172319 71971138

ISBN 978-981-16-8015-1 ISBN 978-981-16-8016-8 (eBook)
https://doi.org/10.1007/978-981-16-8016-8

This Springer imprint is published by the registered company Springer Nature Singapore Pte Ltd.
The registered company address is: 152 Beach Road, #21-01/04 Gateway East, Singapore 189721, Singapore

To my family, especially my loving wife, Dandan Li

—Shaopeng Zhong

To my family and my mom

—Daniel (Jian) Sun

Foreword by Tien Fang Fwa

Traffic congestion and travel delays are perennial problems in practically all major cities in the world. Transportation engineers equipped with modern highway and traffic engineering theories have not been able to provide solutions to tackle these problems effectively to the satisfaction of the general public. With the advances in computer technologies and rapid increase in computing power, coupled with the growth of and innovations in artificial intelligence, information and telecommunication technologies, there evolves a host of new tools and analytics that allows transportation engineers to develop new solutions and gain new insights into the old problems. Big data analytics is one of these exciting and promising tools that have generated much interest among transportation researchers. Like all new technologies, big data analytics comes with many challenges, some of which may not be apparent to general users and even researchers.

In the above-mentioned context, this book by two experienced professionals offers a very timely reading for transportation engineering practitioners and researchers. It helps the reader to gain a deeper understanding of the value of big data applications in solving urban traffic problems, and be better prepared to benefit from the new opportunities and associated challenges. Dr. Zhong has more than 20 years of professional experience in the field of sustainable transportation planning, land use and transportation integration modeling, urban transport network analysis, and logic-driven transport big data analysis. Dr. Sun, my colleague in the College of Future Transportation, Chang'an University, conducts research on smart city and intelligent transportation systems, including risk and resilience of urban transportation infrastructures, transportation environment, transport big data analytics and simulation.

The book focuses on the methods and applications of multi-source traffic big data. It first introduces the theory of land use and transportation integration, and presents the relevant statistical models and methods (Chap. 1). A valuable feature of the book is the informative case studies which are dealt with in necessary detail. They provide useful references for the reader to appreciate the cause and effect relationship between events, and how and why the data-driven method works. Case studies are introduced from the following perspectives, including travel demand analysis (Chaps. 2 and 3), traffic congestion pattern detection and related policy analysis (Chaps. 4 and 5);

travel time estimation (Chaps. 6 and 7); drivers' behavior and road safety (Chaps. 8–10); and urban traffic emission (Chaps. 11 and 12).

The well-structured book presents an ideal reading for graduate research students in the fields of traffic engineering and transportation planning. It would also be a good reference for researchers in the same fields who are interested in big data applications.

T. F. Fwa
Emeritus Professor, National University of Singapore
Distinguished Professor
Dean, College of Future Transportation
Chang'an University
Xi'an, China

Foreword by Feng Xiao

This is an authoritative academic monograph on the mining methods and applications of multi-source traffic big data. Existing research on traffic big data analysis often adopts data-driven methods, which leads us to only know what it is, but not why. This book attempts to link traffic phenomena to the reasons behind it based on traffic big data, which can be a good reference for future research. I believe that every researcher in the field of transportation can get enlightenment from this book.

Feng Xiao
Professor, School of Business Administration
Southwestern University of Finance and Economics
Chengdu, China

Preface

Big data plays an increasingly important role in urban transportation. The core is to provide a new "traffic information environment" and strong data support for a comprehensive and accurate evaluation of the operation of the transportation system. How to overcome the scarcity of a variety of empirical data and limited data sources, diagnosing the bottleneck problems hindering urban transportation development, thus establishing appropriate planning methods under the background of multi-source information is not only the frontier topic in transportation planning, but also the future development direction of the industry.

In addition, the separation of different urban land uses in space is the root of travel demand. In turn, urban transportation system is an important factor affecting urban land use. There is a very complex interaction and restriction relationship between urban transportation and built environment (land use), which constitutes a "spiral" interaction mechanism among urban built environment, accessibility, transportation facilities, and travel demand. However, the existing traffic big data analysis models and methods often ignore the interactive feedback relationship between urban built environment and travel behavior, so that the existing research results often have certain limitations and cannot explore and find the root cause behind the phenomenon.

This book is unique in that it starts from the relationship between urban built environment and travel behavior and focuses on analyzing the origin of traffic phenomena behind the data through multi-source traffic big data, which makes the book different from the previous data-driven traffic big data analysis literatures.

The book focuses on understanding, estimating, predicting, and optimizing mobility patterns. Readers can find multi-source traffic big data processing methods, related statistical analysis models, and practical case applications from this book.

This book bridges the gap between traffic big data, statistical analysis models, and mobility pattern analysis with a systematic investigation of traffic big data's impact on mobility patterns and urban planning.

Academic and practicing planners would be interested in reading the book. It can also be used as a reference for students majoring in traffic engineering, urban and regional planning, and transportation planning and management.

Dalian, China Shaopeng Zhong
Xi'an, China Daniel (Jian) Sun

Acknowledgments

The following fund was obtained by Shaopeng Zhong.

This research has been supported by the National Natural Science Foundation of China (Project No. 71971038 and 71701030), and the Fundamental Research Funds for the Central Universities of China (Project No. DUT20GJ210).

The following fund was obtained by Daniel (Jian) Sun.

This research has been supported by the National Natural Science Foundation of China (Project No. 71971138, 52172319, and 72150410445).

Many people supported and helped in the preparation of this book. Shaopeng Zhong would like to thank the following scholars: Ao Liu, Zhen Li, Jing He, Xiaohan Zhou, Meihan Fan, Jiachao Liu, Yalan Wang, Quanzhi Wang, Jin Guo, Zhong Wang, Xiaoyan Xie, Jianqiang Cui, Fan Yang, Jian Zhang, and Bin Ran.

Daniel (Jian) Sun would like to thank the following scholars: Taha Benarbia, Fangxi Chen, Yizhe Huang, Xueqing Ding, Kaisheng Zhang, Zhiwei Yin, Yingwei Ye, Guangyue Nian, Shaojie Wu, Suwan Shen, Lihui Zhang, Yi Zhu.

Contents

About the Authors

Dr. Shaopeng Zhong is an associate professor in School of Transportation and Logistics at Dalian University of Technology. In 2005, he received his bachelor's degree in transportation engineering from Harbin Institute of Technology, China. In 2010, he obtained his doctorate from Southeast University, China. He is a visiting scholar in urban and regional planning at University of North Carolina at Chapel Hill (2008–2010). He is a guest professor at Technical University of Denmark (2017–2018). He has more than 20 years of professional experience in the field of sustainable urban planning and transportation planning, land use and transportation integration modeling, road congestion pricing, logic-driven transport big data analysis, emergency logistics, and shared autonomous mobility. He has written and published four books and more than 30 scientific papers in the top journals in the field of transportation planning, such as *Transportation Research Part A, C, and E*, *European Journal of Operational Research*, *Journal of Transport Geography*, *Computers, Environment and Urban Systems*, *Journal of Transport & Health*, *Journal of Transport and Land Use*, and *Journal of Transportation Engineering*. He is a member of the Youth Expert Committee of China Intelligent Transportation Systems Association and a member of the Intelligent Transportation Professional Committee of China artificial intelligence society. He is the guest editor of *Journal of Transport and Land Use* and *Journal of Advanced Transportation* and an editorial board member of *Transportation Letters*, *Transportation Management*, *Journal of Civil Engineering Inter*

Disciplinaries, and *Frontiers in Future Transportation*. He is the chairman of the traffic behavior investigation and analysis technical committee of the World Transport Convention. He is in the organizing committee and scientific committee of seven international conferences, such as the International Workshop on Integrated Land Use and Transport Modeling (ILUTM), 6th International Symposium on Travel Demand Management (TDM), and Transportation Research Congress (TRC). Personal website: http://faculty.dlut.edu.cn/2010011103/en/index.htm

Dr. Daniel (Jian) Sun is a professor and an executive dean of College of Future Transportation, Chang'an University. He has been working as the director and the professor of Smart City and Intelligent Transportation (SCIT) Interdisplinary Center, Shanghai Jiao Tong University (2011–2021). He obtained his Ph.D. in Transportation Research Center, University of Florida, in 2009, and has been a senior visiting scholar at ETH-Zurich (2018.9–2019.3). His main research interests include urban transportation planning and land use, traffic control, urban driver behavior and simulation, and urban transportation environment. He has been serving as the committee chair of Smart City and Intelligent Transportation sub-committee in World Transport Convention (WTC), has published more than 60 SCI/SSCI indexed journal papers since 2010, and has more than 30 papers accepted and presented in TRB annual meeting. He has served in editorial committee board of several journals, including *Transportation Research Interdisciplinary Perspectives* (since 2019), *Journal of International Transportation* (since 2012), *Journal of Traffic and Transportation Engineering* (English Version) (since 2014), and has been the chief member of road and traffic engineering sub-committee, Shanghai Society of Civil Engineering (since 2012). Moreover, he has been an expert reviewer for the Transportation Science & Technology Project, Ministry of Transport, China, and the National Science & Technology Award since 2014.

Personal website: http://js.chd.edu.cn/jiaotong/sj2_en/list.htm

Chapter 1
Logic-Driven Traffic Big Data Analytics: An Introduction

Abstract With the acceleration of urbanization, transportation problems have become the main bottleneck limiting the development of modern cities. Traffic congestion, traffic safety, environmental pollution, and other phenomena arise one after another, which have sharply affected the economic construction and operation efficiency of the city. The fundamental way to resolve the urban transportation problem is to integrate urban transportation and land use at the same time, feedback to each other, and achieve mutual coordination. Given this, this chapter firstly analyzes the interaction mechanism between land use and transportation from both macro and micro aspects and proposes the goal of integrated land use and transportation development. Secondly, four classic urban development cases, namely Stockholm, Copenhagen, New York, and Dalian, are selected to compare the process, causes, and social impact of their coordinated development model of transportation and land use. Finally, this chapter summarizes the transportation big data methods for investigating the relationship between transportation and land use. This chapter can not only provide a theoretical basis and technical groundwork for the research of subsequent chapters but also promote readers to have a more comprehensive understanding of the integration of transportation and land use.

Keywords Urban transportation · Land use · Relationship · Integrated planning

1 Integrated Theory of Urban Land Use and Transportation

The acceleration of urbanization, which is associated with the development of politics, economics, culture, and technology, has created new challenges for urban development. The relationship between urban transportation and land use can promote or restrict the development of urban space to guide the sustainable development of cities. To provide readers with a deeper understanding of the relationship between land use and transportation, this section analyzes the fundamental mechanisms of interaction between land use and transportation and discusses the aim of integrating the development of land use and transportation (Zhong & Bushell, 2017a).

S. Zhong and D. Sun, *Logic-Driven Traffic Big Data Analytics*,
https://doi.org/10.1007/978-981-16-8016-8_1

1.1 Introduction of Concepts Related to Transportation and Land Use

1.1.1 Transportation

Urban transportation has a wide range of meanings. One such meaning is that it represents a balance between travel demand and travel supply. Travel demand represents the demand for transportation infrastructure and the tools required for social-economic activities that involve the spatial redistribution of people and goods. Travel supply represents the sum of transportation services provided by transportation producers, which includes the supply of transportation infrastructure, tools, and other related services necessary to meet travel demand. Thus, travel demand and travel supply exist in a dynamic equilibrium: demand generates supply, and supply exists to serve demand.

Key concepts related to urban travel demand include travel volume, travel structure, travel distance, and travel spatial distribution. Travel volume is the number of objectives (e.g., vehicles and pedestrians) crossing a section of road per unit time in a selected period. Travel structure is the proportion of travel volume in a certain area using various travel modes within a certain period of time. Travel distance is the distance between the departure point and destination of a trip. Travel spatial distribution is the expected trajectory of travel volume, as opposed to the spatial distribution of the actual path. Related concepts characterizing urban travel supply include capacity and transportation infrastructure. Capacity refers to the maximum traffic flow that can be accommodated in a transportation infrastructure during a given time period, whereas transportation infrastructure represents the foundational structures and systems for transporting people and goods.

1.1.2 Land Use

Land use is a way of characterizing land according to what can be built on the land and for what purpose the land can be utilized. Thus, land use determines the type of community, environment, or settlement that can exist on a certain type of land; that is, land use is based on the functional dimension of land for different human purposes or economic activities (Zhong et al., 2021b). Countries differ in the methods they apply to characterize urban land use; for example, the United States (Anderson, 1976) divides urban land into six usage categories, whereas China divides urban land into nine usage categories, according to its new national standard, *Code for Classification of Urban Land Use and Planning Standards of Development Land GB50137 (Revised)* (China, 2018).

Related concepts that are used to characterize urban land use are mainly built environment attributes, that is, the geographic attributes of different land use types, such as land use density, diversity, design, destination accessibility, and distance to transit, as shown in Table 1. Land use density is the concentration of different

Table 1 The definition of the five "Ds" built environment attributes

Five Ds	Attributes
Density	Residential density Population density Building density Job density Amenity density
Diversity	Land use mix Jobs-housing balance Retail floor area
Design	Characteristics of street network Intersection/street density Route directness Link to node ratio Sidewalk/footpath continuity Building area and building height Number of entrances and exits for highway and ring road
Destination accessibility	Job within one mile Distance to amenities Distance to stops/stations Density of public transport stops Car parks and setbacks Distance to CBD/city centre Distance to the coast
Distance to transit	Number of bus stops and metro stations per unit area Distance to nearest transit stop

land uses within an area, while diversity is the extent of mixed land use within an area. Design describes the street network characteristics within an area. Destination accessibility is a measure of the ease of access to attractions. Distance to transit is commonly measured as an average of the shortest street routes from the residences or workplaces in an area to the nearest railway station or bus stop in the area (Ewing & Cervero, 2010; Fonseca et al., 2021).

1.2 Mechanism of Interaction Between Urban Transportation and Land Use

1.2.1 Mechanism of Interaction at a Macro Level

Urban land use system and transportation system are inter-influential and interconnected urban subsystems, and thus urban land use patterns form the basis of urban transportation patterns. In particular, a given urban land use pattern leads to a corresponding urban transportation pattern, and a given urban transportation pattern must be supported by a corresponding land use pattern (Giuliano, 2004; Rodrigue et al.,

2016; Vickerman et al., 1999). This results in an interactive feedback loop between urban land use patterns and urban transportation patterns at a macro level, as shown in Fig. 1.

The development of a city primarily relies on its transportation system, which forms the framework of urban space and is, therefore, the city's supply system. Innovations in transportation technology and the optimization of transportation network systems can alter the accessibility of entire areas or local areas, leading to changes in the function and structure of land use and therefore the spatial redistribution of land use. As land supports social and economic activities in cities, the spatial separation of different types of land use is the root cause of travel demand. Consequently, a change in land use will change the travel demand and thereby affect the planning, maintenance, and upgrading of transportation infrastructure and services such as roads and public transportation. These changes will also affect accessibility and trigger a new loop of interactive feedback between land use and transportation (Zhong et al., 2020).

In addition to these internal relationships, changes in external factors such as land use characteristics, technologies, economics, policies, and populations also affect the interaction between land use and transportation. First, changes in land use characteristics, such as the characteristics of the urban circle layer and the supply of available land, alter land users' activity patterns, thus changing travel demand. Second, changes in technologies, economics, and policies alter the overall level of accessibility of a city and the relative accessibility of different locations in the city, resulting in changes in the city's morphological characteristics and development patterns. Finally, as a city's population comprises the main source of travel demand, population growth generates additional travel demand, which creates new requirements for travel supply capacity and results in the re-allocation of transportation resources.

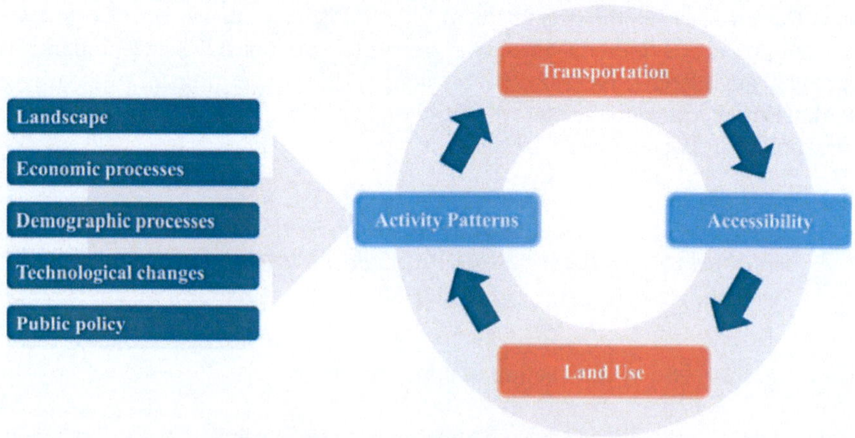

Fig. 1 The relationship between urban transportation and land use systems (Rodrigue et al., 2016)

1.2.2 Mechanism of Interaction at a Micro Level

Land use subsystem and transportation subsystem not only interact at a macro level but also are mutually reinforcing and interdependent at a micro-level (Wegener, 2004; Zhong et al., 2021b). This manifests as four micro-level mechanisms of interaction between urban land use and transportation, as described below and depicted in Fig. 2.

The linking role of accessibility. Accessibility is defined as the ability of an individual to reach his/her destination (e.g., a shop, a school, or a hospital). The planning of transportation and land use systems is often based on the assumption that increased travel capacity will bring value to users, which means that the typical goal of urban planning is to maximize accessibility. However, accessibility is not only related to the mechanism of interaction between land use and transportation; accessibility also affects the attractiveness of land, thereby changing residential location choice, enterprise location choice, and the nature of land use, which causes the redistribution of populations and employment locations. Because people consider the accessibility of a location when selecting locations for work, shopping, leisure (recreation and entertainment), living (residential housing), and construction sites (for houses and commercial buildings), the distribution of land use activities within a city is affected by variations in accessibility. Moreover, the redistribution of urban

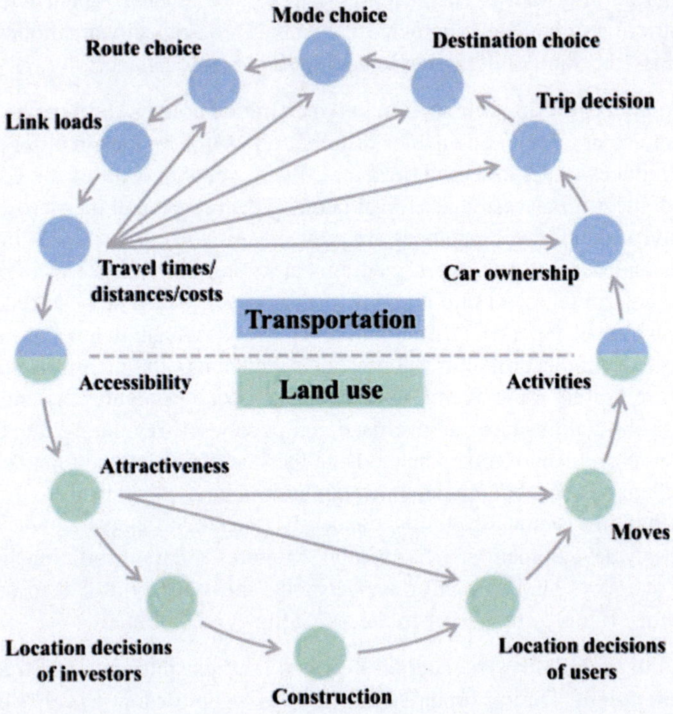

Fig. 2 Feedback loops between urban transportation and land use

economic, cultural, commercial, and other activities redistributes the travel demand of users of an urban transportation network and thereby affects users' travel behavior, such as their travel destinations, travel modes, and travel routes. Thus, the operating efficiency and traffic flow state of an urban transportation network changes in response to users' travel behavior, which alters travel times, distances, and costs and further affects the accessibility of destinations (Rodrigue et al., 2016; Zhong et al., 2015). Accessibility, therefore, plays a key role in connecting an urban land use system with a transportation system at the micro-level.

The mutually regulating interaction between floor-area ratio and road capacity. Land development in a city, which may be commercial, industrial, or residential, increases a city's floor-area ratio, thereby leading to many more trips being taken by urban dwellers. The concomitant increase in travel demand generates an improvement in transportation facilities, an expansion of road capacity, and an improvement in traffic convenience and travel time reliability, thereby increasing land value and attracting investors to pursue further land development and construction activities. Thus, the interaction between road capacity and floor-area ratio enters a new cycle. However, such a cycle cannot continue indefinitely, because there is a limit to how much the capacity of urban transportation facilities can be increased by reconstruction. Therefore, at a certain intensity of land development, traffic congestion beings to occur on road sections in some areas, leading to a decrease in the performance of the transportation system and in the marginal benefit of land use, which in turn results in the restriction of land development in the area. Road capacity and floor-area ratio are therefore constantly interacting to achieve a dynamic balance.

The mutually regulating interaction between mixed land use and travel distance. The mixed use of urban land enables different types of activities in close proximity and thus reduces travel time and distance. When a certain level of mixed land use is reached, most urban residents' travel demand decreases, and their cross-regional travel activities and travel distances are reduced. As a result, increases in the level of mixed land use can transform medium- and long-distance travel into regional travel and horizontal travel into internal vertical travel, which helps to reduce travel distance and travel dependence on urban roads. However, a high level of mixed land use can lead to various problems, such as environmental pollution, traffic congestion, and decreased safety for residents, as urban travel becomes highly concentrated, and transportation facilities become overused and poorly self-regulating. In such situations, the management of travel demand and the spatial allocation of land use cannot meet the demands of urban transportation infrastructure, which leads to the collapse of transportation systems, severe imbalances of urban travel supply and demand, and the disorderly development of urban spatial structures and transportation. Therefore, the level of mixed land use and travel distance must be controlled to achieve an optimal range of outcomes in urban areas (Zhong et al., 2022).

The mutually regulating interaction between land use patterns and transportation development. During urban spatial growth, people adopt travel behaviors to overcome the spatial separation of land that is used for different purposes, which means that the evolution of land use patterns and the spatial distribution of residential and employment areas fundamentally affect the travel distribution of urban transportation. Concomitantly, differences between the capacities of various travel modes lead to variations in these modes' adaptability to the size, intensity, and distribution density of travel demand, such that peoples' travel mode choices vary according to urban scales and land use patterns. Accordingly, the service level and travel supply capacity of urban transportation networks must be adjusted to meet changing travel demands; thus, urban land use patterns determine the development of urban transportation systems. Conversely, the level of urban transportation development affects the distribution of urban land use. For example, a change in urban transportation development can cause the dispersal of the dense population in a city's central area into the city's less central regions; this causes the city's commercial center to become concentrated in a smaller area, which decreases the level of mixed land use. The planning and construction of urban transportation also guide land use and urban development, as land development and utilization is intensified in the vicinity of transportation facilities. Thus, land use patterns and urban transportation development mutually interact, in both positive and negative ways.

1.3 Aim of Integrating the Development of Urban Land Use and Transportation

The aim of integrating the development of urban land use and transportation is to optimize urban development and urban transportation development. This requires a fundamental emphasis on the comprehensive and coordinated development of transportation and land use to optimize urban transportation structures, realize improvements in transportation and land use, and promote sustainable urban development (Wegener, 1996; Zhong & Bushell, 2017b; Zhong et al., 2021a). This aim comprises the following three requirements.

1.3.1 Optimization of the Travel Demand Model to Achieve Reductions in Transportation Development

The optimization of urban land use is the core aim of integrating transportation and land use, as land use directly leads to travel demands, and transportation guides land use. Therefore, the optimal integration of transportation and land use creates a mutually beneficial interaction that promotes compact land use patterns and the development of a healthy urban environment and an efficient public transportation system. If fundamental changes can be made to travel demand, the spatial distribution of travel, and travel mode selection, travel demand patterns and transportation structures can be optimized, thereby enabling reductions in (i.e., more efficient) development of urban transportation and helping to realize the effective management of urban travel demand.

1.3.2 Optimization of Urban Spatial Structure to Achieve a Balance Between Travel Supply and Demand

Inherent limits to the capacities of urban roads and unsuitable patterns and intensities of land use inevitably create an imbalance between travel supply and demand. For example, an urban development model of single-center concentric circles leads to the excessive concentration of urban functions in the central area of a city, with this over intensified land development generating high concentrations of traffic and thus increased pressure on the transportation system. This is accompanied by the imbalanced utilization of urban space resources, resulting in the limitation of urban development. Urban travel supply must therefore be coordinated to ensure the effective growth of travel demand and the rational allocation of construction land resources, thereby realizing a balance between travel supply and demand, urban transportation, and land use. For example, transportation planning departments should recognize the fundamental interaction between land use systems and transportation systems and thus give full play to the leading role of transportation systems (especially public transportation systems) in urban development planning. Such an approach would facilitate the development of a comprehensive model of an urban transportation system and land use, which could then be used to optimize travel supply and demand to realize the coordinated development of transportation and land use. Transportation planning departments should also promote the joint development of traffic zoning and functional zoning, traffic corridor and development axes, passenger transport hubs, and urban center systems, according to the travel modes and land use effects associated with these aspects and the requirements of a given project.

1.3.3 Coordination of Land Use and Transportation to Achieve Sustainable Urban Development

Urban transportation links humans, cultures, cities, and countries, and thus profoundly affects social and economic development. However, the development of urban transportation has various adverse consequences in cities worldwide, such as increased traffic congestion, decreased traffic safety, increased use of energy and land resources, and increased air and noise pollution. For example, the transportation sector in China accounts for 46%, 78%, and 83% of NO_X, CO_X, and hydrocarbon emissions, respectively, in metropolises. Furthermore, China's land resources are becoming increasingly scarce due to rapid increases in urbanization and numbers of motor vehicles, which will inevitably lead to severe restrictions in the availability of land for urban development. This means that fewer land resources will be available to meet essential road traffic requirements in cities.

An additional consideration is that the Chinese government has set strategic goals of achieving peak greenhouse gas emissions before 2030 and carbon neutrality by 2060, in line with global efforts to reduce emissions; these strategic goals will require the restructuring of land use and transportation systems. Various restructuring strategies have been devised and implemented by transportation planning and management departments in different countries; these involve the conversion of existing integrated transportation systems to less polluting modes to create more sustainable cities, as shown in Table 2.

Positive social effects have resulted from strategies such as the bicycle travel mode used in Denmark, the road congestion pricing used in London (UK), the motor vehicle sales tax applied in Singapore, and the traffic guidance screen and bus priority system established in China. However, these strategies have little ability to contribute to the achievement of net-zero emissions in the long term. This is because the transportation system and land use system are essential and mutually interacting components of a city, which means that optimizing only one of these systems leads to short-term improvements in the environment and not long-term, sustainable urban development.

Therefore, urban transportation and land use problems can only be solved by approaches based on integrated planning and coordinated development aimed at maximizing the efficiency of transportation operations and intensive land use and promoting sustainable urban development. That is, an efficient and convenient urban transportation network and organization system is one in which the transportation system facilitates controlled, intensive land use, reduces traffic congestion and environmental pollution, promotes social equity, and ensures the rational exploitation of resources. This enables residents' travel demands to be met and maximum transportation efficiency to be achieved at a minimal social cost, which is the foundation of sustainable urban development.

2 Empirical Analysis of the Interactive Relationship Between Urban Land Use and Transportation Systems

In this section, we compare the processes, causes, and social effects of four typical urban development modes—that of Stockholm, Sweden; Copenhagen, Denmark; New York, United States; and Dalian, China—to highlight the interactive relationship between urban land use and transportation systems.

2.1 Transit-Oriented Development Mode: Stockholm, Sweden

The transit-oriented development (TOD) mode generates a city with a rational and orderly layout and has been applied effectively in many cities worldwide. Stockholm, the capital of Sweden, is considered to be one of the most prosperous cities built

Table 2 Environment-oriented strategies and tactics

Tactics	Strategies				
	Reduce travel demand	Reduce motor vehicle use	Develop green transportation	Improve the operational efficiency of the transportation network	Vehicle technical improvement
Technology Infrastructure; Vehicles; Energy	Transit-Oriented Development (TOD)	Community road	Construction of public transportation, bicycles, and pedestrian infrastructure; New public transportation vehicles	Construction of roads; Parking facilities	New energy vehicles; Alternative energy
Management Planning; Management; Control	Land use planning; Suburbanization management	Traffic restrictions; Parking restrictions; Traffic congestion mitigation	Bus priority measures	Transportation management; Urban transportation planning	Emission regulation; Improvement of fuel standards; Vehicle inspection
Information Communication; Publicity	Remote real-time communication	Publicity	Real-time traffic information broadcasting	Car navigation; Traffic guidance display	Environmental awareness
Policy Tax; Charge	Land tax	Road congestion pricing; Fuel taxes; Motor vehicle sales tax	Public transportation ticket price reduction strategy	Road congestion pricing; Parking charge	Fuel tax; Environmental protection tax

Fig. 3 Stockholm subway system map (Metro of Stockholm, 2011)

on the TOD mode, which was initiated by its transportation planning department after World War II based on similar approaches used in the United States and other countries. As an example of TOD in action, since 1950, Stockholm has built subways consisting of three lines (i.e., the green, red, and blue lines), and as shown in Fig. 3, these lines are arranged in a radial pattern. This pattern was adopted to efficiently connect the central urban area of Stockholm, which is dispersed over 14 islands and a peninsula, with the other parts of the city. Thus, the central urban area is at the center of a star-shaped subway system that provides transportation between satellite towns and the central area of Stockholm (Paulsson, 2020). These satellite towns contain high-density residential areas within walking distance of rail transit stations and represent a more intensive type of urban satellite development than was applied in other countries after World War II.

The construction of Stockholm's new town was undertaken in two slightly overlapping stages. From 1945 to 1975, the first generation of satellite towns (e.g., Vallingby and Farsta) was constructed along the path of the three subway lines; these towns were built based on a functional combination of residential areas, work areas, and service areas to achieve a balance of internal travel between workplaces and residences. This unified urban development strategy meant that all of these first-generation satellite towns followed the same developmental pattern. In the 1970s, the second generation of satellite towns (e.g., Kista and Spagna) was constructed; these were built based on the first-generation satellite towns but included convenient transportation links with the other towns and regions, rather than focusing solely on balancing internal travel.

Overall, Stockholm is a remarkably successful example of the application of the TOD mode. First, the appropriate distances between the central city and the new towns mean that these areas are conveniently connected without the overlapping of urban functions. Second, the radial urban structure helps to protect ecologically rich areas and promotes the formation of urban landscapes (Paulsson, 2020). Third, the judicious integration of the transportation hub and the new city center established a walking-friendly community that enhances pedestrian flow and provides a stable source of customers for businesses. Fourth, the use of private vehicles is greatly reduced, as public transportation accounts for the largest proportion of travel modes in Stockholm. Finally, the public transportation system guides the development of the city, which optimally serves the resulting new travel demand and thereby prevents the initiation of a vicious cycle of urban sprawl.

2.2 *Hand-Shaped City Planning Mode: Copenhagen, Denmark*

Copenhagen is the capital and largest city of Denmark and is home to the country's largest port. Denmark is part of Scandinavia in northern Europe; it has a long coastline, and approximately 10% of its land area is covered in forest. The planning of cities in Denmark has for many years been based on sustainable development and the preservation of natural landscapes. Copenhagen, therefore, uses rail transit to guide urban development, and its hand-shaped city planning is considered one of the most successful examples of urban planning worldwide (Knowles, 2012).

Copenhagen's hand-shaped city plan, the first sketch of which is shown in Fig. 4, was developed in 1947 and implemented in the subsequent decades. As Copenhagen faces the Oresund Strait to the east, the five "fingers" of the city extend to its northern, western, and southern areas, with the "palm" containing the central city, including its medieval historic areas. The fingers, therefore, serve as radial corridors of intensive urban construction, between which are areas of forests, farmland, and open green space in which urban development is prohibited. The development of the central city area has also been supported by the provision of comprehensive public transportation services, as exemplified by Copenhagen's rail transit system (Fig. 5) and well-developed bicycle transportation network. As shown in Fig. 6, there are numerous large bicycle parking lots near the Copenhagen rail transit station. Copenhagen's rail transit system also has dedicated bicycle carriages, which promotes synergies between bicycle and rail transit and further increases the proportion of public transportation trips.

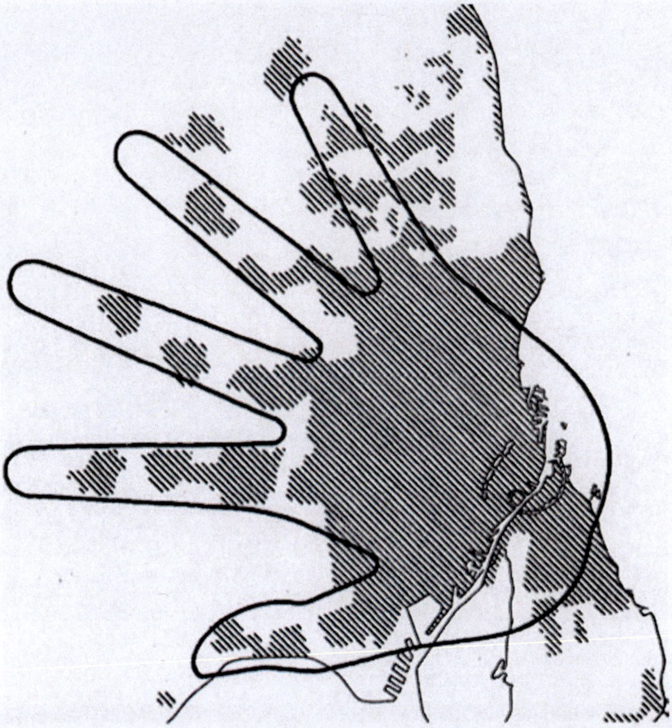

Fig. 4 Sketch from the first "finger" city planning by the regional planning office in 1947

Inevitably, some difficulties were encountered during the construction of Copenhagen. For example, the development of the three fingers to the north was sluggish, which led to disorderly urbanization and the use of part of the green spaces between the fingers. Therefore, in 1961, the two fingers to the south were forcibly extended to decrease the pressure on the central city. In addition, due to the adverse effects of the increase in private vehicle ownership on rail transit and those of the 1970s oil crisis on the economy, the originally planned polycentric urban structure of Copenhagen was not achieved.

Eventually, most of the hand-shaped structure of Copenhagen was completed by 1989. Since then, Copenhagen has continuously adjusted its development plan. The success of its hand-shaped urban layout design was due to the local government's long-term vision based on the natural characteristics of the city and the fact that the design was generally accepted as the standard for regional development (Papa & Bertolini, 2015). This hand-shaped urban design continues to support the functions of the central city and facilitate transportation between the central urban area and outer areas. In particular, Copenhagen closely integrates transportation with land use to achieve a balance between employment and housing areas. Moreover, the local government restricts the use of vehicles in some urban areas, where it promotes walking and cycling between public transportation modes. Overall, the hand-shaped

Fig. 5 The route map of Copenhagen rail transit

Fig. 6 Large bicycle parking lot near Copenhagen rail station

development of Copenhagen has played a positive role in environmental protection and resource conservation.

2.3 Urban Sprawl Development Mode: New York, United States

The examples of Stockholm and Copenhagen illustrate that the TOD and hand-shaped development modes ensure the orderly development of cities. Unscientific urban planning, however, can create various problems in the development of cities, such as the urban sprawl seen in numerous cities, for example, in the United States.

The United States is a highly developed capitalist country that was the first to develop suburbanization; however, it was also the first country to exhibit urban sprawl. New York is the largest city and port in the United States and is one of the most affluent urban areas in the world. However, the urban sprawl caused by disorderly development in New York has had substantial adverse effects on the economy, society, and environment in the area (Bae & Richardson, 2017).

There is contrasting urban spatial density on either side of the Hudson River of New York: the urban spatial density of Manhattan is extremely high, whereas that on the other side of the river is much lower. Figure 7 presents a population density map of New York, created by the London School of Economics. Disordered urban sprawl is particularly evident in the expansive low-density areas in the suburbs. As another example, Fig. 8 depicts urban sprawl in the Bronx, a borough that lies on the outskirts of New York.

New York's urban sprawl is representative of urban sprawls across the United States, which began in the early twentieth century. By this time, the urbanization of

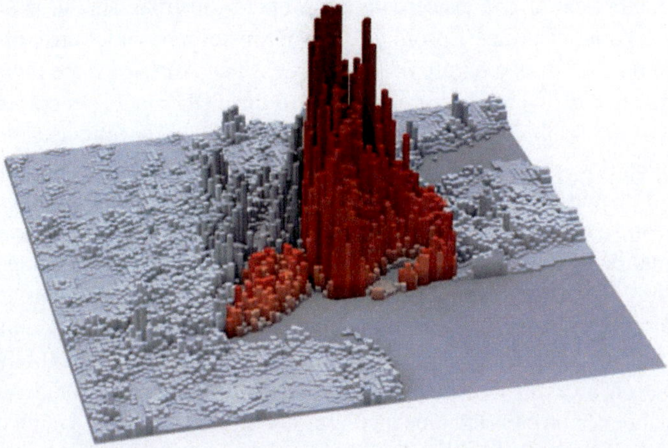

Fig. 7 New York city population density distribution map (Colman, 2019)

Fig. 8 Urban sprawl in the Bronx, New York

New York was largely complete, and the high spatial density of the old part of the city had led to numerous problems that caused a substantial decline in the quality of life. Consequently, residents with sufficient capital began moving out of the central city to develop garden cities, which was the start of suburbanization. After World War II, the economy of the United States flourished, and the number of cars in the country increased significantly, which led to large-scale suburbanization. Disorderly low-density single-use urban areas continued to expand, ultimately resulting in urban sprawl.

Many factors cause and exacerbate the problem of urban sprawl in the United States. First, Americans have a profound love of private cars, which are considered an essential part of American family life. This means that Americans are highly dependent on private cars, which is not conducive to the TOD mode. Second, prolonged development has caused the land rents in central cities to become extremely high, which renders these areas unsuitable for residential use. Second, the government subsidizes the construction and operation of expressways, which facilitates the use of private cars to access outer suburbs, leading to people living in these areas and thus the continuation of suburbanization. Finally, the development of the United States is dominated by the market, which leaves little room for government intervention.

The emergence of urban sprawl has generated economic, social, and environmental problems (Habibi & Asadi, 2011). First, the development of large-scale single-function areas cannot meet residential travel demand, as it increases residents' dependence on cars and thus increases energy consumption, which ultimately becomes a vicious circle. Second, disorderly development wastes land resources and extensively alters the natural landscape. Third, it results in increased pollutant emissions, which further aggravate environmental pollution. Since the 1970s, New York has been continually adjusting its planning strategies to curb urban sprawl.

2.4 Tide Transportation Development Mode: Dalian, China

Although there are considerable ideological, cultural, and social development differences between China and the United States, suburbanization and the geographic separation of people's workplaces from their residences are also problems that afflict Chinese urban development.

Dalian, located at the southern end of Liaodong Peninsula, is an important central city and port on the northern coast of China. The urban development of contemporary Dalian began when Tsarist Russia occupied this large tourist area and built Dalian's port; subsequently, the city gradually developed in parallel with port construction, following a single-center development mode.

Since the reform and opening up of China, economic growth and urban development have accelerated significantly in Dalian, with its central urban area has become a single land use area that houses financial and commercial companies. Increased land prices and traffic congestion led to gradual decreases in the number of residential areas, and thus many residents moved out to establish suburbs. The demand for residential land in the suburbs has continued to increase, while workplaces remain concentrated in the central city area, resulting in the separation of workplaces and residences. In recent years, the establishment of new industrial parks for electronic information, software, and cultural creativity companies has exacerbated this separation, as these industrial parks attract many highly skilled and highly paid people. These people can afford to live in residential areas that are not close to their workplaces and prefer to do so as these farther areas are more pleasant and have better infrastructure. However, this increases their travel time between their homes and workplaces.

This separation of workplace and residential areas in Dalian has led to several urban problems. For example, many people travel into the city center during the morning peak period and travel out of the city during the evening peak period, which induces tide transportation, resulting in uneven distribution of morning and evening peak traffic flow, as shown in Fig. 9. As Dalian is located in a hilly area that is inconvenient for bicycle travel or public transportation connections, its residents are more dependent on private cars for transportation than those in other cities in China, which exacerbates the problem of tide transportation in the urban transportation system in Dalian (Wang et al., 2019). First, tide transportation causes severe congestion during peak hours and wastes road resources, as shown in Fig. 10; unfortunately, these problems cannot be solved solely by rebuilding and expanding transportation facilities. Second, the entry of many private cars into the city center creates a dearth of available parking spaces, which leads to disorderly parking. Overall, the urban layout–driven problem of tide transportation wastes resources, generates environmental pollution,

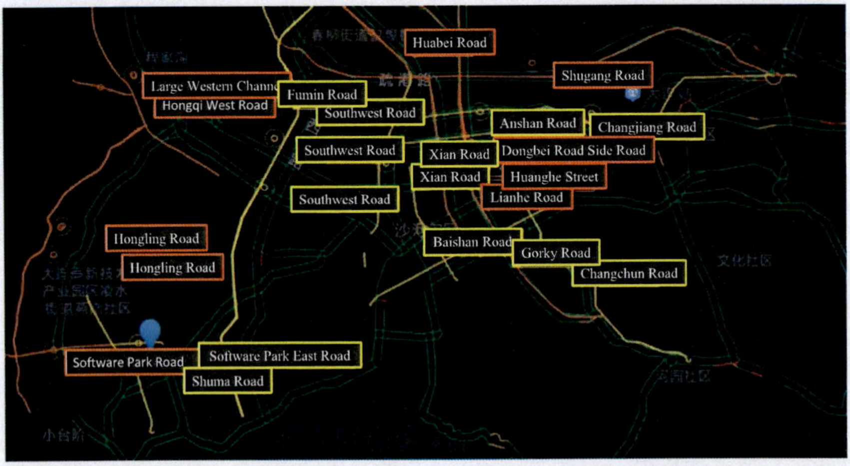

Fig. 9 Congested road sections in the morning and evening peak hours in Dalian

Fig. 10 Morning peak period transportation on Lingshui road in Dalian

and causes social and economic losses (Wang et al., 2019). In response, Dalian is currently trying to alleviate its transportation bottlenecks by creating designated tidal lanes and giving priority to bus services.

In summary, the afore-discussed urban development modes greatly affect all aspects of a city's development. Herein, we conduct a comparative analysis of the above examples to more clearly summarize their characteristics. The results of this analysis are presented in Table 3.

Table 3 Summary of the development and characteristics of each case city

City	Development circumstance	Development characteristics
Stockholm	TOD mode has formed a "Star" urban structure	The orderly and efficient use of urban land is conducive to resource conservation and environmental protection; The corresponding supporting facilities in the residential area are relatively perfect, and the urban transportation is convenient
Copenhagen	TOD mode has formed a "Finger" urban structure	Urban public transportation has developed rapidly, and urban transportation is convenient; The ecological green space has been effectively protected, and the city has expanded in an orderly manner
New York	Due to suburbanization and strong dependence on private cars, there has been a severe phenomenon of urban sprawl	Urban land use efficiency is low, and different land use types are scattered; The increase in residents' travel demand and strong dependence on cars have caused environmental pollution and waste of resources
Dalian	Due to the separation of workplace and residence, there has been a severe phenomenon of tide transportation	The separation of workplace and residence leads to a long commute distance. Tide transportation causes severe traffic congestion and waste of transportation resources at morning-evening rush hours

3 Models and Methods of Integrating Urban Land Use and Transportation

This section briefly introduces the methods used in this book, namely spatial econometrics, spatial statistical analysis methods, multivariate statistical analysis methods, and methods solely or predominantly used to evaluate the effectiveness of policies. The subsequent chapters provide more details on these methods.

3.1 Spatial Econometrics

3.1.1 Spatial Autoregressive Models

The rapid development of data acquisition techniques is causing exponential growth in the amounts of spatial data, because of which research on the processing of spatial data is a focus of many studies. Spatial data is always spatially autocorrelated and thus cannot be analyzed using classical regression models. This means that spatial autoregressive (SAR) models are essential for analyzing spatial data (Anselin, 2010), with SAR models being the most basic and one of the most well-studied types of spatial econometric models.

In transportation applications, the traffic flow in a road network is spatially dependent, and the spatial distribution of the flow is spatially aggregated. That is, if the road near a targeted road has a large flow value in the same time period, then the flow value of the targeted road is also considerable. Thus, a SAR model takes spatial correlation as the essential characteristic of the research variables (Anselin, 2013).

3.1.2 Geographically Weighted Regression Models

Geographically weighted regression (GWR) models are a powerful tool for exploring the characteristics of spatial non-stationarity, which has become a focus of research in recent years due to the burgeoning interest in spatial analysis. The parameters of a given GWR model reflect changes in spatial location (Fotheringham et al., 2003). There are at least three cases in which the model coefficients are not fixed in different regions: if randomization of a research area leads to a certain amount of spatial non-stationarity; if the spatial properties of some relationships in a research area vary between regions; and if ordinary linear regression models cannot adequately measure the spatial interaction such that one or more variables are omitted from the model or incorrect functional relationships are included in the model.

In recent years, GWR models have been widely used in various fields, including in transportation applications for the analysis of the spatial heterogeneity, quantification of the non-stationarity of a traffic network, and for coordinating traffic flows in the modeling of main urban roads. A GWR model incorporates spatial location information on data by assuming that the coefficients in a linear regression model are a function of the geographical location of a research object (Fotheringham et al., 1998). The model then examines the non-stationary spatial characteristics of spatial data to obtain estimated values of the coefficient function at each geographical coordinate to further concretize the model. That is, an estimation model is obtained for each research unit.

3.1.3 Spatial Autoregressive Moving Average Models

Spatial autoregressive moving average (SARMA) model consider the interaction of adjacent individuals and the influence of exogenous variables on a research object. It was first proposed by Anselin in 1996 as improvements of traditional mixed regression SARs and spatial error models.

SARMA model is like spatial moving average models, in that it can explain the spatial correlations between error terms (Ord, 1975) but also like SAR models in that it can consider spatial correlations in the dependent variable.

3.2 *Spatial Statistical Analysis Methods*

3.2.1 Moran's Index (Moran's *I*)

Moran's I is a correlation coefficient that measures the overall spatial autocorrelation of a data set. That is, it measures how similar an object is to its surrounding objects. If objects are attracted to or repelled by each other, this indicates that observations concerning these objects are not independent, i.e., they are autocorrelated. This violates a basic assumption in statistics: the data under analysis are independent (Dray et al., 2006). Thus, the presence of autocorrelation must be determined, as it renders most statistical tests invalid.

Spatial autocorrelation is multidirectional and multidimensional and thus can be used to find patterns in complicated data sets. Moran's I is similar to correlation coefficients, as its value ranges from -1 to 1; however, while correlation coefficients indicate any form of correlation, from a perfect correlation to no correlation, Moran's I uses more complex, spatial calculations to reveal the following aspects:

- if $I = -1$, there is perfect clustering of dissimilar values (a perfect dispersion);
- if $I = 0$, there is no autocorrelation (perfect randomness); and
- if $I = +1$, there is perfect clustering of similar values (a perfect non-dispersion).

3.2.2 Geary's Coefficient (Geary's *C*)

Geary's C is akin to Moran's I, as it is also a global index that can be used to measure the spatial autocorrelation of regional targets. The calculation of Geary's C differs from that of Moran's I—Geary's C is based on the dispersion between observations, whereas Moran's I is based on the cross-multiplication of median dispersions—but they afford similar empirical results (Bivand & Wong, 2018). Furthermore, Geary's C and Moran's I are calculated using the exact same process, and these parameters can be used as (rough) replacements for each other in many analyses.

Geary's C theoretically ranges in value from 0 to 2, but 2 is not the strict upper bound. Overall,

- if $C = 0$–1, there is a positive spatial autocorrelation;
- if $C = 1$, there is no spatial autocorrelation; and
- if $C > 1$, there is a negative spatial autocorrelation.

3.2.3 Geographical Detector (Geodetector)

Spatial differentiation is the spatial manifestation of natural and socioeconomic processes and has been a crucial way for humankind to understand nature since the days of Aristotle (384–322 BC). Geodetector is a new statistical method for the detection of spatial differentiation and its driving factors (Wang et al., 2016) and is based on the assumption that if an independent variable has an essential influence on a dependent variable, then the independent variable and the spatial distribution of the dependent variable should be similar. Geographic differentiation can be expressed using classification algorithms, such as environmental remote sensing classification, or determined based on experience.

Geodetector performs well in the analysis of type quantities and can also be used for the analysis of appropriately discretized sequential quantities, ratio quantities, or interval quantities. Therefore, Geodetector can detect both numerical data and qualitative data, which is a significant advantage. Another unique advantage of Geodetector is its ability to detect the interaction of two factors on the dependent variable, which is typically detected by adding the product of two factors to the regression model to test its statistical significance. However, the interaction of two factors is not necessarily a multiplicative relationship, and Geodetector can be used to determine whether an interaction exists, and if so, its strength, direction, and linearity (or non-linearity). Geodetector achieves this by superimposing the two factors and then calculating and comparing the value of each single-factor q-statistic and the q-value. This approach is effective, as the superposition of two factors includes a multiplicative relationship and other relationships, and if these exist, Geodetector can detect their presence.

3.3 *Multivariate Statistical Analysis Methods*

3.3.1 Cluster Analysis

1. The k-means clustering algorithm

The k-means clustering algorithm is the most fundamental and frequently used partitioning algorithm used in clustering analysis. It is an unsupervised real-time clustering algorithm widely used in science and industry and was first proposed by McQueen in 1967.

The operation process of the k-means algorithm is as follows: First, k initial clusters are identified, the initial values of the centers of these k initial clusters are set, and then the required number of iterations or conditions (e.g., convergence conditions) that eventually stop the algorithm are set (Hartigan & Wong, 1979). Second, the clusters are quantified according to a certain similarity measurement criterion, so that similar data form clusters with commonality. Third, the average vector of each cluster is used as the cluster's center, and then the algorithm and related operations are executed. The process is then repeated and is finally terminated when the previously set conditions are met.

2. Fuzzy c-means (FCM) algorithm

The FCM algorithm is the most widely used fuzzy clustering algorithm based on an objective function (Hartigan & Wong, 1979) and typically treats a clustering problem as a nonlinear programming problem with constraints. The FCM algorithm thereby obtains the optimal fuzzy partition and clustering results of data sets by applying iterative optimization theory.

To use the objective function method to solve a clustering problem, researchers first combined this method with mean-square approximation theory to construct a nonlinear programming function with constraints (Kamel & Selim, 1994; Xie & Beni, 1991). The sum of square errors in this class has since become a general type of clustering objective function. In 1973, Dunn extended the concept of fuzzy partitioning to develop the weighted within-group sum of squared error (WGSS) error function. Bezdek et al., (1984) introduced the parameter m and used it to develop an infinite family of WGSS error functions, which he called the FCM clustering algorithm. Thus, this algorithm can be regarded as an improved hard c-means clustering algorithm.

The classification of traffic data requires the identification of traffic states. As a traffic flow state continuously changes, it is often analyzed using fuzzy concepts such as congestion and unimpeded traffic. However, FCM can accurately reflect the actual distribution of data objects, and thus the FCM clustering algorithm can be used to identify traffic states.

3. Gaussian mixture models (GMMs)

GMMs use the Gaussian probability density function to accurately quantify data by decomposing it into several models (Hinton et al., 2012). That is, a set of observational data can be fitted by a mixture of multiple single Gaussian models, irrespective of the distribution of the set of data and the law obeyed by the data.

The Gaussian distribution, also called the normal distribution, is a probability distribution that is crucial in the fields of mathematics, physics, and engineering, and has a major influence on many aspects of statistics. As the distribution is bell-shaped, it is often called a bell-shaped curve.

GMMs are similar to k-means clustering, as they also use an iterative algorithm for calculations and finally converge to a local optimum. However, GMMs may be more appropriate than the k-means clustering algorithm when clusters vary in size and are correlated. As GMMs exploit the Gaussian probability density function for learning, they are typically used for density detection, in addition to clustering applications. However, while k-means clustering classifies each observation into a particular type of data set, GMMs reveal the probability of an observation being classified into a different type of dataset.

4. Dirichlet-multinomial regression (DMR) models

Text data generated via text mining are usually accompanied by metadata, such as the author, publication location, and date. Many extensions of polynomial mixed-topic models have been developed to take these metadata into account, with DMR models being one example.

DMR models use a collaborative feature vector based on the latent Dirichlet allocation (LDA) model, as this vector generates accurate semantic information by considering text data and its coexisting metadata (Chen & Li, 2013). Therefore, in the fields of text mining and urban functional area recognition, DMR models have afforded results closer to reality than those afforded by LDA models. Furthermore, compared with other models specially designed for analyzing specific data (e.g., the author-topic model and the topic-over-time model), the DMR model exhibits superior topic-extraction performance and yields more concise and effective calculations.

3.3.2 Multivariable Linear Regression Models

Multiple linear regression analysis is a direct generalization of univariate linear regression and is widely used in many fields, such as transportation, medicine, and agriculture. It determines the numerically interdependent linear relationships between multiple variables by, for example, performing regression analysis of a dependent variable to multiple independent variables and the regression analysis of multiple dependent variables to multiple independent variables (Harrell et al., 1996).

There are three main applications for multiple linear regression analysis. First, it can be used to determine the strength of the influence of an independent variable on the dependent variable. Second, it can be used to predict the magnitude of change in the dependent variable in response to a change in an independent variable. Third, it can be used to predicts trends and future values, such as point estimates.

Model fitting is a key consideration when choosing a model for multiple linear regression analysis, as the addition of independent variables to a multiple linear regression model inevitably increases the amount of explanatory variance in the dependent variable. Thus, the addition of too many independent variables without a theoretical basis may cause a model to overfit.

3.3.3 Structural Equation Models

Structural equation models (SEMs) are a multivariate statistical analysis technique for establishing, estimating, and testing causal relationships. They include a series of multivariate statistical analysis methods, such as regression analysis, factor analysis, path analysis, and multivariate analysis of variance. SEMs are thus a very versatile, linear statistical modeling technique that use theory to conduct hypothesis testing (Fornell & Larcker, 1981).

SEMs comprise a measurement model and a structural model; the measurement model determines the relationship between the observed index and the latent variable, and the structural model determines the relationship between the various latent variables. The variables that can be directly measured in a given problem are recorded as observed variables or explicit variables, whereas those that cannot be directly measured are recorded as latent variables. SEMs are also called latent variable causality models, as they represent the relationship between latent variables.

3.4 *Methods for Evaluating Policy Effectiveness*

3.4.1 Difference-In-Differences Method

The difference-in-differences (DID) method is another important tool for causal inference. For example, it is used to analyze the effects of policies, whereby it treats institutional changes and new policies as comprising a natural experiment that is exogenous to an economic system. The key aspect of the DID method is its use of individual data for regression. The first difference was made before and after the beginning of the policy, and the second difference was made between the policy-affected individuals (treatment group) and the policy ineffective individuals (control group). The DID method then examines these differences to determine whether a policy has had a statistically significant effect (Bertrand et al., 2004). The advantage of the DID method is that its way of estimating the effects of exogenous policy shocks avoids estimation bias due to simultaneous error and the missing variable error in an endogenous problem.

The application of the DID method relies on three strict basic assumptions. (1) The randomization of groups caused by policies, which is reflected in both the policy action time and the role of individuals. (2) A relatively clean random grouping, such that the treatment group is influenced by policy but the control group is not. (3) The treatment group and the control group follow the same trend during the sample period; that is, there is a parallel trend.

As urban areas contain a high-level form of human space activities, from a regional perspective the frequency and intensity of policies issued by managers are significantly higher in urban areas than in non-urban areas. Urban economics studies therefore often focus on the effects of classic policy shocks in an area, such as the expansion of a university, the construction of a high-speed rail, subway line, or shopping mall, and the establishment of a large enterprise.

3.4.2 Regression Discontinuity Design Models

Regression discontinuity design (RDD) models are statistical methods for the evaluation of policy effects that determine the size of a disposal effect in non-experimental conditions. These models comprise three types of variables—running variables, outcome variables, and binary disposition variables—and examine the relationships between them (Chen et al., 2013) based on the principle that each experimental individual corresponds to an observable driving variable. Thus, if the value of an individual's driving variable is greater than the value at a known discontinuity, an individual accepts a given treatment, whereas if the value of an individual's driving variable is less than the value at a known discontinuity, the individual does not accept a given treatment. This results in all test individuals being divided into two categories, separated by the discontinuity; the individuals whose corresponding driving variables are greater (less) than the discontinuity are denoted the experimental (control) group. If other control variables that affect the outcome variable are continuous at the discontinuity, the distribution of the experimental individuals near the discontinuity can be regarded as random, which means that the probability of the individuals accepting the treatment is random. Thus, the average difference between the experimental group and the control group near the discontinuity reflects the size of the disposal effect.

4 Summary of Contents

This book is a study of logic-driven traffic big data analytics and is divided into 12 chapters. The main contents and chapter arrangement are shown in Fig. 11. This chapter describes the basis of research in this field and introduces the theory and

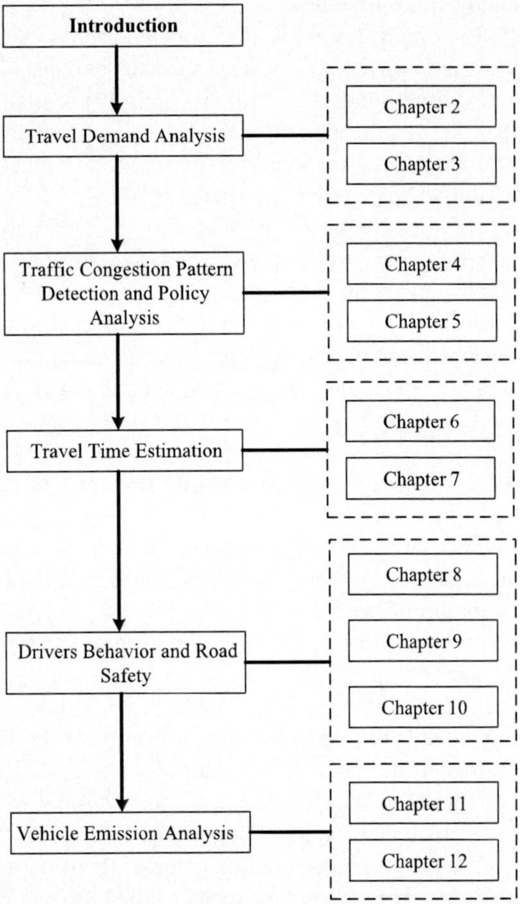

Fig. 11 Main task of the book

analysis of models of land use and transportation integration. Chapters 2 and 3 cover travel demand analysis, which examines the spatiotemporal changes in urban travel demand. In real life, travel demand is often in a prolonged state of spatiotemporal imbalance, which is the main reason for increased traffic congestion. Chapters 4 and 5 describe methods for the detection of traffic congestion patterns and related policy analysis. Traffic congestion reduces travel time reliability and thus makes it difficult for travelers to accurately estimate travel time. Addressing this problem, Chapters 6 and 7 examine travel time estimation, as such estimates directly affect the behavior of travelers during trips (e.g., their travel frequency and travel mode) and further affect road traffic safety. Drivers' behavior and road safety is covered in Chaps. 8, 9, and 10. Finally, Chaps. 11 and 12 incorporate aspects from previous chapters to investigate the long-term environmental effects of a transportation system on a city and society.

In the following paragraphs, the main contents of each chapter are briefly introduced.

Chapter 2 discusses a spatiotemporal distribution model for travelers' origin–destination (OD) demand based on multisource data. Such models are crucial to understanding the spatiotemporal distribution state of travel demand, which enables the formulation of practical transportation planning policies. Data from the center of Chengdu, China, are taken as the research target in this chapter, and multisource data and topic models are used to identify the functional attributes of each area in the sampled region. Then, the *k*-means clustering algorithm is used to divide Chengdu into functional zones. Finally, a SAR model is applied to examine the relationship between urban built environment attributes and the spatiotemporal distribution of travel OD demand and to analyze the spatiotemporal distribution pattern of travelers' OD demand.

Chapter 3 is a case study of the spatiotemporal evolution of ridesourcing markets in Shanghai under the new restriction policy. Urban built environment attributes are closely related to the spatiotemporal distribution of urban residents' travel, which in turn affects the spatiotemporal distribution of the demand for online ride-hailing services. It is therefore vitally important for online hailing systems to analyze and identify the main factors affecting demand in the ridesourcing market to function in the constantly changing market. To this end, this chapter applies a two-level growth model to examine the effect of multilevel factors, such as land use, accessibility, and weather conditions, on the demand for online ride-hailing and the market share of Didi Express and Didi Taxi services. The results will assist urban planners and transportation operators to formulate policies to improve transportation systems.

Chapter 4 details the effect of exclusive bus lanes on the average speed of buses and cars in a city in China. The creation of such lanes is one method to reduce traffic congestion, which causes enormous economic losses in cities. Traffic congestion is caused by factors such as imbalanced land distribution, poor transportation structure, and insufficient transportation facilities, in addition to spatiotemporally imbalanced travel demand (as discussed in Chaps. 2 and 3). Public transportation can alleviate urban traffic congestion, as it is a large-volume, high-speed, high-efficiency, and low-cost travel mode that maximizes the efficiency of transportation by establishing and maintaining a balance between supply and demand. A transit priority policy has therefore been deemed an essential part of transportation development strategies in many countries. Thus, based on empirical data, this chapter proposes and develops a method for quantitatively evaluating the effect of exclusive bus lanes on buses and other multi-person vehicles.

In addition to congestion mitigation strategies such as the transit priority policy mentioned in Chap. 4, it is crucial to identify the spatiotemporal distribution of traffic congestion, because such identification enables the factors causing and influencing traffic congestion to be determined and optimized to alleviate traffic congestion. Therefore, Chap. 5 analyzes spatiotemporal congestion patterns on urban roads based on geographical positioning system (GPS) data from taxis. The FCM clustering algorithm is used to perform a clustering analysis of taxi GPS data from Shanghai, which in turn enables a probability based analysis of urban transportation patterns. Then, taking probabilistic classification as the dependent variable, a mixed SARMA model is used to evaluate the influence of environmental factors on congestion patterns.

Chapter 6 examines travel time estimation based on urban built environment attribute data and low-frequency floating vehicle data, as urban transportation systems are subject to real-time dynamic changes. For example, changes in a traffic flow state may cause dramatic fluctuations in urban transportation systems, such that travel time cannot be accurately estimated. Although some studies have estimated travel time based on traffic flow theory or data-driven methods, the effect of land use on transportation has not been considered. Therefore, this chapter develops a novel method that uses floating car data, which contains only location information, for road segment travel time estimation based on probability statistics, considering surrounding built environment attributes. This also enables the distribution of road segment travel time to be obtained by replacing the distribution of road section length with the distribution of vehicle number as the proportional coefficient of travel time distribution.

Chapter 7 explores the spatially heterogeneous effects of urban built environments on road travel time variability. Road segment travel time is an important index of traffic flow state that can be used to evaluate the traffic conditions on road segments and thus reflect real-time traffic congestion. Accurate road segment travel time estimation is thus a key part of transportation planning, transportation control, and transportation management. To this end, this chapter applies a global regression model and a GWR model to quantitatively analyze the effect of the urban built environment on road travel time, based on a comprehensive set of taxi GPS data, land use data, and road network data (e.g., speed).

Chapter 8 focuses on the who, when, where, and how of taxi-driver speeding, as a way of analyzing road safety, in the form of a comparative study of Shanghai and New York. Speeding is one form of road users' travel behavior, and a comprehensive analysis of such behavior is needed for the correct prediction of users' travel demand and to guide users to travel safely and efficiently. This also enables the formulation of practical transportation management policies and ensures the smooth development of urban transportation planning. Speeding is the leading cause of traffic accidents, which result in significant injury, loss of life, and property damage. Therefore, in this chapter, floating taxi data from Shanghai and New York is analyzed using the driver–road–environment identification model to determine the relationship between the characteristics of driving behavior, road attributes, environmental factors, and the speeding behavior of taxi drivers. This analysis reveals what transportation control measures should be taken to decrease speeding by drivers of taxis and other vehicles.

Chapter 9 determines the effects of congestion on drivers' speed choice by assessing the mediating role of state aggressiveness from floating taxi data, as traffic congestion is the main cause of negative emotions (e.g., tension, disgust, and anger). Travel delays caused by traffic congestion create time pressure tension in drivers given the uncertain travel time, which worsens the travel experience and triggers aggressive driving behavior, leading to a decrease in traffic safety. Therefore, based on the method discussed in Chap. 8, this chapter examines the influence of traffic congestion on drivers' choice of travel speed in terms of drivers' aggressiveness state and analyzes the mediating effect of driver type, value of time, and working hours on aggressiveness state.

Chapter 10 comprises an analysis of the spatiotemporal distribution of traffic incidents based on urban built environment attributes and microblog data. Traffic accidents severely threaten the life and property of road travelers. To reduce traffic accidents, their distribution law must be understood; based on this information, effective travel safety measures can be developed. The discussion and findings in Chaps. 2–9 confirm that the urban built environment closely interacts with transportation systems, and thus the urban built environment affects the spatiotemporal distribution of traffic accidents by affecting the travel behavior of travelers and the distribution of urban travel demand. Accordingly, Chap. 10 uses a GWR model and microblog data with continuous spatiotemporal distribution characteristics to investigate the relationship between the attributes of the built environment and the spatiotemporal distribution of traffic accidents and to identify the main factors affecting this distribution.

In Chap. 11, taxi-hailing choice behavior and the economic benefits of emission reduction are analyzed based on multi-mode travel big data, as the transportation sector is the primary contributor to greenhouse gas emissions worldwide. Effective means to reduce traffic emissions include technological innovations that enhance the energy efficiency of vehicles, economic methods that increase travel costs, and practical methods that encourage and develop urban residents' use of sustainable travel options. This chapter, therefore, examines the relationship between population, land use attributes, transportation attributes, and users' travel mode choices, based on taxi, metro, and shared bicycle data. Then, based on the main factors that influence travel mode choices, various transportation-related energy-saving and emission-reduction measures are proposed. The economic cost–benefits of each measure are also analyzed from an environmental perspective, which affords information to guide transportation planning authorities in their development of congestion-mitigation, energy-saving, and emission-reduction policies.

Chapter 12 is an analysis of spatiotemporal traffic line source emissions based on massive DiDi online car-hailing service data, as such services are key to reducing severe traffic pollution problems associated with increased vehicle ownership, increased travel demand, and the aggravation of traffic congestion. Traffic pollution is deleterious to the health of urban residents and is environmentally unsustainable. A comprehensive analysis of the spatiotemporal distribution characteristics of urban traffic emissions will facilitate sustainable urban development. To this end, this chapter takes the urban road network in Shanghai as its research object, and calculates urban traffic emissions from floating vehicle data, analyzes these emissions' distribution characteristics, and explores how emissions are affected by the built environment and land use. This analysis lays a foundation for sustainable urban transportation planning and development.

References

Anderson, J. R. (1976). *A land use and land cover classification system for use with remote sensor data* (Vol. 964). US Government Printing Office.

Anselin, L. (2010). Thirty years of spatial econometrics. *Papers in Regional Science, 89*(1), 3–25.

Anselin, L. (2013). *Spatial econometrics: Methods and models* (Vol. 4). Springer Science & Business Media.

Bae, C. H. C., & Richardson, H. W. (2017). *Urban sprawl in western Europe and the United States.* Routledge.

Bertrand, M., Duflo, E., & Mullainathan, S. (2004). How much should we trust differences-in-differences estimates? *Quarterly Journal of Economics, 119*(1), 249–275.

Bezdek, J. C., Ehrlich, R., & Full, W. (1984). FCM: The fuzzy c-means clustering algorithm. *Computers & Geosciences, 10*(2–3), 191–203.

Bivand, R. S., & Wong, D. W. (2018). Comparing implementations of global and local indicators of spatial association. *TEST, 27*(3), 716–748.

Chen, J., & Li, H. (2013). Variable selection for sparse Dirichlet-multinomial regression with an application to microbiome data analysis. *The Annals of Applied Statistics, 7*(1).

Chen, Y., Ebenstein, A., Greenstone, M., & Li, H. (2013). Evidence on the impact of sustained exposure to air pollution on life expectancy from China's Huai River policy. *Proceedings of the National Academy of Sciences, 110*(32), 12936–12941.

China, P. R. (2018). Code for classification of urban and rural land use and planning standards of development land. Ministry of Housing and Urban-Rural Development.

Colman, M. S. M. S. C. (2019). See how NYC's urban density stacks up against other major cities. 6sqft. https://www.6sqft.com/see-how-nycs-urban-density-stacks-up-against-other-major-cities/

Dray, S., Legendre, P., & Peres-Neto, P. R. (2006). Spatial modelling: A comprehensive framework for principal coordinate analysis of neighbour matrices (PCNM). *Ecological Modelling, 196*(3–4), 483–493.

Ewing, R., & Cervero, R. (2010). Travel and the built environment: A meta-analysis. *Journal of the American Planning Association, 76*(3), 265–294.

Fonseca, F., Ribeiro, P. J., Conticelli, E., Jabbari, M., Papageorgiou, G., Tondelli, S., & Ramos, R. A. (2021). Built environment attributes and their influence on walkability. *International Journal of Sustainable Transportation,* 1–40.

Fornell, C., & Larcker, D. F. (1981). Evaluating structural equation models with unobservable variables and measurement error. *Journal of Marketing Research, 18*(1), 39–50.

Fotheringham, A. S., Brunsdon, C., & Charlton, M. (2003). *Geographically weighted regression: The analysis of spatially varying relationships.* Wiley.

Fotheringham, A. S., Charlton, M. E., & Brunsdon, C. (1998). Geographically weighted regression: A natural evolution of the expansion method for spatial data analysis. *Environment and Planning A, 30*(11), 1905–1927.

Giuliano, G. (2004). Land use impacts of transportation investments. *The Geography of Urban Transportation, 3*, 237–273.

Habibi, S., & Asadi, N. (2011). Causes, results and methods of controlling urban sprawl. *Procedia Engineering, 21*, 133–141.

Harrell, F. E., Jr., Lee, K. L., & Mark, D. B. (1996). Multivariable prognostic models: Issues in developing models, evaluating assumptions and adequacy, and measuring and reducing errors. *Statistics in Medicine, 15*(4), 361–387.

Hartigan, J. A., & Wong, M. A. (1979). Algorithm AS 136: A k-means clustering algorithm. *Journal of the Royal Statistical Society Series C-Applied Statistics, 28*(1), 100–108.

Hinton, G., Deng, L., Yu, D., Dahl, G. E., Mohamed, A. R., Jaitly, N., et al. (2012). Deep neural networks for acoustic modeling in speech recognition: The shared views of four research groups. *IEEE Signal Processing Magazine, 29*(6), 82–97.

Kamel, M. S., & Selim, S. Z. (1994). New algorithms for solving the fuzzy clustering problem. *Pattern Recognition, 27*(3), 421–428.

Knowles, R. D. (2012). Transit oriented development in Copenhagen, Denmark: From the finger plan to Ørestad. *Journal of Transport Geography, 22*, 251–261.

Metro of Stockholm. (2011). U-Bahn von Stockholm. Mapa-Metro.Com. https://mapa-metro.com/de/Schweden/Stockholm/Stockholm-Tunnelbana-Karte.htm. August 27, 2021.

Ord, K. (1975). Estimation methods for models of spatial interaction. *Journal of the American Statistical Association, 70*(349), 120–126.

Papa, E., & Bertolini, L. (2015). Accessibility and transit-oriented development in European metropolitan areas. *Journal of Transport Geography, 47*, 70–83.

Paulsson, A. (2020). The city that the metro system built: Urban transformations and modalities of integrated planning in Stockholm. *Urban Studies, 57*(14), 2936–2955.

Rodrigue, J. P., Comtois, C., & Slack, B. (2016). *The geography of transport systems*. Routledge.

Vickerman, R., Spiekermann, K., & Wegener, M. (1999). Accessibility and economic development in Europe. *Regional Studies, 33*(1), 1–15.

Wang, G. M., Li, Y. Q., & Xu, M. (2019). Integrating the management and design of urban road network to alleviate tide traffic. In *2019 IEEE Intelligent Transportation Systems Conference (ITSC)* (pp. 708–713). IEEE.

Wang, J. F., Zhang, T. L., & Fu, B. J. (2016). A measure of spatial stratified heterogeneity. *Ecological Indicators, 67*, 250–256.

Wegener, M. (1996). Reduction of CO_2 emissions of transport by reorganisation of urban activities. In *Transport, land-use and the environment* (pp. 103–124). Springer.

Wegener, M. (2004). *Overview of land use transport models*. Emerald Group Publishing Limited.

Xie, X. L., & Beni, G. (1991). A validity measure for fuzzy clustering. *IEEE Transactions on Pattern Analysis and Machine Intelligence, 13*(8), 841–847.

Zhong, S., & Bushell, M. (2017a). Built environment and potential job accessibility effects of road pricing: A spatial econometric perspective. *Journal of Transport Geography, 60*, 98–109.

Zhong, S., & Bushell, M. (2017b). Impact of the built environment on the vehicle emission effects of road pricing policies: A simulation case study. *Transportation Research Part A: Policy and Practice, 103*, 235–249.

Zhong, S., Cheng, R., Li, X., Wang, Z., & Jiang, Y. (2020). Identifying the combined effect of shared autonomous vehicles and congestion pricing on regional job accessibility. *Journal of Transport and Land Use, 13*(1), 273–297.

Zhong, S., Gong, Y., Zhou, Z., Cheng, R., & Xiao, F. (2021a). Active learning for multi-objective optimal road congestion pricing considering negative land use effect. *Transportation Research Part C: Emerging Technologies, 125*, 103002.

Zhong, S., Wang, Z., Wang, Q., Liu, A., & Cui, J. (2021b). Exploring the spatially heterogeneous effects of urban built environment on road travel time variability. *Journal of Transportation Engineering, Part A: Systems, 147*(1), 04020142.

Zhong, S., Jiang, Y., & Nielsen, O. A. (2022). Lexicographic multi-objective road pricing optimization considering land use and transportation effects. *European Journal of Operational Research, 298*(2), 496–509.

Zhong, S., Wang, S., Jiang, Y., Yu, B., & Zhang, W. (2015). Distinguishing the land use effects of road pricing based on the urban form attributes. *Transportation Research Part A: Policy and Practice, 74*, 44–58.

Chapter 2
A Spatio-temporal Distribution Model for Determining Origin–Destination Demand from Multisource Data

Abstract A scientific understanding of the spatio-temporal distribution of road travel demand is a prerequisite for formulating effective countermeasures to traffic congestion. Accordingly, this chapter analyzes the relationship between urban built environment attributes and origin–destination (OD) demand in the specific spatial structure of a city, thereby guiding decision-makers on how to solve traffic congestion problems. Multisource data and a Dirichlet multinomial regression model are used to reveal the functional zones and spatial structure of a city. A spatial autoregressive model is then applied to reveal the relationship between urban built environment attributes and the spatio-temporal distribution of OD demand. Finally, data from the downtown area of Chengdu (China) are used to validate the model and method and analyze their performance.

Keywords Spatio-temporal distribution of OD demand · Built environment · Dirichlet multinomial regression model · Spatial autoregressive model

1 Introduction

The relationship between built environment attributes and origin–destination (OD) demand can be analyzed in specific spatial structures. Urban built environment is a product of urban spatial change and profoundly affects residents' travel behavior, such as the distance, frequency, route, and mode choice of their travel. Thus, urban built environment attributes ultimately affect the spatio-temporal distribution of residents' OD demand. The quantification of the spatio-temporal patterns of OD demand and the determination of the dependence of OD demand on the built environment attributes is therefore fundamental to understanding and predicting the dynamics of OD demand. The interactions between different regions' OD demand can cause the geographical concentration of spatio-temporal distributions of OD demand between different regions. Consequently, determining the effect of urban built environment on OD demand is a spatial correlation problem that is inseparable from the spatial structure of a city. However, in conventional OD spatio-temporal distribution analysis, the intrinsic influence of urban built environment is not sufficiently considered.

S. Zhong and D. Sun, *Logic-Driven Traffic Big Data Analytics*,
https://doi.org/10.1007/978-981-16-8016-8_2

Furthermore, studies have rarely considered the effects of a city's functional zones, despite these being the most natural spatial manifestation of urban layouts.

Different urban functional zones provide different functions that enable various social and economic activities. These functional zones are artificially designed by urban planners or are naturally generated by residents' lifestyles and reflect the spatial structure of a city. The advent of multisource data and machine learning has made it possible to determine and extract urban spatial structures; thus, mining a large body of rich data can reveal the potential and real spatial structure of a city, which was previously a near-impossible task.

This chapter has two objectives. The first objective is to extract urban functional zones using a Dirichlet multinomial regression (DMR) model based on multisource data. The second objective is to establish a spatial econometric model of urban built environment attributes and OD demand for a specific spatial structure.

The main contributions of this chapter are as follows:

1. A DMR model (a type of topic model) based on multisource data is adopted to divide urban spatial structure.
2. The spatio-temporal distribution of OD demand is analyzed, considering the effects of the built environment attributes.
3. A hybrid mechanism is developed, based on the DMR model and a spatial econometric method, for analysis of the spatio-temporal distribution of OD demand.

2 Literature Review

2.1 *Research on the Application of Topic Models in Transportation*

Current research on topic models focuses on two applications: (1) the detection or prediction of travel behavior and (2) the extraction of data for traffic analysis or policy research from a large amount of textual information. Examples of the former include work by Zhang et al. (2018), who used topic models based on the latent Dirichlet allocation (LDA) model to detect traffic accidents in social media data; by Hasan and Ukkusuri (2014), who classified urban activity patterns using the LDA model; and Markou et al. (2019), who applied the LDA model to predict taxi demand hotspots. Examples of the latter include the use of the LDA model to extract textual information related to traffic accidents from the Internet for subsequent research (e.g., Pereira et al., 2013, 2015). More specifically, Sun and Kirtonia (2020) devised a regional theme model based on the LDA model to determine the theme characteristics of a transportation research area; Das (2021) used the LDA model to understand pedestrians' and cyclists' perceptions of autonomous vehicle safety; and Qi et al. (2019) used the modified LDA model and the modified hierarchical LDA model to identify different driving styles.

2.2 Research on the Built Environment

Attributes of the built environment, such as density, land use diversity, street design, destination accessibility, and distance from public transportation, are closely related to travel behavior (Cervero & Kockelman, 1997; Ewing & Cervero, 2001; Zhong & Bushell, 2017b; Zhong et al., 2015). Bhat and Guo (2007) found that people are less likely to choose to travel by private car in high-density areas with land use diversity, destination and public accessibility, and better street design. Zhong et al. (2015) and Zhong and Bushell (2017a) have confirmed that built environment attributes have a significant influence on the land use effects of road tolls. Maat and Timmermans (2009) found that the built environment attributes of residential areas and workplaces have a strong influence on whether commuters drive to and from work. Zhao (2014) analyzed randomly sampled data from Beijing and reported that rapid changes in urban built environment have contributed to the gradual decline in the rate of bicycle use in China. Chen et al. (2018) explored the relationship between built environment attributes and bicycle theft and found that theft incidents increase in dense areas with more bicycles. Sabouri et al. (2020) analyzed data from 24 different regions in the United States to determine the effect of built environment attributes on demand for the ride-hailing service Uber and revealed that the demand is negatively correlated with intersection density and destination accessibility. Wang and Zhang (2021) identified that shared autonomous vehicles are more efficient and generate fewer vehicle miles traveled in denser cities with more connected networks and diverse land use development patterns than in less dense cities. Finally, Chan et al. (2021) showed that built environment attributes affect walking behavior.

Through the above literature review, we can draw the following conclusions,

1. There has been little research on the effect of urban built environment on the spatio-temporal distribution of OD demand; and
2. Few studies have considered the effect of spatial structure on OD demand.

3 Discovery of Activities in a Region

3.1 Map Segmentation

A road network naturally segments a city into areas. In this chapter, each of these areas is regarded as the basic unit within which the functions of a city are performed, and thus these areas together form an urban functional zone. Within each area are points of interest (POI), where people are provided with basic facilities for social and economic activities. People are therefore attracted by POIs to perform such activities (e.g., recreation or work) in or between various regions, which results in their taking many trips. Conversely, people's travel shapes the functional attributes of these areas, such that these areas fully meet people's needs.

Chengdu, a sub-provincial city and the capital of Sichuan province, China, is used here as a case study. A series of morphological operations (Figs. 1, 2, 3 and 4) is applied to divide the road network of Chengdu (comprising freeways, national highways, provincial highways, railways, and other roads) into 329 areas, which serve as the spatial structures in the subsequent analysis.

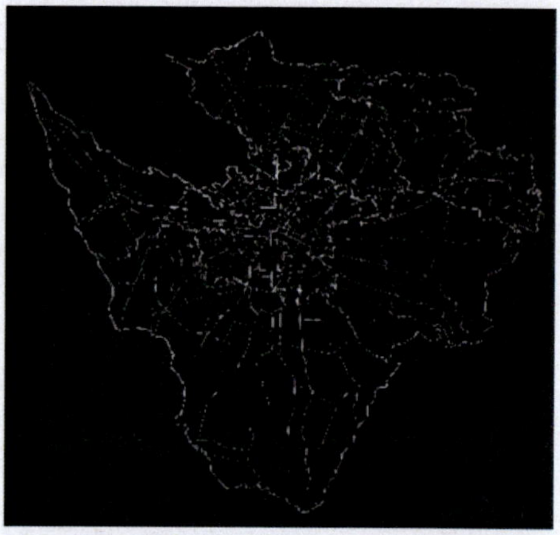

Fig. 1 Binarization

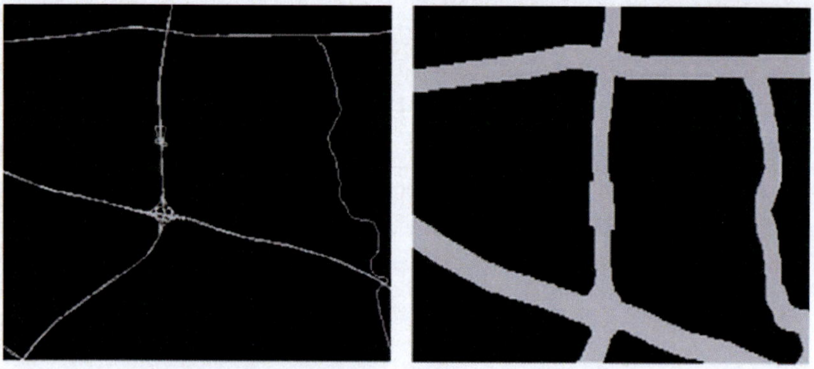

(a) (b)

Fig. 2 **a** Before dilation, **b** after dilation

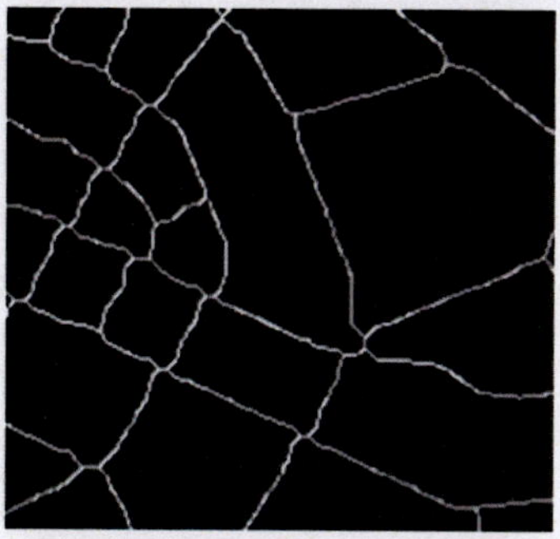

Fig. 3 Thinning

Fig. 4 Connected component labeling

3.2 *Topic Model*

3.2.1 Preliminary

The LDA model is a topic model proposed by Blei et al. (2003) for use in text mining. It considers each document in a corpus as a mixture of multiple topics and that each word in a document is derived from a certain topic. When all the words of a corpus have been given, the distribution of topics in each document and the distribution of words corresponding to each topic can be obtained by mathematical derivation. Therefore, the LDA model can extract all of the latent semantic information of a sentence or a word (Wei & Croft, 2006).

However, the LDA model cannot determine document-level semantic information. Therefore, improved topic models based on the LDA model have been proposed, such as DMR models, which incorporate a collaborative feature vector that contains feature information unique to each document. This vector is used to better identify the semantic information of each document, thereby affording more accurate classification results. Moreover, DMR models can fuse multisource information, such as the metadata of a document, via the collaborative feature vector. This means that in the

Fig. 5 Generation process of the DMR model

(1) For each topic *t*,

1) $\lambda_t : \mathrm{N}(0, \sigma^2 I)$
2) $\beta_t : \mathrm{Dir}(\eta)$

(2) For each document *d*,

1) For each topic t of document d, $\alpha_{dt} = \exp(x_d^T \lambda_t)$
2) $\theta_d : \mathrm{Dir}(\alpha_d)$

(3) For the nth word in document *d*,

1) $Z_{dn} : \mathrm{multi}(\theta_d)$
2) $W_{dn} : \mathrm{multi}(\beta_{Z_{dn}})$

context of functional zone division, a DMR model combines information from environmental and location semantics to determine the functional attribute of a region. In this case study, environmental semantics are expressed in the built environment attributes, whereas location semantics are expressed in the road network, which suits the needs of our research.

The generation process of the DMR model is shown in Fig. 5 as follows.

For each topic t, λ_t is a vector with the same length as the collaborative feature vector; N(·) is the Gaussian distribution with σ^2 as the variance of the prior parameter values and I is the standardized matrix; θ_d and β_t denote the document-topic distribution and topic-word distribution respectively, while η and α are their respective hyperparameters, and Dir(·) is the Dirichlet distribution. In the DMR model, each document d corresponds to a hyperparameter vector α_d. Different from the LDA model, the DMR model introduces x_d as the feature vector that encodes metadata values and supports any feature attributes, such as author, mailbox, and other information (Mimno & McCallum, 2008). Here, multi(·) is the multinomial distribution; Z_{dn} is the topic number of the nth word in document d, we can get the probability distribution of Z_{dn} from document-topic distribution θ_d; W_{dn} is the probability distribution of the nth word in document d with the Z_{dn}^{th} topic. The central problem of the topic model is to estimate θ_d and β_t, which respectively denote the document-topic distribution and topic-word distribution. In general, this problem can be accomplished by Gibbs sampling and variational inference (Sun & Yin, 2017).

3.2.2 Collaborative Feature Vector

For each area $r = 1, 2, \ldots, R$ (where R is the total number of areas), we compute the collaborative built environment attribute feature vector, which represents the distribution characteristics of the built environment attributes in area r. As shown in Table 1, the variables of the built environment attributes comprise variables for traffic-related built environment attributes (for five types of road attributes) and for

Table 1 Built environment attribute variables

	Attributes		Formula
Traffic built environment attributes	Freeway		$V_{ri} = \frac{length_{ri}}{LENGTH_r} * \log \frac{R}{NUM_i}$
	National highway		
	Provincial highway		
	Railway		
	Other roads		
Other built environment attributes	Edifice	Auto service	$B_{rj} = \frac{p_{rj}}{AREA_r}$
	Expressway service	Toll station	
	Police	Bus passenger station	
	Shopping mall	Parking lot	
	Railway station	Entertainment	
	Airport	Medical service	
	Gas station	Government agency	
	Finance service	Accommodation	
	Science education	Others	
	Tourism	Bus stop	

Note V_{ri} is the TF $*$ IDF value of the ith road category of area r; B_{rj} is the jth built environment variable value of area r; $length_{ri}$ represents the length of the ith road category of area r; $LENGTH_r$ denotes the total length of all the roads in area r; R is the total number of areas; NUM_i represents the number of areas with the ith road category in R areas; p_{rj} represents the number of jth built environment attributes in area r; $AREA_r$ is the acreage of area r

other built environment attributes (for 20 types of built environment facilities). The former variables are calculated using the term frequency-inverse document frequency (TF-IDF) weighting technique, which is commonly used to evaluate the importance of a word in a document of a corpus. TF-IDF assumes that the importance of a word increases (decreases) in direct (inverse) proportion to its frequency of use in a document. Therefore, the more frequently a word appears in a certain document and the less it appears in other documents, the more representative it is of that specific document.

In this study, the original data is 329 $\times$ 25 dimensions. After factor analysis, 10 factors are obtained, and x_i is a row vector of 1 $\times$ 10. Let $x_r = (F_{r1}, F_{r2}, ..., F_{r\xi}, ..., F_{r10,})$, where $F_{r\xi}$ is the ξth built environment factor value of area r.

In this study, the problem of identifying the functions in an area can be analogized to the problem of discovering the latent topics of a document. Table 2 depicts the analogy relations.

Table 2 Analogy from region-activities to document-topics

Area-functional zones		Document-topic
Area r	→	Document d
Functional zones k	→	Topic t
POI i, j	→	Word n
Collaborative POI feature vector x_r	→	Metadata of a document x_d

3.2.3 Application of the DMR Model

In the DMR model used in the case study, α is specified by the cooperative feature vector x_r with η set to 0.01, λ_t obeying a normal distribution, and variance σ^2 set to 0.5. The total number of topics T is set to 10 and the model is trained for 1000 iterations, with α optimized and the model saved after every 50 iterations. The parameter settings used in this case study are those proposed by Mimno and McCallum (2008).

3.2.4 Identification of the Functional Zones

The DMR model is used to determine the function distribution probability of different areas. The topic distribution for area r is a K-dimensional vector $\theta_r = (\theta_{r,1}, \theta_{r,2}, \ldots, \theta_{r,K})$, where $\theta_{r,k}$ is the probability that area r has a function k. Thus, k-means clustering is used to aggregate 329 regions into K clusters, each of which is denoted a functional zone. Next, multiple cross-validations are performed for different K values, and the validation with the maximum average Silhouette coefficient is selected. This yields nine functional zones, as shown in Fig. 6, which are annotated with respect to the following three aspects.

1. POI configuration in functional zones. We rank POI categories in a functional zone (denoted internal ranking, IR) and in all functional zones for each POI category (denoted external ranking, ER) according to the density value of each POI category in each functional zone. These rankings are detailed in Tables 3 and 4.
2. Function intensity. We feed OD demand points into the kernel density estimation model to define function intensity. For m points, $g_1, g_2, \ldots, g_m$, located in a two-dimensional space, the function intensity at area r of any function zone is defined as:

$$\Lambda_{(r)} = \frac{1}{mb}\sum_{\varphi=1}^{m} K\left(\frac{d_{\varphi,r}}{b}\right) \tag{1}$$

where $d_{\varphi,r}$ is the distance from the φth travel OD point g_φ to area r, b is bandwidth, and $K(\cdot)$ is kernel function that decays as $d_{\varphi,r}$ increases. In our study, we

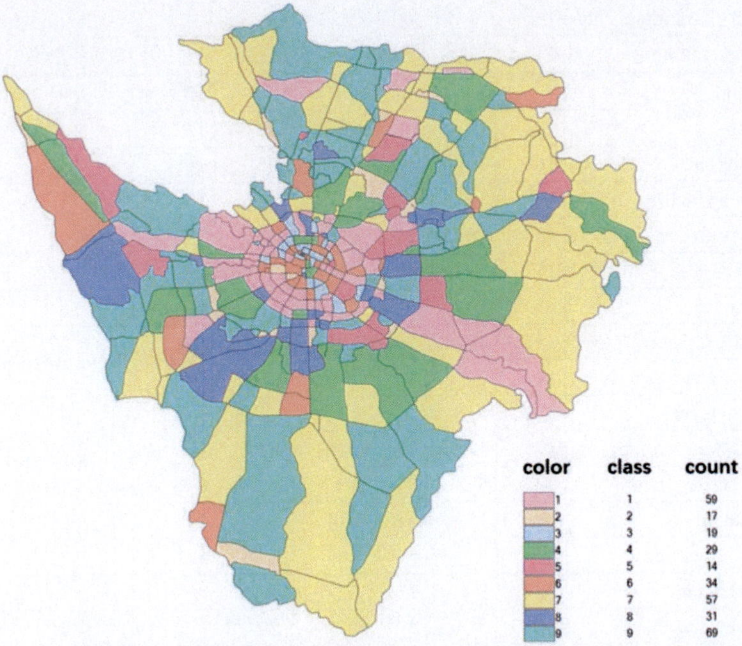

Fig. 6 Functional zones

use the Gaussian function as the kernel function. Figure 7 shows the function intensity distribution of functional zone Z9, which is the most widely covered cluster.

3. The hourly arrival/departure distribution of functional zones. This is determined for all nine functional zones on weekdays and weekends, as depicted in Fig. 8.

This annotation procedure reveals that functional zones Z1, Z2, Z3, and Z6 have complete functional coverage, relatively complete infrastructure, and a high distribution intensity of OD demand, which shows that the development of these four regions is relatively mature. In contrast, the annotation reveals that zones Z4, Z5, Z8, and Z9 are areas requiring development. However, the IR and ER of each POI in zone Z7 are difficult to determine using the existing data. These annotation results are listed in Table 5.

4 Spatio-temporal Distribution Mode of OD Demand

Our second objective is to establish a spatial econometric model of urban built environment attributes and OD demand. This is necessary as the built environment attributes vary between regions and they are spatially correlated. Therefore, it can be

Table 3 Ranking results of IR

Internal ranking	Z1	Z2	Z3	Z4	Z5	Z6	Z7	Z8	Z9
Bus stop	7	14	14	4	7	12	1	9	4
Tourism	14	12	16	13	14	13	8	13	15
Toll station	16	13	17	15	15	16	15	15	14
Shopping mall	1	1	1	1	1	1	2	1	1
Railway station	18	16	17	19	18	18	16	18	16
Police	15	17	12	16	17	15	17	16	17
Parking lot	8	6	6	10	9	6	11	8	11
Others	11	10	10	11	11	10	13	12	10
Bus passenger station	17	19	15	16	19	16	17	16	18
Medical service	3	3	4	3	4	3	4	4	2
Expressway service	19	17	17	16	15	19	19	20	19
Government agency	9	9	8	9	3	8	9	11	9
Gas station	13	15	13	14	13	14	12	14	12
Financial service	5	7	5	6	8	5	7	5	7
Entertainment	2	2	2	2	2	2	3	2	3
Science education	6	5	3	7	5	4	6	7	6
Auto services	4	8	11	5	6	9	5	6	5
Edifice	12	11	9	12	12	11	14	10	13
Airport	19	19	17	20	19	19	19	19	20
Accommodation	10	4	7	8	10	7	10	3	8

assumed that the built environment attributes have a significant spatial effect on OD demand.

4.1 Extraction of the Basic Factors of Urban Built Environment Attributes

Given the purpose of this study and the availability of data, the POI densities and road densities in each functional zone are selected as the initial built environment variables. In addition, the classification result of the DMR model is combined with eight 0–1 variables, namely $\Omega_1, \Omega_2, \ldots, \Omega_8$. For example, if area r is within functional zone Z1, the 0–1 variable of area r is $(1, 0, 0, 0, 0, 0, 0, 0)$; similarly, if area r is within functional area Z9, the 0–1 variable of area r is $(0, 0, 0, 0, 0, 0, 0, 0, 1)$. The 0–1 variables represent the spatial structure in the model, which can further reveal the influence of urban environment attributes on the spatio-temporal distribution of OD demand. Overall, we measure the urban built environment attributes using 33 raw

Table 4 Ranking results of ER

External ranking	Z1	Z2	Z3	Z4	Z5	Z6	Z7	Z8	Z9
Bus stop	1	9	8	3	7	6	2	5	4
Tourism	5	1	8	4	7	3	6	2	9
Toll station	6	1	9	5	4	7	8	3	2
Shopping mall	3	6	1	4	8	2	9	5	7
Railway station	7	1	9	8	6	4	5	2	3
Police	3	7	1	8	5	2	9	4	6
Parking lot	3	5	1	6	7	2	9	4	8
Others	3	7	1	5	8	2	9	4	6
Bus passenger station	5	8	1	6	8	3	7	2	4
Medical service	3	5	1	7	8	2	9	6	4
Expressway service	6	2	6	4	1	6	6	5	3
Government agency	3	7	1	6	4	2	9	5	8
Gas station	2	6	1	8	4	7	9	5	3
Financial service	3	6	1	5	8	2	9	4	7
Entertainment	2	7	1	6	5	3	9	4	8
Science education	3	5	1	7	8	2	9	4	6
Auto services	2	7	1	6	8	3	9	4	5
Edifice	4	6	1	5	7	2	9	3	8
Airport	2	2	2	2	2	2	2	1	2
Accommodation	4	5	1	6	8	2	9	3	7

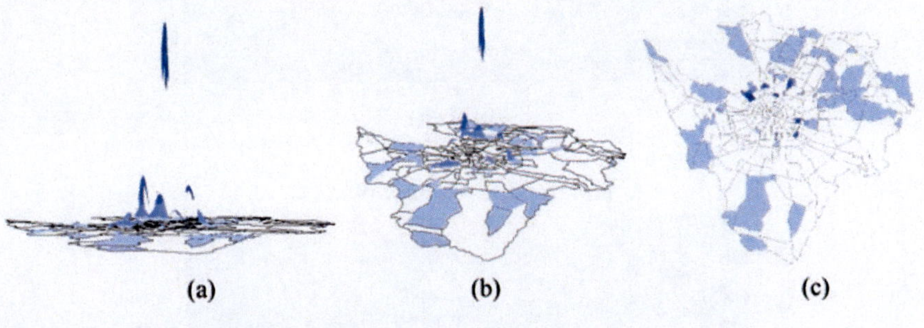

Fig. 7 Function intensity distribution of functional zone Z9

variables that are composed of 25 built environment attribute variables (listed in Table 1) and eight 0–1 variables.

Understandably, there may be correlations between some raw variables. Thus, to aggregate data, reduce dimensionality, and ensure that variables are independent of each other, using factor analysis, we extract key and effective built environment factors from the (normalized) raw variables, subject the extracted factors to analysis,

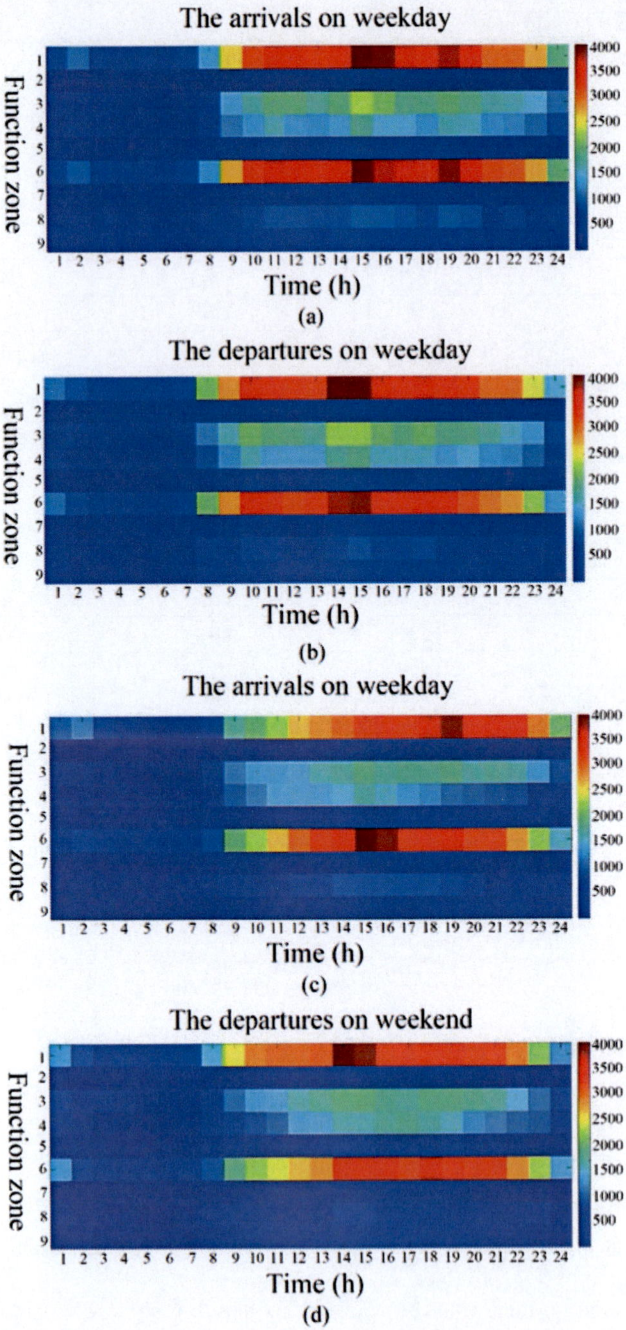

Fig. 8 Hourly distribution of arrival/departure amount of functional zones

Table 5 Annotation results of functional zones identification

Number	Function	Number	Function
Z1	Educational research area, financial and insurance area	Z5 Z8 Z9	Transportation hub area
Z2	Historical and cultural area	Z6	Entertainment area
Z3	Entertainment area, Administrative office area	Z7	Unidentified area
Z4	Rural resort area		

and then linearly combine all factors that are correlated. This affords 14 factors (Table 6), which serve as the basic input variables of the spatial econometric models. Combining the 0–1 explanatory variables and data information, the meanings of the factors can be interpreted as follows:

Table 6 Factors analysis results

Number	Built environment semantic information	Component
F1	Basic service infrastructure	Bus station, shopping, medical service, government, finance, entertainment, education, building, accommodation, provincial highway, other roads, other POIs
F2	Car service infrastructure	Gas station, car service, Ω_1
F3	Highway distribution	Freeway, toll station, Ω_2
F4	Municipal office infrastructure	Police, Ω_3
F5	Agricultural and sideline infrastructure	Ω_4
F6	Highway service infrastructure	Highway service, Ω_5
F7	Entertainment infrastructure	Ω_6
F8	National highway distribution	National highway, Ω_7
F9	Ω_8	Ω_8
F10	Railway	Railway
F11	Passenger transportation station	Passenger transportation station
F12	Travel	Travel
F13	Railway station	Railway station
F14	Airport	Airport

Table 7 Result of diagnostic test

Spatial error	Moran test	2.7762	[0.005]
	LM-error test	5.5165	[0.019]
	Robust LM-error test	7.5764	[0.006]
Spatial autocorrelation	LM-lag test	23.8002	[0.000]
	Robust LM-lag test	25.8601	[0.000]

Note p-value is in the square brackets

4.2 Model Selection

GeoDa is designed to introduce non-experts to spatial data analysis, developed under the auspices of the National Science Foundation funded Center for Spatially Integrated Social Science (Anselin et al., 2010). In this research, we perform testing in GeoDa to identify a suitable model, which reveals that a spatial autoregressive (SAR) model gives the best fit. Specifically, we perform an ordinary least squares test, and the result is shown in Table 7. According to a model selection method (Anselin, 1988), LM-lag and robust LM-lag tests should be given more weight than LM-error and robust LM-error tests; accordingly, we select the SAR model.

Moran's I is used to perform global autocorrelation and local autocorrelation tests to detect the spatial dependence of regression residuals. Real-world data, which represents the O demand in each area at the morning peak on November 1, 2016 (Tuesday), denoted tueAMO, are tested; the results are shown in Figs. 9 and 10.

As can be seen from Fig. 9, Moran's I is 0.191 and Z is 15.546, and these are significant at the $\alpha = 0.01$ level. This implies that tueAMO is significantly correlated to the global geographical distribution. Similarly, as shown in Fig. 10, the local indicators of spatial association illustrate the spatial agglomeration of specific values within regional units. These results show that the effect of the built environment attributes on the spatio-temporal distribution of OD demand in an area is determined by the surrounding areas and by the internal characteristics of the area.

4.3 SAR Modeling

The SAR model focuses on the interdependence of the decision-making behavior of agents, which is represented as the following formula:

$$Y = \tau + \mu WY + \rho X + \varepsilon \tag{2}$$

Here, Y is the observed value, τ is the constant term, μ is the spatial autocorrelation parameter to be estimated, reflects the spatial correlation inherent in sample data, and measuring the influence of the adjacent regions of Y. WY is the spatial autoregressive

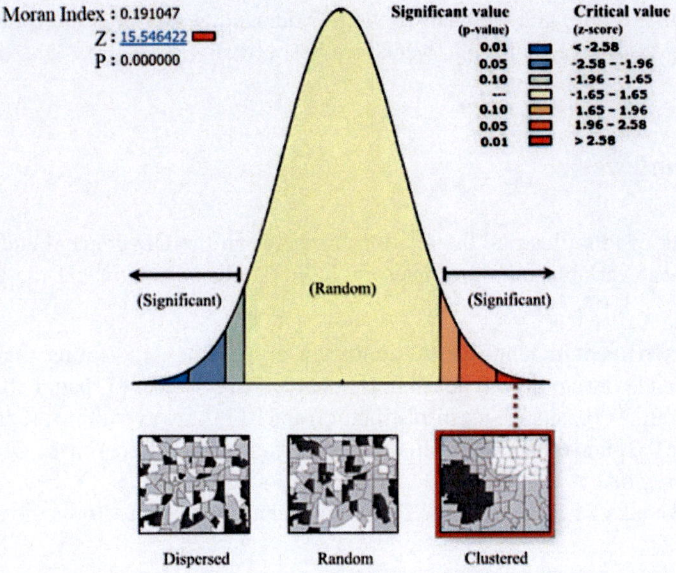

Fig. 9 Result of global Moran test

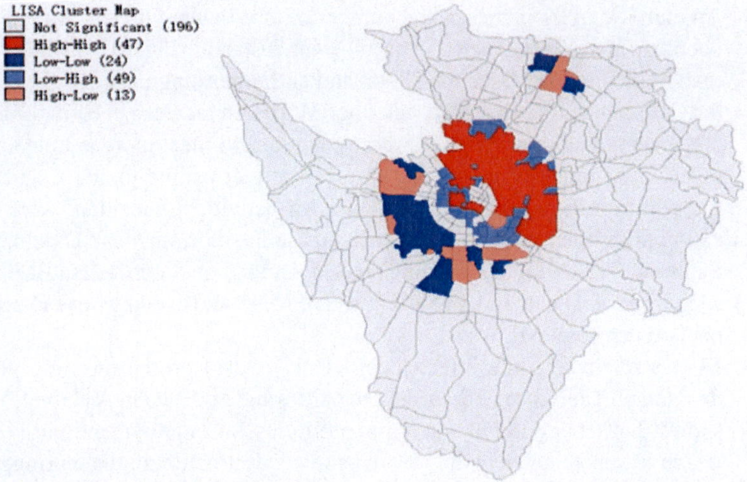

Fig. 10 Cluster map of LISA

dependent variable. is the regression parameter, X is the explanatory variable, and ε is the error term.

As mentioned, the factors obtained from the factor analysis serve as the explanatory variables. The OD demand in a specific area at a specific time is regarded as the dependent variable, and we analyze the OD demand at morning and evening peak usage periods on weekends and weekdays, respectively.

4.4 Analysis

The fitting results illustrate the relationship between the OD demand and the built environment variables and are listed in Table 8. The results reveal the following trends.

1. The SAR coefficient WY is significant at all times, indicating that the OD demands of regions are non-independent and are subject to spatial effects.
2. Overall, F1 (basic service infrastructure) and F13 (railway station) are significant at the 0.01 and 0.05 levels at all times and have positive effects on the OD demand of a region.

   We take F1 as an example for detailed analysis, which affords the following results.

   - F1 has the most positive effect on the spatio-temporal distribution of OD demand in Chengdu. This means that the addition of each unit of basic service facilities increases tueAMO by 13.11%.
   - An increase in F1 increases the number of jobs in a region and attracts more people to arrive in the region in the morning rush hour and leave in the evening rush hour [i.e., increases the D demand at the morning peak (tueAMD) and the O demand at the evening peak (tuePMO)]. An increase in F1 in the region also means that the region has more functions to meet more people's needs for a wider range of activities. This will attract people to arrive and leave the region during non-working hours for activities other than work (e.g., eating and shopping) and will thus increase the morning peak D demand on Saturday (satAMD), the evening peak O demand on Saturday (satPMO), the evening peak D demand on Saturday (satPMD), and evening peak D demand on Tuesday (tuePMD).
   - F1 has relatively little effect (coefficient < 20%) on the morning peak O demand on Tuesday and Saturday (tueAMO and satAMO, respectively). This shows that Chengdu has many opportunities for employment and activity, which means that Chengdu will be people's destination in the morning peak period rather than their origin.

3. The significances of the effects of other built environment attributes on OD demand, represented by F2, F4, F7, F8, F9, F11, F12, F13, and F14, vary over time. We take F11, F12, and F13 as examples for detailed analysis, which reveals the following trends.

   - F11 (passenger transportation station) has a significantly positive effect on morning peak D demand and evening peak O demand on Tuesday and Saturday (tueAMD, tuePMO, satAMD, and satPMO), which is consistent

Table 8 Fitting results of the SAR model

	tueAMO	tueAMD	tuePMO	tuePMD	satAMO	satAMD	satPMO	satPMD
WY	0.621***	0.429***	0.502***	0.507***	0.627***	0.472***	0.485***	0.451***
ε	9.663***	14.771***	17.353***	16.259***	10.098***	13.946***	22.713***	22.626***
F1	13.107***	28.165***	29.136***	19.228***	11.383***	23.968***	35.754***	29.104***
F2	5.179*	−5.982**	−3.554	5.41	6.756**	−2.271	−2.82	4.362
F3	−3.363	−3.322	−4.244	−4.84	−3.165	−3.195	−5.483	−6.005
F4	6.268**	4.003	7.124*	6.425**	6.204**	3.809	8.062	5.484
F5	−0.422	0.625	0.562	0.019	1.053	0.119	2.845	0.864
F6	−3.27	−0.034	−1.878	−3.567	−3.064	−1.003	−2.641	−4.574
F7	2.083	4.406*	5.245	3.724	3.037	5.25**	4.483	3.251
F8	−7.012***	−4.438*	−6.094	−3.102	−7.197***	−4.981*	−7.134	−5.529
F9	0.177	1.46	−1.032	−0.484	0.191	3.188	−1.625	−1.386
F10	1.636	3.141	3.86	3.955	1.421	3.402	6.314	4.485
F11	1.142	8.367***	11.009***	3.133	2.018	7.905***	14.015***	3.445
F12	1.901	1.691	2.39	3.735	2.068	7.271***	5.301	5.294
F13	8.315***	9.859***	12.376***	10.001***	8.308***	8.865***	15.997***	16.685***
F14	2.537	2.336	3.741	6.952**	2.86	3.616	3.332	7.2

Note: (*, **, ***) means significant at $\alpha = (0.10, 0.05, 0.01)$; "tue" and "sat" mean Tuesday (weekday) and Saturday (weekend) respectively; "AM" and "PM" mean the morning peak and the evening peak respectively; "O" and "D" mean the departure amount and the arrival amount respectively with reality. That is, people generally arrive at a transportation hub from their homes early of day, start their day's travel, travel back to the transportation hub at a later time of day, and return to their homes to end their day's travel (i.e., in this context, the transportation hub is their D and their home is their O).

- F12 (travel) only has a significant effect on the morning peak D demand in the weekend, which is consistent with our prediction. That is, people typically travel to areas offering recreational activities in the early part of the weekend. However, unlike on weekdays, people do not necessarily leave an area during the evening peak O demand period.
- F13 (railway station) has different effects on O demand and D demand at different times, especially on satPMO and satPMD. This is also consistent with real-life situations: people are more likely to travel on weekends, meaning that many vehicles are present near a railway station on weekends to pick up or drop off passengers; this significantly affects satPMO and satPMD.

4. The same five factors (F1, F2, F4, F8, and F13) significantly affect morning peak O demand on weekends and weekdays. We attribute this to the low elastic demand for travel in the morning peak times on weekends (to leisure and entertainment travel activities) and the high rigid travel demand in the morning peak times on weekdays (for transit to work and related activities).

5 Conclusion

This chapter describes a case study that outlines a framework for exploring the effect of urban built environment on OD demand in the context of specific spatial structures. The following conclusions are drawn.

1. The effect of urban built environment on OD demand displays significant spatio-temporal variability. That is, the magnitude of the effect of urban built environment attributes vary with geographical location and time. This highlights that when decision-makers are planning and designing a city, they must consider the varying effects of built environment attributes on OD demand. Scientific principles should be used in the design of urban built environment, as this will enable effective regulation of the distribution of traffic volume at peak hours, thereby reducing traffic congestion.
2. Basic service facilities and railway-station service facilities have a positive effect on the spatio-temporal distribution of OD demand. Accordingly, when government departments are planning and utilizing land, they must be cognizant of the effect of these facilities on OD demand. Such cognizance will enable them to devise strategies to prevent large volumes of passengers from gathering at such facilities, as this can lead to traffic congestion and social problems.

References

Anselin, L. (1988). *Spatial econometrics: Methods and models*. Kluwer Academic Publishers.

Anselin, L., Syabri, I., & Kho, Y. (2010). GeoDa: An introduction to spatial data analysis. In *Handbook of applied spatial analysis* (pp. 73–89). Springer.

Bhat, C. R., & Guo, J. Y. (2007). A comprehensive analysis of built environment characteristics on household residential choice and auto ownership levels. *Transportation Research Part B: Methodological, 41*(5), 506–526.

Blei, D. M., Ng, A. Y., & Jordan, M. I. (2003). Latent Dirichlet allocation. *The Journal of Machine Learning Research, 3*, 993–1022.

Cervero, R., & Kockelman, K. (1997). Travel demand and the 3Ds: Density, diversity, and design. *Transportation Research Part D: Transport and Environment, 2*(3), 199–219.

Chan, E. T., Schwanen, T., & Banister, D. (2021). The role of perceived environment, neighbourhood characteristics, and attitudes in walking behaviour: Evidence from a rapidly developing city in China. *Transportation, 48*(1), 431–454.

Chen, P., Liu, Q., & Sun, F. (2018). Bicycle parking security and built environments. *Transportation Research Part D: Transport and Environment, 62*, 169–178.

Das, S. (2021). Autonomous vehicle safety: Understanding perceptions of pedestrians and bicyclists. *Transportation Research Part F: Traffic Psychology and Behaviour, 81*, 41–54.

Ewing, R., & Cervero, R. (2001). Travel and the built environment: A synthesis. *Transportation Research Record, 1780*(1), 87–114.

Hasan, S., & Ukkusuri, S. V. (2014). Urban activity pattern classification using topic models from online geo-location data. *Transportation Research Part C: Emerging Technologies, 44*, 363–381.

Maat, K., & Timmermans, H. J. (2009). Influence of the residential and work environment on car use in dual-earner households. *Transportation Research Part A: Policy and Practice, 43*(7), 654–664.

Markou, I., Kaiser, K., & Pereira, F. C. (2019). Predicting taxi demand hotspots using automated internet search queries. *Transportation Research Part C: Emerging Technologies, 102*, 73–86.

Mimno, D. M., & McCallum, A. (2008, July). Topic models conditioned on arbitrary features with Dirichlet-multinomial regression. In *UAI* (Vol. 24, pp. 411–418).

Pereira, F. C., Rodrigues, F., & Ben-Akiva, M. (2013). Text analysis in incident duration prediction. *Transportation Research Part C: Emerging Technologies, 37*, 177–192.

Pereira, F. C., Rodrigues, F., Polisciuc, E., & Ben-Akiva, M. (2015). Why so many people? Explaining nonhabitual transport overcrowding with internet data. *IEEE Transactions on Intelligent Transportation Systems, 16*(3), 1370–1379.

Qi, G., Wu, J., Zhou, Y., Du, Y., Jia, Y., Hounsell, N., & Stanton, N. A. (2019). Recognizing driving styles based on topic models. *Transportation Research Part D: Transport and Environment, 66*, 13–22.

Sabouri, S., Park, K., Smith, A., Tian, G., & Ewing, R. (2020). Exploring the influence of built environment on Uber demand. *Transportation Research Part D: Transport and Environment, 81*, 102296.

Sun, L., & Yin, Y. (2017). Discovering themes and trends in transportation research using topic modeling. *Transportation Research Part C: Emerging Technologies, 77*, 49–66.

Sun, Y., & Kirtonia, S. (2020). Identifying regional characteristics of transportation research with Transport Research International Documentation (TRID) data. *Transportation Research Part A: Policy and Practice, 137*, 111–130.

Wang, K., & Zhang, W. (2021). The role of urban form in the performance of shared automated vehicles. *Transportation Research Part D: Transport and Environment, 93*, 102744.

Wei, X., & Croft, W. B. (2006). LDA-based document models for ad-hoc retrieval. In *Proceedings of the 29th Annual International ACM SIGIR Conference on Research and Development in Information Retrieval* (pp. 178–185).

Zhang, Z., He, Q., Gao, J., & Ni, M. (2018). A deep learning approach for detecting traffic accidents from social media data. *Transportation Research Part C: Emerging Technologies, 86*, 580–596.

Zhao, P. (2014). The impact of the built environment on bicycle commuting: Evidence from Beijing. *Urban Studies, 51*(5), 1019–1037.

Zhong, S., & Bushell, M. (2017a). Built environment and potential job accessibility effects of road pricing: A spatial econometric perspective. *Journal of Transport Geography, 60*, 98–109.

Zhong, S., & Bushell, M. (2017b). Impact of the built environment on the vehicle emission effects of road pricing policies: A simulation case study. *Transportation Research Part A: Policy and Practice, 103*, 235–249.

Zhong, S., Wang, S., Jiang, Y., Yu, B., & Zhang, W. (2015). Distinguishing the land use effects of road pricing based on the urban form attributes. *Transportation Research Part A: Policy and Practice, 74*, 44–58.

Chapter 3
Spatiotemporal Evolution of Ridesourcing Markets Under the New Restriction Policy: A Case Study in Shanghai

Abstract Identifying and understanding factors that influence the demand of ridesourcing market is essential for online hailing systems to improve the quality of service. This chapter proposes a two-level growth model (GM) to identify the potential multi-level factors that may affect online ride-hailing service demand. By using the massive datasets from Didi Chuxing, Inc., including both Didi Express and Didi Taxi services, the order number fluctuations at different urban circle zones after the implementation of restrictions on ridesourcing in Shanghai, 2016 were analyzed, to assess the competition and mutual complementarities between Express and Taxi, the two major services provided by Didi Chuxing. The relative market share of Express was estimated to reveal the possible related spatial and temporal factors, which further demonstrates significant positive associations between ridesourcing demand and built environment factors, such as commercial/residential land use, public transport accessibility, as well as weather conditions. Metro service availability and rainy weather were found correlated with a relatively higher market share of Express service. Additionally, compared to the regular road transit service, the metro system was found to have a stronger correlation with the ridesourcing demand. Findings of this study may provide guidelines for urban planning and traffic operations, which in turn assists to achieve high-quality ridesourcing service for travellers.

Keywords Car-hailing service · Growth model (GM) · Multi-level factors · Demand and market share change · New restrictions on ridesourcing

1 Introduction

Online ridesourcing service has experienced a rapid growth in recent years (SAE International, 2018), which provides travellers a convenient and efficient way to schedule rides through smartphone-based applications instead of waiting and hailing taxis on-road. This feature has made it an increasingly popular travel mode, especially in major metropolitan cities. As the largest ridesourcing platform in China, Didi Chuxing completed 7.43 billion rides for more than 450 million passengers in over 400 cities during 2017, accounting for 85.6% of the overall annual car-on-demand services. Among these, the taxi-hailing service—Didi Taxi has about 2

S. Zhong and D. Sun, *Logic-Driven Traffic Big Data Analytics*,
https://doi.org/10.1007/978-981-16-8016-8_3

million registered drivers, providing more than 1.1 billion rides across the country (Analysys Qianfan, 2016).

Despite the growing popularity of ridesourcing service, governments as well as researchers continue to contemplate the equity and efficiency issues, raising critical debates on social welfare, safety, privacy, energy consuming, potential congestion and related environmental impact (Contreras & Paz, 2018; Rogers, 2015; Sun et al., 2018; Toner, 2010). Cohen and Shaheen (2016) demonstrated that the shared mobility directly influenced and was influenced by most facets of urban planning, which may be grouped into four categories: travel behavior, environmental, land use, and social affairs. D'Orey and Ferreira (2014) found taxi sharing employment assists to reduce individual travel fare, total travel distance, as well as the operational costs. Cramer and Krueger (2015) ascribed the higher capacity utilization of Uber comparing with the traditional taxis to four aspects: advanced driver-customer matching technologies, large scale, flexible pricing/labor supplying strategies and regulations on taxis. Hall and Krueger (2017) analyzed and discussed the social impact from an economic perspective and found that the hourly earnings for Uber drivers are almost invariant to work duration.

In the context of taxi fleet management, the existing studies revealed that no entry control may lead to an oversupply of taxi in creating excessive of destructive competition, thus to deteriorate the service quality (Dempsey, 1996), while entry control may cause deficiencies of service availability (Schaller, 2007) and increase in medallion prices or license values (Çetin & Eryigit, 2011). The similar situations also occurred in the ridesourcing services, which aroused the attentions from both governments and academia. A survey of 380 ridesourcing (Uber/Lyft/Sidecar) users in San Francisco indicated that compared with taxis, ridesourcing passengers are generally with shorter wait time, and the mode is consequently more reliable. At least half of ridesourcing trips replaced the modes other than taxi, including public transit and driving (Rayle et al., 2016). In China, Shanghai initiates regulations to ban the online hailing during peak hours in 2014. Since then, enforcement actions against the illegal operation of ridesourcing were released in other Chinese major cities, including Jinan, Nanchang, Urumqi, and so on. On Dec. 21, 2016, the local municipal government of Shanghai issued Order No. 48, specifying restrictions toward ridesourcing service market, which mandated the level of vehicle dimension parameters and required the operational vehicles must be registered with a local license plate and satisfy certain emission standard (i.e. China Stage 5). The registered drivers should have household registration (Hukou) permits and with fewer than five traffic violations during the past 12 months. Overall, the government intends to regulate the ridesourcing market to ensure operational safety and to protect the rights and benefits of both passengers and drivers.

In addition to the potential outcomes on the supply side, researchers have also investigated the factors related to the demand, as well as the forecasting methods. Studies on taxi demand were first carried out using aggregated indicators, such as regional average income and household vehicle ownership (Gilbert et al., 1976). Schaller (2005) found that the number of metro commuters, the number of households without vehicle and the demand of taxi trip to airport as the primary factors to estimate the demand of taxis at city level. Additionally, weather was also found to potentially affect the route and mode choice of travellers and impact the overall trip duration (Guo et al., 2007; Kamga et al., 2015; Khattak & Palma, 1997; Singhal et al., 2014). For example, vehicle usage was generally higher during summer comparing to winter (Bergstrom & Magnusson, 2003). In recent years, with the widely applied positioning technologies (Location Based Service of smart phones as well as other devices), position and trajectory data were obtained to delve into the operation dynamics, such as short term demand forecasting (Chang et al., 2010; Zhao et al., 2016), real-time traffic condition estimation (Yao et al., 2017), and route choice preference for particular drivers, e.g. taxi drivers (Sun et al., 2014). At the same time, demographic, socioeconomic, and land use data were incorporated to reveal the factors that influence taxi demand (Yang & Gonzales, 2014; Yang et al., 2018). To this end, the operational behavior and travel patterns were investigated (Gao et al., 2013; Liu et al., 2015; Yao & Lin, 2016).

In summary, the existing studies on ridesourcing, taxi management, and influencing factors, haven't incorporated spatial and temporal control variables simultaneously, not to mention their potential interactions. Moreover, still other factors, including lane use, weather, and transport accessibility attributes, may be either time varying or constant (time-independent) within a certain period but are important for policy related analyses. This study attempts to reveal factors that may affect ridesourcing demand. In comparison with the existing studies, the time series longitudinal data (e.g. time-varying and time independent explanatory variables) was incorporated and combined to identify the potential factors related to the demand of ridesourcing service. Specifically, the grid-based analysis was used to create urban and suburban subject areas, referred to as cells in this study. Then, a two-level growth model with random intercept/slope coefficients and auto regression covariance structure [RIS-AR(1)] was introduced to analyze the trip data provided by Didi. Built environment measures and weather conditions were introduced to capture the across-subject and within-subject variation respectively. By incorporating both location (different city circle zones) and time as control variables, the model evaluated longitudinal change in ridesourcing service before and after the implementation of Shanghai's new restriction policy to shed lights on the relationship between built environment, weather condition and the demand of the service. Statistical tests were carried out to assess the outcomes of measurement at different city circle layers. The mutual complementarity between the services of Express and Taxi was also revealed not only at the municipal city level but also among circle layers.

2 Data and Methods

2.1 Data Description and Pre-process

The Express and Taxi order dataset used in this study were randomly sampled and extracted from November 2016 to February 2017 in Shanghai, China. The entire study area includes eight central districts and seven suburb districts, with the average numbers of weekday orders at about 20,000 and 33,500 for Taxi and Express, respectively. Each order includes order ID, driver ID, vehicle ID, longitude and latitude location, timestamps when the trip starts and ends, in which the order ID, driver ID and vehicle ID were purposely encrypted to avoid privacy issues.

Based on the order data, the spatial distribution of pickup locations was obtained and are presented in Fig. 1. Although Express had comparable larger order volume, the taxi service was more prevalent at certain suburban areas (e.g. 1 and 2), which may be partially explained by the restrictions on suburban taxis (License Plate initialized with C) to enter into the central urban areas. This study used a grid-cell of 1 km × 1 km combining with expressway rings of Shanghai for the analysis. All cells were divided by the three ring expressways into four groups of zones—the Core group, the Mid circle group, the Outer circle group and the Suburban group. Cells located right on the ring roads were purposely removed to ensure: (1) clear identification of cell type and cell area in each group, and (2) obvious spatial separation between groups.

On Dec. 21, 2016 (Wednesday), Shanghai municipal government released a new restriction policy on online ridesourcing service, which mandatorily requires both the driver residence (Hukou) and vehicle registration location at Shanghai. Consequently, this study chose order data from eleven weekdays, including Nov. 30, Dec. 7, 14, 21, 28 of 2016 and Jan. 4, 11, 18, Feb. 15, 22, 28 of 2017, in which the dates within Spring Festival period were purposely left out. When selecting the sample days, four days before Dec. 21 (including Dec. 21, as the policy was released in the afternoon, drivers on Dec. 21 were considered as not affected), were intentionally chosen. Then, another four days (Dec. 28, Jan. 4, 11, 18, all Wednesday) were chosen right after the

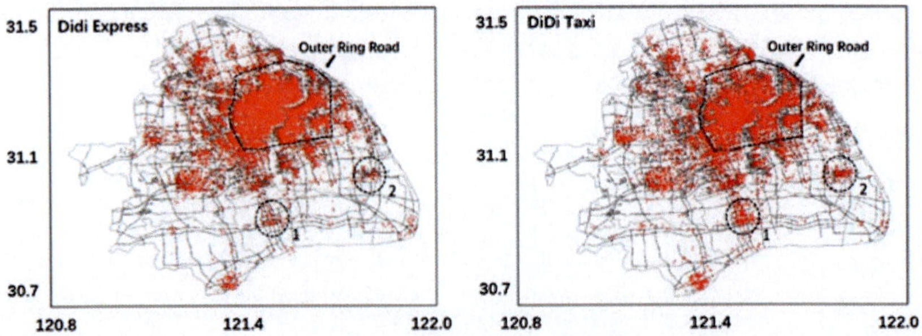

Fig. 1 Pickup location of Didi Express and Didi Taxi

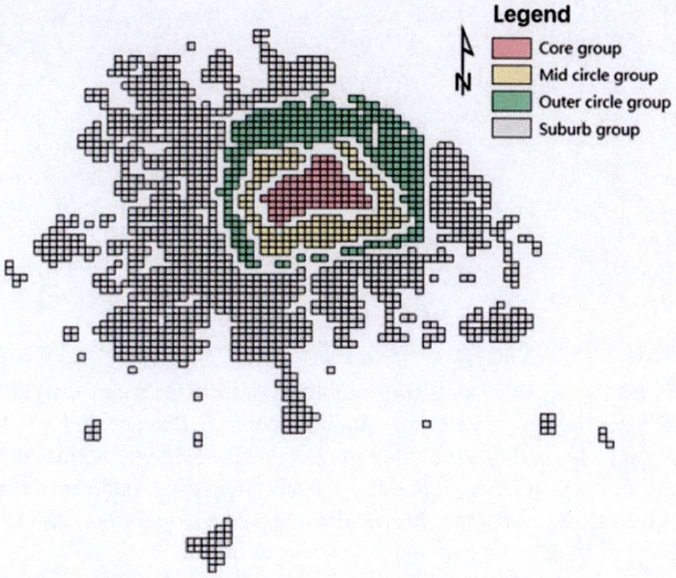

Fig. 2 Grid-based division at different city circle layer

policy releasing date, to investigate the immediate effect of the policy. As the Chinese Spring Festival is on Jan 28, and many residents who are non-local Shanghai native (more than 70%) were leaving the city to their hometown between Jan. 21 (1 week before the festival) and Feb. 11 (2 weeks after the festival), the last three days were chosen as Feb. 15, 22, and 28 to avoid the effect of Spring Festival. For each cell, the order amount of both Taxi and Express service were calculated for the selected days. All records were aggregated into a cell-by-week data structure in which longitudinal counts of orders were nested within cells. Cells with less than three records, or with daily number of orders less than 20 were removed to avoid the sample data sparsity at suburb area. Finally, as represented in Fig. 2, all cells were divided into four circle layers, by the three urban elevated circle roads, in which the cell number at Core group, Mid circle group, Outer circle group and Suburb group was 86, 125, 239, 947, respectively. To this end, the order numbers in different cells may be further investigated from both time and spatial perspectives.

2.2 Two-Level Growth Model

This study proposed a two-level growth model, in which Level 1 units are repeated measures at various time for each cell, and Level 2 units are individual subjects—cells from different circle layer groups. The two-level model, within and between-subject, can be described as follows:

Level 1:

$$y_{ij} = \pi_{0j} + \pi_{1j} t_{ij} + \varepsilon_{ij}, \varepsilon_{ij} \overset{i.i.d}{\sim} N(0, \sigma_{\varepsilon}^2) \tag{1}$$

Level 2:

$$\begin{aligned} \pi_{0j} &= \beta_{00} + \beta_{01} x_j + u_{0j} \\ \pi_{1j} &= \beta_{10} + \beta_{11} x_j + u_{1j} \end{aligned}, \begin{pmatrix} u_{0j} \\ u_{1j} \end{pmatrix} \overset{i.i.d}{\sim} N\left(\begin{pmatrix} 0 \\ 0 \end{pmatrix}, T_{\beta}\right) \tag{2}$$

$$T_{\beta} = \begin{pmatrix} \sigma_{u0}^2 & \sigma_{u01}^2 \\ \sigma_{u01}^2 & \sigma_{u1}^2 \end{pmatrix} \tag{3}$$

where, in Level 1, y_{ij} represents the number of orders (order amount) at time i for cell j, and t_{ij} is the time, while π_{0j} (the intercept coefficient) and π_{1j} (the slope coefficient) are assumed to have random effects with bivariate normal distribution, as shown in Eq. (2). In Level 2, β_{00} and β_{10} are the average intercept value and the slope for given x_j, while β_{01} and β_{11} are the regression slopes for x_j, interpreting the order amount difference for π_{0j} and π_{1j} across each cell.

The general format of the multilevel growth model can be expressed with matrixs as below:

$$Y = X\beta + ZU + e \tag{4}$$

where, Y denotes the repeated measurements of individual gd cell, β is the vector of fixed effects with matrix X containing values for certain independent variables. U is the vector of random effects with matrix Z containing values for other explanatory variables, and e is the vector of residuals.

To examine the change of relative market structure, for each cell the ratio of Express order to the total order of both Express and Taxi services was calculated to reveal the relative market share of Express service. As the market share is bounded between 0 and 100%, which violates the assumptions needed for linear regression, log-odds of the dependent variable was introduced, so that the transformed quantity is unbounded both from below and above. In other words, a logistics regression was used here for the relative market share of Express service, instead of the simple linear regression, as presented below:

$$\log\left(\frac{Y}{1 - Y}\right) = X\beta + ZU + e. \tag{5}$$

The results, along with its original cell-based order data, were used for model estimation. Since the order data was randomly drawn from the operational datasets, the number of orders for some suburban cells at certain time points may be significantly lower than the common level due to the data sparsity, which may in turn cause outliers in the derived market share ratios and affect the final result. Thus, for each cell, records with 50% difference above or below the average value of the studied 11 sample days were replaced with null values. Since the growth model can handle missing value in longitudinal records and repeated measures with unequal time intervals, cells with incomplete records were saved for model estimation. Finally, the total number of valid order records within the 11 sample days turns out to be 316,171 (117,151 for Taxi and 199,020 for Express).

To reduce the skewness of cell-based demand, a natural log-transformation was carried out for the order amount in each cell. The distribution of transformed measurements approximated the normal form and thus could be used for the implementation of the growth model.

2.3 Modeling Variables

The overall regional weather data were obtained from the Meteorological Bureau of Shanghai (2017). Five of the eleven days were with light rain, and the temperature of the sampling days was stable, with wind speed under Scale 3 (about 4 m/s). The weather state was denoted as categorical variables, 1 for rainy and 0 for the others. The temperature variables including both average daytime and night temperature degrees were treated as continuous variables. For the cell-based time-invariant data, two groups of variables were used. The first was obtained to evaluate public transport coverage of each cell (Yu et al., 2012). For metro service, a two-level categorical coding was used to represent whether any metro station exists. For bus service, since the number of bus stations varies greatly between different cells, bus-line serving index (BSI) was introduced to evaluate the level of service at each cell, which can be calculated as follows:

$$BSI_i = \sum_j x_{ij}\alpha_j \tag{6}$$

where BSI_i represents bus-line serving index at cell i; x_{ij} is defined as percentage of area in cell i that is within 500 m of bus station j; α_j denote the number of bus lines at station j.

The second time-invariant variable category contains the proportion of different type of land use within each cell. In this study, the proposed land uses include commercial, residential, transportation, industrial, educational, governmental and green space (Zhang et al., 2017; Zhong & Bushell, 2017).

T variables chosen and used in the model are as follows, with the data source and descriptive statistics presented in Table 1:

1. Outcome variables:
   (a) Longitudinal records of **order amount** for both Taxi and Express services and their summation as total;
   (b) **Market share** of Didi Express during the study period.

Table 1 Symbol and descriptive statistics for variables in the growth model

Variables	Symbol	Data source	Min	Mean	Max
Circle	γ_{01}	ArcGIS file	0	–	3
Week	γ_{10}	Didi order date	0	5.82	13
TA: Metro	γ_{02}	Metro network	0	–	1
TA: BSI	γ_{03}	Transit network	0	0.96	7
LU: Com	γ_{04}	Land use file	0	0.12	1
LU: Res	γ_{05}	Land use file	0	0.19	1
LU: Trans	γ_{06}	Land use file	0	0.18	1
LU: Ind	γ_{07}	Land use file	0	0.21	1
LU: Edu	γ_{08}	Land use file	0	0.09	1
LU: Gov	γ_{09}	Land use file	0	0.07	1
LU: Gre	γ_{010}	Land use file	0	0.13	1
W: Rain	γ_{011}	Meteorological bureau	0	–	1
W: Temp_day	γ_{012}	Meteorological bureau	3	10.42	19
W: Temp_night	γ_{013}	Meteorological bureau	−2	5.64	11

2. Level 1 explanatory variables:
   (a) **Week** (γ_{10}): coded as 0, 1, 2, 3, 4, 5, 6, 7, 11, 12, 13, representing the eleven sample days;
   (b) **Rain** (γ_{011}): coded 0 for no rain, and 1 for moderate rain since only moderate rain has occurred over the studying time period;
   (c) **Temp_day** (γ_{012}): daytime temperature degree;
   (d) **Temp_night** (γ_{013}): night temperature degree;
3. Level 2 explanatory variables:
   (a) **Circle** (γ_{01}): coded as categorical variable 0, 1, 2, 3 for Core group, Mid circle group, Outer circle group and Suburban group, respectively;
   (b) **Metro** (γ_{02}): dummy variable (1—cells with metro station, 0—without);
   (c) **BSI** (γ_{03}): continuous measurements representing the quality of bus service within a cell;
   (d) **Land use pattern**: the proportion of different land use within the cell, including commercial (γ_{04}), residential (γ_{05}), transportation (γ_{06}), industrial (γ_{07}), educational (γ_{08}), governmental (γ_{09}) and green space (γ_{010}).

In this study, both the longitudinal data of order amount and market share were treated as outcome measures. The effect of individual background was assumed as constant over time, thus not included in the level-2 equation. The analyses were conducted using MIXED Procedure in SAS software package, in which different circle layers were treated as classification variables. A RIS-AR(1) (Random Slope-Intercept) structure was introduced: the cell level intercept and slope were modelled with random coefficients, while AR(1) was used to model the within-cell covariance structure of time series data.

3 Results and Discussions

3.1 Descriptive Results

Descriptive statistics were obtained to provide a straightforward understanding of the daily order amount change and the change of relative market share of Express service at different zones. The studying period was divided into three phases, corresponding to four weeks before the implementation of restriction, four weeks after the implementation of restriction (right before the Spring Festival), and three weeks after the Spring Festival. As presented in Fig. 3a, b, the daily cell order amount of Express service decreased along with the observed three phases, while a complementary rise of taxi service orders was found during corresponding periods. As a result, the total cell-based order amount only suffered a slight decrease, as presented in Fig. 3c. The change from Phase 1 to Phase 2 was significant, but became much smaller from Phase 2 to Phase 3. This means the market experienced a significant change process after the implementation of the New Restriction Policy, and then gradually stabilized after the Spring Festival. Moreover, different levels of changes were identified across the circle groups, in which the two inner circle groups had more decreases comparing with the other two.

For more intuitive reflection of the complementarity between the two services, the average order numbers within two periods (i.e. Dec. 1 to 14, 2016, and Feb. 1 to 14, 2017) were calculated for different circle groups. In addition to the changes of **cell order amount** for different circle/zone areas across time (phases 1, 2 and 3), the percentage change of **circle order amount** (order amount at different circle groups) between the two periods is presented in Fig. 4. For each circle group, the loss of the order amount of the Express service was accompanied by a corresponding similar but opposite rise of the taxi service. Moreover, the Core and Mid circle groups suffered comparable larger drop of total orders (both were about 9%) while the Outer circle and Suburb groups decreased only by 1% and 2%, respectively.

For the Express, as presented in Fig. 5, the market share change was not evident in suburban group, while enduring evidently decrease at the core and intermediate groups in Phase 2, and then slightly decreased during Phase 3. The outer circle group, however, experienced a constant decrease of market share during the observation period. The descriptive results indicated potential interaction between different city circle layers and time, which is worth further investigating as an interactive effect in the model.

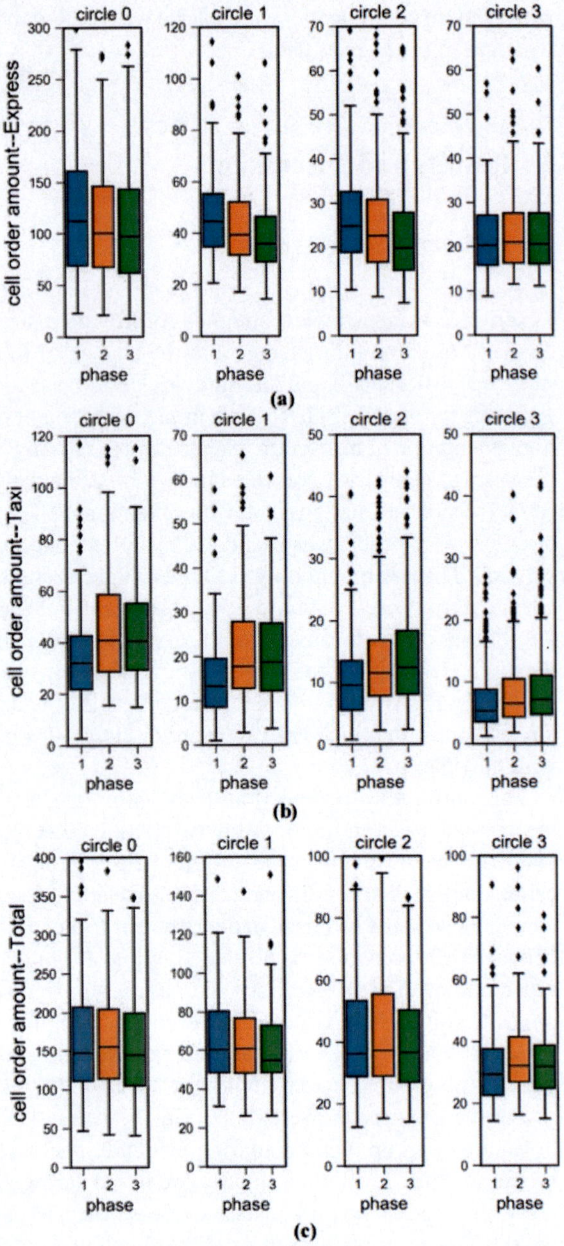

Fig. 3 Daily cell order amount for various ridesourcing services, **a** Didi Express, **b** Didi Taxi, **c** sum of both services

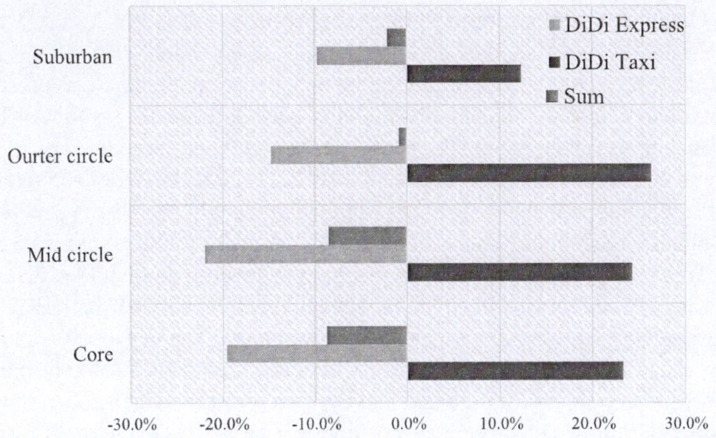

Fig. 4 Percentage change of order amount at different circle layers between Dec. 2016 and Feb. 2017

Fig. 5 Box plot of cell market share of Didi Express

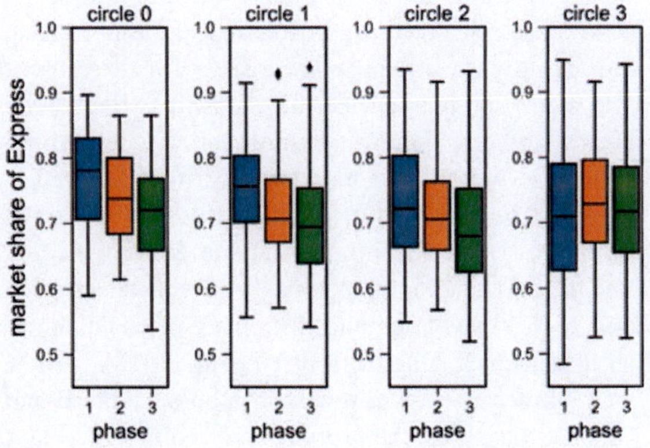

3.2 *Analytic Results*

Three models (Model A, Model B and Model C) were formulated in a stepwise process based on the obtained trip datasets. First, a growth model (Model A) with random intercept and slope was formulated as a linear function of time to test the difference at the circle group level. In Model B, the individual background independent variables, including transport accessibility and land use patterns were added at Level 2 to explain the between-subject variability (Kim & Mokhtarian, 2018). Since the time-varying covariates may also have impacts on the outcome, weather and temperature conditions were added to model the in-subject variability in Model C (Bocker & Jan Faber, 2016). For each model, the fit statistics -2LL (-2 log likelihood), AIC (Akaike information criterion) and BIC (Bayesian Information Criterion) were introduced to validate the model improvement.

3.2.1 Order Amount Generated in Study Area

Results of the three models for both Express and Taxi were obtained, as presented in Table 2. It was found that during the studying period, the order amount of Express decreased with time ($\hat{\gamma}_{10} = -0.023, p < 0.001$, Model A), accompanied by a complementary growth of overall taxi orders ($\hat{\gamma}_{10} = 0.022, p < 0.001$, Model A). Meanwhile, the total order amount in the study area suffered only a slight decrease ($\hat{\gamma}_{10} = -0.009$), which suggests that some Express demands may transfer to the taxi service or other modes. This is consistent with Fig. 3c, in which the overall amount of order actually didn't change much across three phases. The complementarity between Express and taxi services were also evident at the circle group level. The negative result of fixed effect of urban circle ($\hat{\gamma}_{01} < 0, p < 0.001$) indicates that cells at outer groups have comparable fewer orders for both taxi and Express services, which coincide with the sparse distribution of the service at the suburban area. The fixed effect estimation of cross-level interactions of circle and time (Circle × Week) for both Express and total orders have a significant positive effect on the outcomes ($\hat{\gamma}_{11} > 0, p < 0.001$), standing for a slower decreasing trend of orders from outer groups. Correspondingly, the negative interaction between circle and time (Circle × Week) of taxi service ($\hat{\gamma}_{11} = -0.002, p < 0.001$) indicates that the outer groups have less increase in orders comparing with the inner groups. As for the effect of other transportation modes, both metro and bus services have positive effect on the order amount, indicating that areas with better public transport services tend to have a higher demand for ridesourcing service. The demand was partially derived from those transferred to taxi or Express from public transport. For this reason, taxi stands are often located close to the metro station to improve transfer convenience. Moreover, from the perspective land use and development, areas with well-developed commercial facilities generally have higher travel demand, which are often accompanied by better public transport services (Sun & Guan, 2016; Yu et al., 2018).

A similar effect was revealed in both Model B and Model C for land use pattern. Of these variables, the proportion of commercial land showed a significant positive statistical result with the highest rate of change ($\hat{\gamma}_{04} = 1.641, p < 0.001$ for Express, in Model C; $\hat{\gamma}_{04} = 0.474, p < 0.05$ for taxi, in Models B and C), and the proportion of residential land also demonstrated significant positive effect on the outcome ($\hat{\gamma}_{05}$. $= 0.217, p < 0.1$ for Express; $\hat{\gamma}_{05} = 0.046, p < 0.1$ for taxi and $\hat{\gamma}_{05} = 0.246, p < 0.01$ for total, all in Model C). Transportation land use was revealed to have significant negative effect on the outcome for Express ($\hat{\gamma}_{06} = -0.202, p < 0.1$ in Model C) while an insignificant positive effect on taxis. This indicated that regions with higher proportion of transportation-related land use tend to produce fewer trip demands. The influence also appeared on green land proportion, where the significant positive effect was found for Express service. Moreover, weather conditions also demonstrated a significant effect on the order amount. For example, the order amounts on rainy days appeared higher as expected ($\hat{\gamma}_{011} > 0, p < 0.001$ for both Express and taxi), and the higher daytime temperature was generally associated with higher order amount ($\hat{\gamma}_{012} = 006, p < 0.001$ for Express and $\hat{\gamma}_{012} = 0.014, p < 0.1$ for taxi). For each case (Express, Taxi, and Total) all the fit statistic indexes (-2LL, AIC and BIC) show a

Table 2 Model results of order count (fixed effect)

Variable	Didi Express						Didi Taxi					
	Model A		Model B		Model C		Model A		Model B		Model C	
	Estimate	Sig	Estimate	Sig	Estimate	Sig	Estimate	Sig	Estimate	Sig	Estimate	Sig
Intercept γ_{00}	4.493	***	4.022	***	4.101	***	3.352	***	2.68	***	2.564	***
Circle γ_{01}	−0.531	***	−0.426	***	−0.425	***	−0.572	***	−0.382	***	−0.382	***
Week γ_{10}	−0.023	***	−0.023	***	−0.025	***	0.022	***	0.022	***	0.022	***
Circle × Week γ_{11}	0.006	***	0.006	***	0.006	***	−0.002	***	−0.002	***	−0.002	***
TA: Metro γ_{02}			0.09	*	0.09	*			0.028		0.028	
TA: BSI γ_{03}			0.002	**	0.002	**			0.004	***	0.004	***
LU: Com γ_{04}			1.639	***	1.641	***			0.474	*	0.474	*
LU: Res γ_{05}			0.217	~	0.216	~			0.046	~	0.046	~
LU: Trans γ_{06}			−0.201	~	−0.202	~			0.073		0.072	
LU: Ind γ_{07}			0.005		0.005				−0.155	**	−0.155	**
LU: Edu γ_{08}			−0.251		−0.253				−0.016		−0.015	
LU: Gov γ_{09}			−0.131		−0.134				0.688		0.689	
LU: Gre γ_{010}			0.371	*	0.37	*			0.115		0.116	
W: Rain γ_{011}					0.038	***					0.122	***
W: Temp_day γ_{012}					0.006	***					0.014	***
W: Temp_night γ_{013}					−0.001	~					−0.012	***
-2LL	13,331.7		13,008.4		12,575.1		17,468.8		17,350.5		17,038.8	
AIC	13,411.7		13,088.4		12,675.1		17,474.8		17,356.5		17,048.8	
BIC	13,575.5		13,252.2		12,879.7		17,489.8		17,371.5		17,073.8	

Variable	Total					
	Model A		Model B		Model C	
	Estimate	Sig	Estimate	Sig	Estimate	Sig
Intercept γ_{00}	4.900	***	3.322	***	3.323	***
Circle γ_{01}	−0.993	***	−0.501	***	−0.509	***
Week γ_{10}	−0.009	***	−0.009	***	−0.012	***

(continued)

Table 2 (continued)

Variable	Total					
	Model A		Model B		Model C	
Circle × Week γ_{11}	0.006	***	0.006	***	0.006	***
TA: Metro γ_{02}			0.138	**	0.139	**
TA: BSI γ_{03}			0.01	***	0.010	***
LU: Com γ_{04}			0.745	~	0.746	~
LU: Res γ_{05}			0.246	**	0.247	**
LU: Trans γ_{06}			−0.043		−0.043	
LU: Ind γ_{07}			−0.412	***	−0.412	***
LU: Edu γ_{08}			−0.081		−0.08	
LU: Gov γ_{09}			0.92		0.81	
LU: Gre γ_{010}			0.154		0.155	
W: Rain γ_{011}					0.079	***
W: Temp_day γ_{012}					0.002	***
W: Temp_night γ_{013}					−0.006	***
-2LL	23,361.1		22,871.4		22,753.9	
AIC	23,369.1		22,879.4		22,763.9	
BIC	23,390.5		22,900.9		22,790.7	

Note ~$p \leq 0.10$; *$p \leq 0.05$; **$p \leq 0.01$; ***$p \leq 0.001$
TA Transport Accessibility; *LU* Land Use; *W* Weather
Com Commercial; *Res* Residential; *Trans* Transportation; *Ind* Industrial; *Edu* Educational; *Gov* Governmental; *Gre* Green space

continuous decline, indicating that the goodness of fit was improved with the increase of explanatory variables.

3.2.2 Relative Market Share of Didi Express

The negative results of fixed effect of circle ($\hat{\gamma}_{01} = -0.008$, $p < 0.05$) stands for a significantly lower level of market share at cells far from the city center, as shown in Table 3. The main negative effect of time and interactive effect "Circle × Week" shows a significant decline of market share and less market share loss at the outer circle. This stands for a diversion of market lose at different circle layers, which is in line with the former observation of the order change of Express service.

The level 1 time-invariant covariate—metro, shows a positive slightly significant effect on the outcome ($\hat{\gamma}_{02} = 0.016$, $p < 0.05$), indicating that the market shares

Table 3 Model results of relative market share of Didi Express (fixed effect)

Variable	Model A		Model B		Model C	
	Estimate	Sig	Estimate	Sig	Estimate	Sig
Intercept γ_{00}	0.77	***	0.748	***	0.742	***
Circle γ_{01}	−0.008	*	−0.003		−0.004	
Week γ_{10}	−0.01	***	−0.01	***	−0.009	***
Circle × Week γ_{11}	0.002	***	0.002	***	0.002	***
TA: Metro γ_{02}	–	–	0.016	*	0.016	*
TA: BSI γ_{03}	–	–	−0.0001		0.0001	
LU: Com γ_{04}	–	–	0.034		0.037	
LU: Res γ_{05}	–	–	0.005		0.006	
LU: Trans γ_{06}	–	–	−0.026		−0.026	
LU: Ind γ_{07}	–	–	−0.023	~	−0.024	~
LU: Edu γ_{08}	–	–	0.013		−0.011	
LU: Gov γ_{09}	–	–	−0.18		−0.179	
LU: Gre γ_{010}	–	–	0.113		0.114	*
W: Rain γ_{011}	–	–	–	–	0.005	*
W: Temp_day γ_{012}	–	–	–	–	−0.001	
W: Temp_night γ_{013}	–	–	–	–	0.002	
-2LL	−7094.1		−7047		−7065.8	
AIC	−7088.1		−7041		−7057.8	
BIC	−7074.4		−7027.3		−7039.6	

Note ~$p \leq 0.10$; *$p \leq 0.05$; **$p \leq 0.01$; ***$p \leq 0.001$
TA Transport Accessibility; *LU* Land Use; *W* Weather
Com Commercial; *Res* Residential; *Trans* Transportation; *Ind* Industrial; *Edu* Educational; *Gov* Governmental; *Gre* Green space

of Express service near metro stations are higher than other areas. In other words, compared with the Taxi service, the Express service has a larger percent of orders attributed to the metro stations nearby. This can be explained by the differences between hailing modes of the two services, as taxis have a natural feature in cruising and are easily identified. Meanwhile, Express service is mainly consisted of private cars, which have to be hailed through the online platform, and consequently seldom cruise on roads. Thus, a better strategy for Express drivers is to park close to certain areas with high demand, such as metro stations nearby.

The market share of Express on rain days during the studying period also appears to be higher than with no rains. The previous research had shown that on rainy days, taxi drivers tend to make more frequent but slightly shorter trips in order to increase their income, and off duty earlier when reaching certain income target (Singhal et al., 2014). Here two possible reasons are: (1) Express generally has flexible pricing mechanism, and charges more during raining days, which may attract more on-duty drivers; (2) Express drivers have flexible work time, and some part-time Express drivers may extend their service time to earn more, which consequently enlarges the market share.

3.2.3 Variance Explanation

The proportion of explained variance by each model compared with the unconditional model is represented using pseudo R^2, proposed by Raudenbush and Bryk (2002) as an analogue to R^2:

$$pseudo\_R^2 = \frac{\hat{\sigma}^2(\text{unconditional model}) - \hat{\sigma}^2(\text{specified model})}{\hat{\sigma}^2(\text{unconditional model})} \tag{7}$$

As presented in "p_R^2" column under "Combined Order" in Table 4, after specifying circle and cross-level effects in Model A, 19.5% between cell variance instead of level-1 variance was explained. The between-subject independent variables added in Model B further assisted to explain 60.6% of the level-2 variation, indicating that about 40% of the variation was accounted for by the land use and transport accessibility predictors. For within cell variance (level-1 variance), only 2% was explained in Model C, which may be attributed to the fact there was no drastic weather change during the sample days. The variance under the column "Market Share of Didi Express" by the models showed a similar trend at level-1. However, both Models B and C only slightly further explained the level-2 variance, which is consistent with the fact that few explanation variables added in Models B and C appeared statistically significant as shown in Table 2.

Table 4 Proportion of variance explained

Models	Combined order		Market share of Didi Express	
	Variance	p_R^2	Variance	p_R^2
Level-1 variance				
Unconditional model	0.1623	–	0.0094	–
Model A	0.1618***	0.33%	0.0093***	1.59%
Model B	0.1612***	0.68%	0.0091***	3.61%
Model C	0.159***	2.03%	0.0089***	5.21%
Level-2 variance				
Unconditional model	1.5132	–	0.0067	–
Model A	1.2185***	19.5%	0.0061***	9.52%
Model B	0.5957***	60.6%	0.0060***	11.16%
Model C	0.5824***	61.5%	0.0058***	13.24%

Note ~$p \le 0.10$; *$p \le 0.05$; **$p \le 0.01$; ***$p \le 0.001$

4 Conclusions

This study proposed a two-level growth model (GM) to investigate the effects of multilevel factors including land use, transport accessibility and weather condition on the online ride-sourcing demand and the relative market structure of Didi Express and Taxi services. The results demonstrated significant positive associations between the ridesourcing demand and factors, such as commercial and residential land use as well as weather conditions (e.g. rain and daytime temperature). The significant positive effect of transport accessibility implies its strong correlation with demand. The comparison of the relative effect of bus and metro service availability indicated that metro could be a more complementary with online ridesourcing demand than bus. In addition, market shares were used to investigate the higher penetration of Express service in cells where metro stations exist and lower penetration in areas where more industrial land use presents. The higher market share of Express in rainy days was attributed to the flexibility of work time and dynamic pricing mechanisms of Didi Express service. These findings strengthen the intuitive opinion that the private-car-consisted online ridesourcing service is more flexible and demand-oriented. The local government, particular the Shanghai municipal bureau of transportation commission may regulate the coupon issuing strategies from the ridesourcing service providers, so that the ridesourcing demand and supply may be coordinated with public transportation system under certain situations, e.g. rainy day, or hot summer (high temperature). Meanwhile, as different land use types also affect the demand significantly (e.g. γ_{04} for commercial land, γ_{07} for in industrial land, and γ_{08} for in educational land), the service provider, such as Didi Inc., may stipulate their vehicle dispatching strategies based on the results from this study, which may help to improve the efficiency and reduce the demand–supply matching time. Or even additional dispatching strategies and pricing schemes may be proposed based on the four divided circle layers in Shanghai, or other cities if applicable.

Results of the study also identified a significant change in outcome measures during the observation period. The results indicated a decrease of Express order amount as well as the relative market share, which was more significant in urban than in suburban areas. Meanwhile a complementary rise in Taxi service was found during the same period. The overall demand only suffered a slight decrease, which was comparably larger in the inner city than in suburb areas. The findings suggested that the Express service was suppressed after the implementation of new regulations on ridesourcing service, which can be attributed to the fact that the inner city has better public transit service to attract the car-hailing passengers. However, this study mainly deals with the ridesourcing services, Taxi, and Express, without touching too much about the traditional offline taxi service. It was figured out by official statistics that the ratio of Didi Taxi orders to the tradition taxis orders is about 7:3, which didn't change much during the investigation period. To this end, the findings for Didi Taxi (the geospatial changes of order number and market share) in this study may reflect the changing of traditional taxi services. As the spatial and temporal changes of two ridesourcing services (Didi Express and Didi Taxi) in Shanghai were modeled and analyzed, results from this study should be able to be extended to other Chinese cities, such as Beijing, or Chengdu. Instead of dividing the city area into four zones, i.e. Core group, Mid circle group, Outer circle group, and Suburb group, different number of zones may be used, but the general research methodology should be similar, as most Chinese cities have the CBD in the central city area. Moreover, if using the order and trajectory data from Uber for a similar analysis in cities in the United States, the zone divisions should be determined differently considering the discrepancy in city land use patterns in China and the United States. As a result, the circle layer division in this study won't be effective any longer, and additional strategies have to be introduced.

Although the study revealed a correlation between the proposed factors and ridesourcing demand and identified inner connections between weather conditions as well as transport accessibility to metro and market share of Express service, it is believed that more detailed hourly based analysis may assist further understanding and validation if such effect. For instance, exploring the interactive effect of weather and time that influences the service demand in a relatively short period. Moreover, future studies may further refine the model by incorporating social-economic, and demographic factors as explanatory variables. An efficient and practical method is to use smart phone data to estimate the dynamically changed population density of each zone, in which the synchronized fluctuations in ridesourcing demand and population density may be identified. Additionally, although the model reflected the longitudinal change of the market, long term observation is needed for further insights. Continuous studies are also required to assess the re-balance of market structure as well as other related issues.

References

Analysys Qianfan. (2016). Accelerated internationalization process of car-on-demand service in China and diversified service development in Q1 2016. http://www.analysyschina.com/view/viewDetail-163.html. Accessed on November 10, 2017.

Bergstrom, A., & Magnusson, R. (2003). Potential of transferring car trips to bicycle during winter. *Transportation Research Part A, 37*(8), 649–666.

Bocker, L., & Jan Faber, M. D. (2016). Weather, transport mode choices and emotional travel experiences. *Transportation Research Part A, 94*, 360–373.

Çetin, T., & Eryigit, Y. (2011). Estimating the effects of entry regulation in the Istanbul taxicab market. *Transportation Research Part A, 45*(6), 476–484.

Chang, H. W., Tai, Y. C., & Hsu, Y. J. (2010). Context-aware taxi demand hotspots prediction. *International Journal of Business Intelligence & Data Mining, 5*(1), 3–18.

Cohen, A., & Shaheen, S. (2016). *Planning for shared mobility*. American Planning Association. ISBN: 978-1-61190-186-3

Contreras, S. D., & Paz, A. (2018). The effects of ride-hailing companies on the taxicab industry in Las Vegas, Nevada. *Transportation Research Part A, 115*, 63–70.

Cramer, J., & Krueger, B. (2015). Disruptive change in the taxi business: The case of Uber. *American Economic Review, 106*(5), 177–182.

Dempsey, P. S. (1996). Taxi industry regulation, deregulation, and reregulation: The paradox of market failure. *Transportation Law Journal, 24*(1), 73–120.

D'Orey, P. M., & Ferreira, M. (2014). Can ride-sharing become attractive? A case study of taxi-sharing employing a simulation modelling approach. *IET Intelligent Transport Systems, 9*(2), 210–220.

Gao, M., Zhu, T., Wan, X., & Wang, Q. (2013). Analysis of travel time patterns in urban using taxi GPS data. In *Proceedings of the 2013 IEEE International Conference on Green Computing and Communications and IEEE Internet of Things and IEEE Cyber, Physical and Social Computing* (pp. 512–517), Beijing, China, August 20–23.

Gilbert, G., Bach, O., Dilorio, F. C., & Fravel, F. D. (1976). *Taxicab user characteristics in small and medium-size cities*. Technical report (UMTA-NC-11-0003-76-1). Urban Mass Transportation Administration.

Guo, Z., Wilson, N. H. M., & Rahbee, A. (2007). Impact of weather on transit ridership in Chicago, Illinois. *Transportation Research Record, 2034*, 3–10.

Hall, J., & Krueger, A. (2017). An analysis of the labor market for Uber's driver-partners in the United States. *ILR Review, 71*(3), 705–732.

Kamga, C., Yazici, M. A., & Singhal, A. (2015). Analysis of taxi demand and supply in New York City: Implications of recent taxi regulations. *Transportation Planning & Technology, 38*(6), 601–625.

Khattak, A. J., & Palma, A. D. (1997). The impact of adverse weather conditions on the propensity to change travel decisions: A survey of Brussels commuters. *Transportation Research Part A, 31*(3), 181–203.

Kim, S. H., & Mokhtarian, P. L. (2018). Taste heterogeneity as an alternative form of endogeneity bias: Investigating the attitude-moderated effects of built environment and socio-demographics on vehicle ownership using latent class modelling. *Transportation Research Part A, 116*, 130–150.

Liu, X., Gong, L., Gong, Y., & Liu, Y. (2015). Revealing travel patterns and city structure with taxi trip data. *Journal of Transport Geography, 43*, 78–90.

Meteorological Bureau of Shanghai. Shanghai historical weather record. http://lishi.tianqi.com/shanghai/index.html. Accessed on November 10, 2017 (in Chinese).

Raudenbush, S. W., & Bryk, A. S. (2002). *Hierarchical linear models: Applications and data analysis methods* (2nd ed.). Sage.

Rayle, L., Dai, D., Chan, N., Cervero, R., & Shaheen, S. (2016). Just a better taxi? A survey-based comparison of taxis, transit, and ride-sourcing services in San Francisco. *Transport Policy, 45*, 168–178.

Rogers, B. (2015). The social costs of Uber. *University of Chicago Law Review Dialogue, 82*, 85–102.

SAE International. (2018). Taxonomy and definitions for terms related to shared mobility and enabling technologies J3163_201809. https://www.sae.org/standards/content/j3163201809/. Accessed on June 8, 2019.

Schaller, B. (2005). A regression model of the number of taxicabs in US cities. *Journal of Public Transportation, 8*(5), 63–78.

Schaller, B. (2007). Entry controls in taxi regulation: Implications of US and Canadian experience for taxi regulation and deregulation. *Transport Policy, 14*(6), 490–506.

Singhal, A., Kamga, C., & Yazici, A. (2014). Impact of weather on urban transit ridership. *Transportation Research Part A, 69*, 379–391.

Sun, D. J., & Guan, S. (2016). Measuring vulnerability of urban metro network from line operation perspective. *Transportation Research Part A, 94*, 348–359.

Sun, D. J., Zhang, C., Zhang, L., Chen, F., & Peng, Z.-R. (2014). Urban travel behavior analyses and route prediction based on floating car data. *Transportation Letters, 6*(3), 118–125.

Sun, D. J., Zhang, K., & Shen, S. (2018). Analyzing spatiotemporal traffic line source emissions based on massive Didi online car-hailing service data. *Transportation Research Part D, 62*, 699–714.

Toner, J. P. (2010). The welfare effects of taxicab regulation in English towns. *Economic Analysis & Policy, 40*(3), 299–312.

Yang, C., & Gonzales, E. (2014). Modeling taxi trip demand by time of day in New York City. *Transportation Research Record, 2429*, 110–120.

Yang, Z., Franz, M. L., Zhu, S., Mahmoudi, J., Nasri, A., & Zhang, L. (2018). Analysis of Washington DC taxi demand using GPS and land-use data. *Journal of Transport Geography, 66*, 35–44.

Yao, B., Chen, C., Cao, Q., Jin, L., Zhang, M., Zhu, H., & Yu, B. (2017). Short-term traffic speed prediction for an urban corridor. *Computer-Aided Civil and Infrastructure Engineering, 32*(2), 154–169.

Yao, C. Z., & Lin, J. N. (2016). A study of human mobility behavior dynamics: A perspective of a single vehicle with taxi. *Transportation Research Part A, 87*, 51–58.

Yu, B., Wang, H., Shan, W., & Yao, B. (2018). Prediction of bus travel time using random forests based on near neighbors. *Computer-Aided Civil and Infrastructure Engineering, 33*(4), 333–350.

Yu, B., Yang, Z. Z., Jin, P. H., Wu, S. H., & Yao, B. Z. (2012). Transit route network design-maximizing direct and transfer demand density. *Transportation Research Part C, 22*, 58–75.

Zhang, K., Sun, D. J., Shen, S., & Zhu, Y. (2017). Analyzing spatiotemporal congestion pattern on urban roads based on taxi GPS data. *The Journal of Transport and Land Use, 10*(1), 675–694.

Zhao, K., Khryashchev, D., Freire, J., Silva, C., & Vo, H. (2016). Predicting taxi demand at high spatial resolution: Approaching the limit of predictability. In *Proceedings of the IEEE International Conference on Big Data* (pp. 833–842), Washington, DC, United States, December 5–8.

Zhong, S., & Bushell, M. (2017). Impact of the built environment on the vehicle emission effects of road pricing policies: A simulation case study. *Transportation Research Part A, 103*, 235–249.

Chapter 4
A Regression Discontinuity-Based Approach for Evaluating the Effect of Exclusive Bus Lanes on Average Vehicle Speeds

Abstract An exclusive bus lane (EBL) policy is the most common and effective bus priority policy, and therefore EBL policies have been implemented in many cities worldwide. This implementation leads to the redistribution of road resources and thus affects buses and other social vehicles to various extents. It is crucial to determine the magnitudes of these effects to enable optimal EBL implementations, but most previous methods that do so are based on simulation experiments and theoretical models, which cannot accurately reproduce real-world road situations. Thus, in this chapter, we propose an EBL evaluation method based on regression discontinuity (RD) analysis, which quantitatively evaluates the effect of an EBL on the speeds of two common types of vehicles: buses and social vehicles. As an example, we analyze the effect of an EBL implementation by examining bus and taxi global positioning system data from Shennan East Road in Shenzhen, China. The results show that after the EBL opening time, the average speed of buses increases by 8.2% and the average speed of taxis decreases by 9.9%, whereas after the EBL closing time, the average speed of buses decreases by 10.0%, and the average speed of taxis increases by 25.9%. Based on these results, we conclude that the closing time of the EBL on Shennan East Road should be delayed to maximize bus speeds at peak usage times.

Keywords Regression discontinuity · Exclusive bus lane · Policy assessment · Floating car data

1 Introduction

The rate of urbanization and the construction of transportation infrastructure are accelerating worldwide. However, the increase in private car ownership in cities means that existing road transportation resources are insufficient to meet travel demand in cities. This has led to large cities experiencing increasingly severe traffic problems, such as traffic congestion, excessive traffic pollution, and traffic accidents. Compared with other travel modes, public transportation is more efficient, has a higher capacity, and lower emissions per capita; thus, it is an important mode of urban transportation that is promoted by researchers and governments. Consequently,

S. Zhong and D. Sun, *Logic-Driven Traffic Big Data Analytics*,
https://doi.org/10.1007/978-981-16-8016-8_4

the development of public transportation is often the first approach taken to attempt to solve urban traffic problems (Diakaki et al., 2015; Viegas & Lu, 2004).

To encourage more travelers to choose public transportation, improve the operational efficiency of public transportation, and give travelers a more comfortable travel experience, public transportation priority policies have been proposed. These include public transportation signal priority and the expansion of public transportation road resources, with the most common and cost-effective type of the latter being the exclusive bus lane (EBL). An EBL is a section of road reserved for use only by buses during a specified period and is a supporting piece of infrastructure in urban transportation networks. EBLs were designed to enable bus networks to manage road traffic problems during peak usage times or emergencies and thereby improve the operational efficiency of bus systems. However, an EBL decreases the availability of road resources, which lowers the traffic efficiency of private cars (Gao et al., 2019; Szarata & Olszewski, 2019). Therefore, the use of EBLs must be examined to better understand its adverse effects and facilitate improved EBL design.

Many diverse and complex factors influence urban road traffic conditions. Thus, it is difficult to incorporate all of these factors into a complete model and then separate and analyze the effect of a single factor. Consequently, most modeling studies have considered only a few influencing factors, because of which these studies have struggled to comprehensively evaluate the influence of an EBL (Zhao & Zhou, 2018; Zheng et al., 2020). However, the generation of big data, which comprises diverse, accurate, and rich data (e.g., global positioning system (GPS) data) and the development of related novel data processing methods have enabled powerful approaches for new studies.

Accordingly, in this chapter, an EBL policy evaluation method based on regression discontinuity (RD) design is proposed. Shennan East Road in Shenzhen, China, is used as a case study, and the effect of an EBL on the speed of buses and social cars on this road is quantitatively evaluated using GPS data. The RD design has the advantage of not requiring the effect of all control variables on the variables to be known; instead, it focuses on the mutation of the result variables at the discontinuous point.

The remainder of this chapter is structured as follows. Section 2 reviews the literature and details the EBL and RD design. Section 3 details the methodology, and Sect. 4 introduces the case study. Section 5 presents the main results and analysis. Finally, Sect. 6 summarizes the findings and discusses their policy implications.

2 Literature Review

This literature review is divided into two parts. The first part reviews the results and methods reported in studies on the effects of EBLs and analyzes the limitations of these studies. The second part reviews studies on RD design, summarizes the development process and application fields of RD design, and explains the feasibility and reliability of the method.

2.1 *Exclusive Bus Lane*

The conventional methods for evaluating the effect of an EBL can be divided into three categories: experience-based methods, simulation-based methods, and theoretical-model-based methods. Early researchers in this field have evaluated EBLs based on their experience and have summarized the methods of designing an EBL (Cox, 1975; Feather et al., 1973). As these empirical studies often did not include quantitative analyzes, other researchers have since used improved traffic assignment models to evaluate the effect of an EBL (e.g., Yu et al., 2015). In addition, researchers have proposed evaluating the effect of an EBL via a multi-mode dynamic stochastic user equilibrium model by assuming a fixed road travel time and via a bi-modal user equilibrium model incorporating travelers' risk-averse behavior (Li & Ju, 2009; Yao et al., 2015). However, the assumptions that form the basis of theoretical models often differ from the bases of an actual road network. Thus, given the wide application of traffic simulation in traffic engineering, some researchers have used computer simulations to study motor vehicle behavior in the presence of an EBL in realistic settings. These studies evaluate the effect of an EBL by comparing traffic flow operation indicators in the absence and presence of an EBL. Shalaby (1999) used TRANSYT-7F to simulate and analyze the behavioral changes of buses and other motor vehicles in the absence and presence of an EBL. The results showed that the EBL significantly improved bus operation, but it somewhat negatively affected nearby non-bus vehicle operation. Currie et al. (2007) proposed a trade-off model for implementing an EBL when road space is limited. They used a traffic microsimulation model and a travel behavior model to evaluate the effect of road resource redistribution and a change in travel mode, and proposed an overall effect assessment method based on social cost-effectiveness. Arasan and Vedagiri (2010) constructed a new microsimulation model for mixed traffic flow, named HETEROSIM, to simulate the behavior characteristics of motor vehicles in real situations and to evaluate the effect of an EBL; they found that the speed of buses would be improved if an EBL was applied in Indian cities with highly mixed traffic flows.

In essence, traffic simulation methods can be regarded as random tests, and their results have certain practical significance. However, because simulations cannot fully reproduce the real state of a road system, the results of simulation-based studies can only be used as theoretical information to aid in implementing an EBL policy and do not accurately reflect post-implementation effects. In addition, the factors affecting urban road traffic states are complex, difficult to identify, and challenging to simulate; thus, the utility of traffic simulation methods is limited by their inability to faithfully reproduce real-life situations. Therefore, an accurate and comprehensive study of the effect of EBL implementation requires field measurements (Zhao et al., 2019).

2.2 Regression Discontinuity

The RD design was first proposed by Campell, a psychologist at Northwestern University (United States), in 1958. Subsequently, the advantages of the RD design were highlighted, and RD methods were gradually improved and accepted by the academic community (Hahn et al., 2001; Lee & Lemieux, 2010; Sacks & Ylvisaker, 1978). Most early research on the application of RD designs was in the field of pedagogy. Thislethwaite and Campell (1960) first applied RD design to pedagogy by analyzing the effect of students' scholarship on future academic achievements. Angrist and Lavy (1999) and Van der Klaauw (2002) have used RD design to evaluate the effect of class size and financial assistance on students' educational outcomes, respectively. Black (1999) first used discontinuous points in a geography study to analyze the effect of school district boundaries on parents' willingness to pay for high-quality schools. Over the past two decades, increasing numbers of researchers in different fields have used RD designs to solve problems, especially in the fields of sociology and economics. For example, RD-based studies have examined aspects of the labor market, medical and health policy, policy economics, crime, and environmental problems (Ebenstein et al., 2017; Imbens & Lemieux, 2008).

In recent years, researchers have increasingly used RD designs to evaluate the effects of traffic policies, such as to analyze the effects of the cost of roadside parking on the retail industry (Hymel, 2014); the effects on driving of policies prohibiting the use of mobile phones while driving (Burger et al., 2014); the effects of urban road pricing policies on surrounding rents (D'Arcangelo & Percoco, 2015); the effects of traffic restriction policies on air pollution and labor supply (Viard & Fu, 2015); the effects of the cancellation of road toll zones on traffic flow and vehicle emissions (Ding et al., 2021); the effects of free bus fares on the travel behavior of older adults (Shin, 2021); and the effects of pricing policy changes on-street parking demand and user satisfaction (Mo et al., 2021). Although researchers have confirmed the applicability of RD designs to traffic policy analysis, no studies have used RD designs to analyze vehicle GPS data to evaluate the effect of an EBL.

3 Methodology

3.1 Introduction to RD and Description of Variables

RD design is a statistical method for the evaluation of policy effects that examines a treatment effect under non-experimental conditions (Liu et al., 2018). Thus, RD studies examine the relationship between three types of variables: running variables, outcome variables, and binary treatment variables. Each experimental individual corresponds to an observable running variable: if the value of this running variable is greater than a known discontinuity point, the individual will accept the treatment; conversely, if the value of the running variable is less than the discontinuity point, the individual will not accept the treatment. In this way, all of the test individuals are divided into two categories with respect to the discontinuity point: those with running variable values greater (less) than the discontinuity point are denoted the experimental (control) group. If other control variables that affect the outcome variable are continuous at the discontinuity point, the distribution of the experimental individuals near the discontinuity point can be regarded as random, i.e., the probability of their accepting the treatment is random. Thus, the average difference between the experimental group and the control group near the discontinuity point reflects the size of the treatment effect.

We use an RD design to estimate the relationship between the average speed of vehicles (buses or taxis) and the EBL, based on EBL operation time. Specifically, we test whether the EBL strategy causes discontinuous changes in the average speed of a vehicle (bus or taxi) after the opening or closing of the EBL. The necessary assumption is that any unobserved determinants of the EBL change smoothly when the EBL is in operation. If the relevant assumption is valid, the adjustment of a sufficiently flexible polynomial or the application of linear regression on the operation time of the EBL will remove all potential sources of bias and allow a causal inference to be made. The RD treatment effects (τ) of the EBL based on average speed can be depicted as in Fig. 1, and derived directly by the following equation:

$$\tau = \lim_{\delta \to 0}\{E[S|t_c - \delta < t < t_c] - E[S|t_c < t < t_c + \delta]\} = E[S(\text{after}) - S(\text{before})] \tag{1}$$

where S is the outcome variable, i.e. the average speed of the vehicle (bus or taxi). t is the running variable, i.e. the time, which is a continuous variable. t_c is the discontinuous point, which is either the opening or closing time of the EBL. $S(\text{after})$ and $S(\text{before})$ are the average speeds of the vehicle (bus or taxi) after and before the

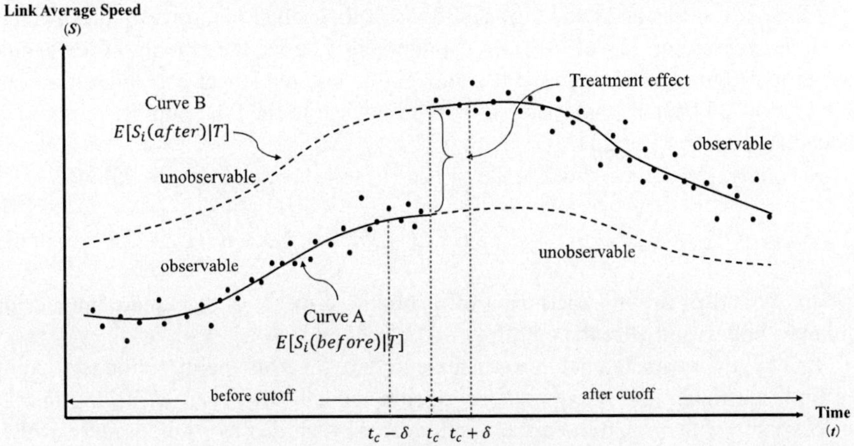

Fig. 1 Sketch of the RD treatment effects of the EBL based on average speed

opening or closing of the EBL. δ is the time gap to the cutoff for the intuition that the RD estimates should use observations "close" to the cutoff (e.g. in this case, at points $t_c - \delta$ and $t_c + \delta$).

3.2 RD Model

The research object of this chapter is an EBL during peak hours. At specific times of the day, an EBL is open only to buses, and there will be discontinuities at its opening time and closing time. Before and after the discontinuous point, only the road resource allocation changes, so an RD model can be used to evaluate the effect of policy discontinuities. For the convenience of regression, the basic model is typically a linear model.

Firstly, the overall regression model describing the vehicle speed on the EBL is proposed as follows:

$$S_t = \alpha + \beta_0 \cdot M_t + \beta_1 t + \beta_2 t \cdot M_t + \gamma X_t + \varepsilon_t \tag{2}$$

where S_t is the average speed of vehicles. t is the number of minutes before and after the cutoff, and t is equal to zero at the cutoff. M_t is a binary treatment variable, which equals to zero when the EBL is open to private vehicles, and equals to one when the EBL is closed and only can be used by buses. X_t is a vector of other control variables. ε_t is the white noise. α is a constant. The coefficient of each variable reflects the degree of influence of each variable on the dependent variable. β_0 represents the treatment effect. β_1 and β_2 reflect the trend of speed over time. γ is a vector combined by the coefficients of other control variables. This paper uses ordinary least squares (OLS) to get the regression results. In addition, the linear model can be extended to a polynomial regression model to test its robustness.

Models of buses and taxis differ in terms of their other control variables, which for a bus regression model includes the number of buses, the number of taxis, and the ratio of buses to taxis. In the taxi model, the taxi passenger carrying ratio (i.e., the proportion of taxis with passengers with respect to the total number of taxis) is added as a control variable.

Secondly, the second-order model is used to test its robustness, as follows:

$$S_t = \alpha + \beta_0 \cdot EBL_t + \beta_1 t + \beta_2 t \cdot EBL_t + \beta_3 t^2 + \beta_4 t^2 \cdot EBL_t + \gamma X_t + \varepsilon_t \tag{3}$$

where β_3 and β_4 are the coefficient of quadratic terms. t^2 is the square number of minutes before and after the cutoff.

Finally, the estimated value is sensitive to data far from the discontinuous point in both the linear and polynomial regression models. However, according to RD theory, only data around the discontinuity can be regarded as a random test, which means that the points close to the discontinuity are regarding as having more weight than points not close to the discontinuity. This target can be achieved by a partially

Table 1 Expressions of the three kernel functions

Kernel function type	Function expression
Rectangular	$K(u) = \frac{1}{2} \quad \lvert u \rvert \leq 1$
Triangular	$K(u) = 1 - \lvert u \rvert \quad \lvert u \rvert \leq 1$
Epanechnikov	$K(u) = \frac{3}{4}(1-u^2) \quad \lvert u \rvert \leq 1$

weighted regression model with kernel function K. As t_c is the cutoff, data with the bandwidth of h on both sides of the discontinuity is separated into two groups. The data before the cutoff is the control group data, and the data after the cutoff is the experimental group data. The linear regression model of the experimental group is expressed as follows:

$$S_i = a_r + b_r \cdot t_i + \varepsilon_r \tag{4}$$

where S_i is the average speed of the vehicle in the minute i of the experimental group (to the right of the discontinuity point). t_i is the number of minutes i after the discontinuity point. ε_r is the white noise of the experimental group. a_r is a constant of the experimental group. b_r is the regression coefficient of the experimental group.

Using a partially weighted regression model, we can obtain (a_r, b_r) satisfying the following formulation, where function K is one of rectangular, triangular, and Epanechnikov kernels, as shown in Table 1 (Mora-Garcia et al., 2015).

$$u = (t_i - t_c)/h \tag{5}$$

$$\left(a_r, b_r\right) = \arg\min \sum_{i=1}^{N_r} [S_i - a_r - b_r(t_i - t_c)]^2 \cdot K(u) \cdot 1 \quad (t_i > t_c) \tag{6}$$

where u is the ratio of the time from the discontinuity point to the ith minute to the bandwidth. N_r is the total number of minutes in the experimental group.

Then the estimated value at the discontinuous point in the experimental group a_r can be calculated by

$$S_r(t_c) = a_r + b_r(t_c - t_c) = a_r \tag{7}$$

Using a similar method, we can obtain the estimated value at the discontinuous point in the control group a_l. Finally, we can get the treatment effect τ by

$$\tau = a_r - a_l \tag{8}$$

To test the robustness of the partially weighted regression model, we change the type of kernel function K and the bandwidth h of the data.

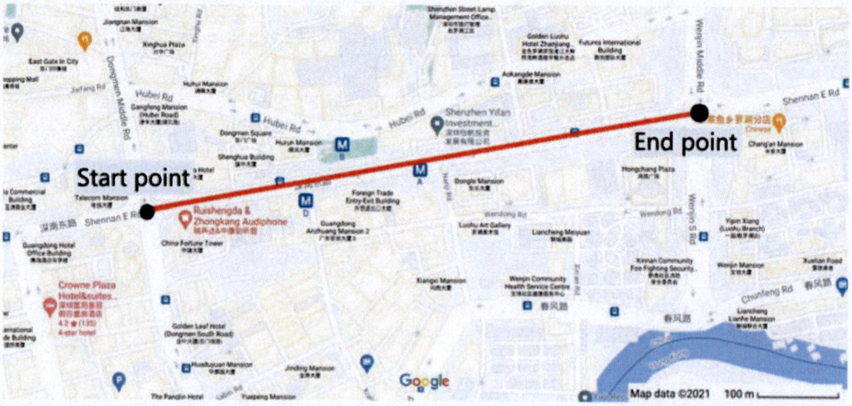

Fig. 2 Schematic diagram of the study road section—Shennan East Road (© Google)

4 Case Study

4.1 Study Region

Our research object is a section of Shennan East Road, a two-way eight-lane arterial road in the center of Shenzhen, China. The section starts at the intersection of Jinwen Middle Road and Shennan East Road and ends at the intersection of Dongmen South Road and Shennan East Road. This section is 900 m long and has two signalized intersections and two curbside bus stops served by 16 bus lines (Fig. 2). The EBL is set at the outermost side of the road, and is open daily during the morning rush hour (7:30–9:30 a.m.) and the evening rush hour (5:30–7:30 p.m.).

4.2 Data Collection

As it is difficult to collect the data of all private vehicles, taxi data are used as a proxy to represent private vehicles' data in this case study. The average speeds per minute of buses and taxis are derived from GPS records every 15 s for 5 h on 5 days (from 6:00 a.m. to 11:00 a.m. on June 9–13, 2014). Shenzhen taxi and bus operation data have different data structures. Taxi GPS data comprises a taxi's serial number, license plate number, direction, speed, recording time, the latitude and longitude coordinates, and the presence of passengers. Bus GPS data comprises a bus's license plate, route number, arrival and departure time, speed, direction angle number, the recording time, and the latitude and longitude coordinates.

4.3 Analysis Process

The process for RD analysis of the EBL based on the GPS data of buses and taxis is introduced in Fig. 3. It consists of the following five steps:

Step 1: Data processing, including screening repeated records, eliminating abnormal records, and correcting the coordinate offset of GPS data.
Step 2: Map-matching of filtered GPS data to the target road using ArcGIS software.
Step 3: Calculation of the time-dependent velocities of buses and taxis using SPSS software.
Step 4: RD design and analysis, including fitting images, solving regression models, and testing the model robustness.
Step 5: Discussion of the results and findings to provide suggestions for system improvement.

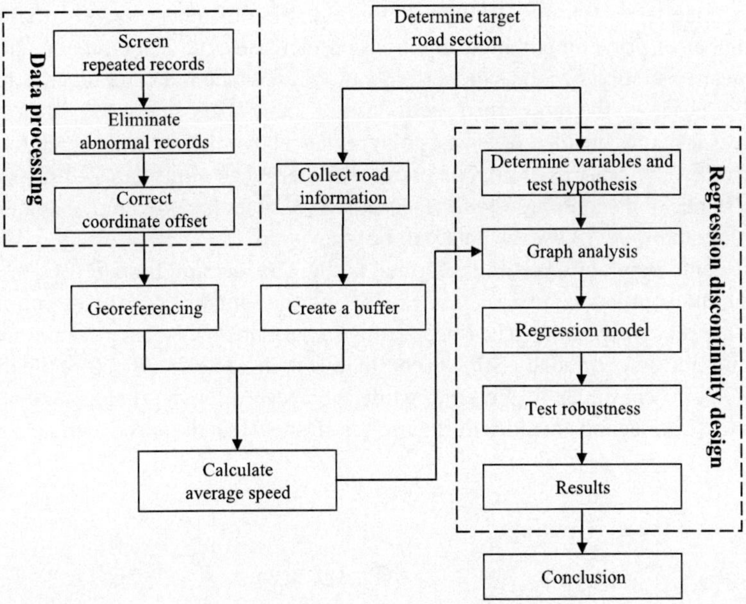

Fig. 3 The RD analysis of BEL based on the GPS data of buses and taxis

5 Results and Analysis

5.1 *Bus Regression Results*

5.1.1 Graph Analysis of Buses

As the opening of the EBL provides additional road resources for buses, the average speed of buses increases significantly after the EBL opens. This means that a discontinuous point of speed is formed, and the mutation value of the discontinuous point is positive. To graphically depict the occurrence of discontinuous points, an image of the entire data is drawn before regression, and the data points are then fitted. The three images in Fig. 4 show the change in the average bus speed over time under various polynomial fitting conditions. Each point in the figure represents the average bus speed for the corresponding number of minutes, and the line in the figure is the fitted time trend line. The quadratic and cubic polynomial fitting graphs show that at the EBL opening time (7:30 a.m.), the average speed of buses decreases by approximately 0.5–0.8 m/s. Without the introduction of other control variables, as the number of polynomials used in the fitting increases, the discontinuous point of the upward mutation becomes increasingly obvious, which indicates that the average speed of buses on the target road section is not linear with time when other control variables are not considered. Using polynomials above the second degree to fit the data creates obvious discontinuous points, but the image alone cannot fully explain the existence of discontinuous points, because the higher the polynomial degree used for fitting, the more likely the data will be over-fitted.

The same graph analysis is performed at the EBL closing time (9:30 a.m.). The discontinuous points can be seen in Fig. 5, and the mutations at the discontinuous points are relatively stable. The linear fitting image shows the change in the slope of the trend line before and after the discontinuous point. The speed before the discontinuous point continues to increase, while the speed after the discontinuous point stabilizes. The second and third fitting images all show that the speed increases before

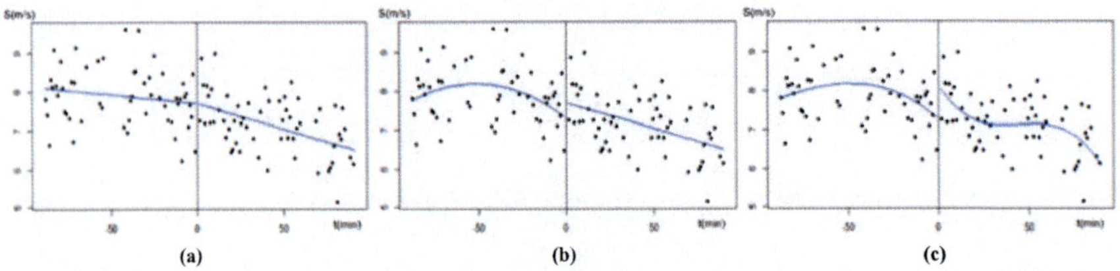

Fig. 4 (**a**) Change of average bus speed overtime during 6:00–9:00 a.m. (1st order global polynomial). (**b**) Change of average bus speed overtime during 6:00–9:00 a.m. (2nd order global polynomial). (**c**) Change of average bus speed overtime during 6:00–9:00 a.m. (3rd order global polynomial)

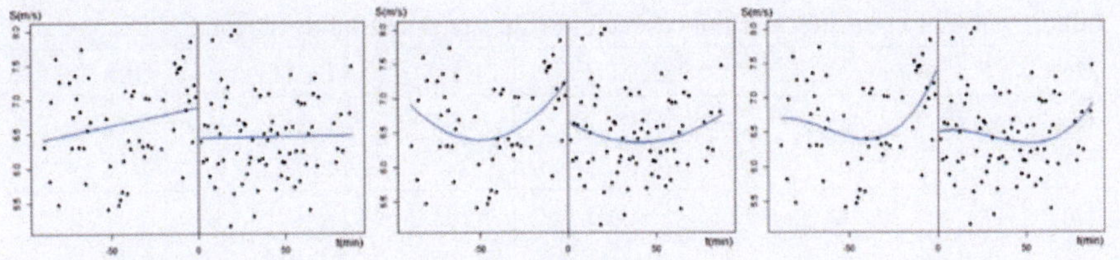

Fig. 5 (**a**) Change of average bus speed overtime during 8:00–11:00 a.m. (1st order global polynomial). (**b**) Change of average bus speed overtime during 8:00–11:00 a.m. (2nd order global polynomial). (**c**) Change of average bus speed overtime during 8:00–11:00 a.m. (3rd order global polynomial)

the discontinuous point and that the speed slightly decreases after the discontinuous point.

5.1.2 Regression Model Analysis of Buses

To further examine whether there is a treatment effect at the discontinuous point, four full-data ordinary least-squares (OLS) regression experiments are conducted in the EBL opening and ending time periods; the regression results are shown in Tables 2 and 3, respectively. OLS(1) (Table 2) is a linear regression of the OLS method that only uses binary treatment variables and time running variables. The obtained treatment effect value is 0.006, which means that after the discontinuous point, the average speed of the bus section increases by 0.006 m/s. As this value is close to zero, and the standard error of the estimate is large (0.238), this treatment effect may not be real. OLS(2) regression is based on OLS(1) and uses three variables: the number of buses and taxis per minute and the ratio of these vehicles on the target road section. The treatment effect after considering other control variables is found to be −0.03, which is too small, and the standard error remains large (0.242); thus, the existence of a treatment effect is again not proven. However, this result confirms the results of image analysis: linear regression fitting cannot reveal obvious discontinuous points. In the regression of OLS(3), a quadratic term (time t) is introduced, based on OLS(2). The treatment effect is 0.34, which is significantly better than that of OLS(2); in addition, the relative value of the standard error is lower (0.347) and the R^2 value of the overall regression (0.363) is higher than those of OLS(2). This shows that regression of the quadratic polynomial is better than linear regression for analyzing the entire dataset. The OLS(4) regression increases the polynomial for time t to the third degree, and the result (0.212) is not markedly different from that of the OLS(3). Thus, it can be considered that the estimate of the disposal effect has stabilized.

The same regression analysis is carried out at the EBL closing time (9:30 a.m.) These results (Table 3) also prove the existence of the discontinuous point; the average speed before and after the discontinuous point decreases by approximately 0.6 m/s. Moreover, the closer the data is to the discontinuous point, the greater the absolute

Table 2 Overall regression result of bus (cutoff: 7:30 a.m.; bandwidth: 180 min)

Model	OLS(1)	OLS(2)	OLS(3)	OLS(4)
Treatment effect	0.006	−0.03	0.34	0.212
	(0.238)	(0.242)	(0.347)	(0.311)
Time	−0.004	−0.01	−0.037	−0.026
	(0.003)	(0.005)	(0.014)	(0.01)
Time × treatment variable	−0.009	−0.005	0.023	0.014
	(0.005)	(0.006)	(0.019)	(0.013)
Constant	7.689	7.345	6.668	6.744
	(0.170)	(0.384)	(0.486)	(0.475)
Number of taxi per min	–	0.003	0.0003	0.001
	–	(0.005)	(0.005)	(0.005)
Number of bus per min	–	0.04	0.037	0.037
	–	(0.029)	(0.029)	(0.029)
Number ratio taxi/bus per min	–	0.02	0.043	0.042
	–	(0.053)	(0.054)	(0.054)
Average speed before the cutoff	7.89	7.89	7.89	7.89
Polynomial order	Linear	Linear	Quadratic	Cubic
Sample size	116	116	116	116
R-square	0.3	0.34	0.363	0.363
Cutoff	7:30 a.m	7:30 a.m	7:30 a.m	7:30 a.m

Note Robust standard errors are in parentheses

value of the treatment effect, with the maximum absolute value being approximately 0.8. This is similar to the results obtained for the EBL opening time (7:30 a.m.) analysis. However, the treatment effect at the EBL closing time is larger than that at the EBL opening time, indicating that the effect of the EBL is greater in the period around its closing time.

By comparing the regression results for the foregoing two discontinuous points, we can determine the effect of the EBL on buses. Before and after the opening time of the EBL, the overall average speed of buses on the target road section increases by approximately 8.2%. Before and after the closing time of the EBL, the overall average speed of buses on the target road section decreases by 9.9%. As the increase in bus speed after the opening time is less than the decrease in bus speed after the closing time, the effect of the opening time is not as obvious as that of the closing time.

Table 3 Overall regression result of bus (cutoff: 9:30 a.m.; bandwidth: 180 min)

Model	OLS(1)	OLS(2)	OLS(3)	OLS(4)
Treatment effect	−0.450	−0.345	−0.636	−0.55
	(0.207)	(0.204)	(0.283)	(0.253)
Time	0.0004	0.002	−0.006	−0.003
	(0.003)	(0.003)	(0.011)	(0.007)
Time × Treatment variable	0.005	0.003	0.042	0.026
	(0.004)	(0.004)	(0.016)	(0.001)
Constant	6.438	5.454	5.602	5.605
	(0.137)	(0.419)	(0.464)	(0.451)
Number of taxi per min	–	0.001	0.0008	0.001
	–	(0.007)	(0.007)	(0.007)
Number of bus per min	–	0.056	0.055	0.053
	–	(0.026)	(0.026)	(0.026)
Number ratio taxi/bus per min	–	0.034	0.037	0.031
	–	(0.091)	(0.089)	(0.089)
Ave. speed before cutoff	6.66	6.66	6.66	6.66
Polynomial order	Linear	Linear	Quadratic	Cubic
Sample size	132	132	132	132
R-square	0.050	0.119	0.177	0.174
Cutoff	9:30 a.m	9:30 a.m	9:30 a.m	9:30 a.m

Note Robust standard errors are in parentheses

5.1.3 Robustness Test for the Bus Treatment Effect

To further test the robustness of the estimated treatment effect, we conduct regression experiments with different data ranges. The regression results of bus data in different bandwidths of the EBL opening time period are shown in Table 4. The results show that when the data range is 60 min before and after the discontinuous point, the difference between the treatment effect estimates obtained through the quadratic regression equation is the smallest; this indicates that the treatment effect is not affected by the regression times when the data bandwidth is 60 min. However, when the data bandwidth is greater or less than 60 min, the linear regression results are lower in magnitude than the quadratic regression results, and the quadratic regression results are closer to the results for the 60-min bandwidth. This once again shows that when performing OLS regression, the results obtained from high-order polynomials are better than those obtained from linear regression, as the former results are barely affected by the data range. Thus, the OLS regression estimation is robust and credible. Compared with the results of regression on all data, the treatment effect estimate in the small data range is larger. This is unsurprising because the estimate obtained using polynomial regression is more affected (than the estimate obtained by linear

Table 4 Regression result of bus in different bandwidth (cutoff: 7:30 a.m.)

Bandwidth	30 min	30 min	60 min	60 min	120 min	120 min
Treatment effect	0.288	0.678	0.627	0.689	0.13	0.64
	(0.601)	(0.956)	(0.355)	(0.547)	(0.285)	(0.418)
Time	0.034	−0.133	−0.013	0.049	−0.023	−0.03
	(0.054)	(0.233)	(0.017)	(0.068)	(0.008)	(0.027)
Time × Treatment variable	−0.071	0.129	−0.025	−0.175	0.011	−0.026
	(0.082)	(0.440)	(0.022)	(0.096)	(0.009)	(0.037)
Constant	7.718	7.586	6.662	6.517	6.559	6.527
	(1.448)	(1.558)	(0.931)	(0.987)	(0.487)	(0.681)
Number of taxi per min	0.009	0.017	−0.018	−0.021	0.001	−0.0003
	(0.038)	(0.041)	(0.022)	(0.022)	(0.013)	(0.014)
Number of bus per min	−0.012	−0.033	0.039	0.065	0.049	0.05
	(0.096)	(0.107)	(0.063)	(0.064)	(0.042)	(0.045)
Number ratio taxi/bus per min	−0.031	−0.190	0.326	0.410	0.051	0.053
	(0.536)	(0.603)	(0.302)	(0.303)	(0.184)	(0.202)
Ave. speed before cutoff	7.47	7.47	7.66	7.66	7.86	7.86
Polynomial order	Linear	Quadratic	Linear	Quadratic	Linear	Quadratic
Sample size	26	26	47	47	80	80
R-square	0.119	0.148	0.253	0.303	0.243	0.285
Cutoff	7:30 a.m	7:30 a.m	7:30 a.m	7:30 a.m	7:30 a.m	7:30 a.m

Note Robust standard errors are in parentheses

regression) by data far away from the discontinuous point. Thus, only by continuously narrowing the data range to a certain interval around the discontinuous point can we determine the true treatment effect value.

Next, we perform a partial regression under different conditions (different kernel functions and different bandwidths), with the target value of the regression being the treatment effect. The results of bus partial regression are shown in Table 5. When the bandwidth is too large, the estimated treatment effect is almost non-existent. As the bandwidth decreases, the estimate increases and stabilizes in the range of 0.55–0.70, similar to the results of the foregoing OLS regression. This indicates that the results obtained by parametric regression and non-parametric regression are similar and are thus reliable.

The same robustness test is performed at the EBL closing time (9:30 a.m.), and the results are shown in Table 6. When the bandwidth is 120 min before and after the discontinuous point, the difference between the treatment effect estimates obtained by the two regression equations is the smallest; this indicates that the treatment effect is not affected by the regression times in this bandwidth.

The results of partial regression are shown in Table 7. As the bandwidth decreases, the estimate increases and then stabilizes in the range of −0.62 to −0.81; this is

Table 5 Estimated treatment effect of bus partial regression (cutoff: 7:30 a.m.)

Kernel function type	Bandwidth			
	90 min	45 min	30 min	15 min
Rectangular	−0.037	0.592	0.689	0.546
	(0.225)	(0.356)	(0.403)	(0.621)
Triangular	0.115	0.64	0.731	0.602
	(0.258)	(0.382)	(0.485)	(0.822)
Epanechnikov	0.035	0.605	0.76	0.549
	(0.244)	(0.362)	(0.453)	(0.772)

Note Robust standard errors are in parentheses

Table 6 Regression result of bus in different bandwidth (cutoff: 9:30 a.m.)

Bandwidth	30 min	30 min	60 min	60 min	120 min	120 min
Treatment effect	−0.697	−0.808	−0.510	−0.866	−0.660	−0.584
	(0.472)	(0.785)	(0.359)	(0.584)	(0.234)	(0.348)
Time	0.019	−0.004	−0.006	0.062	−0.003	−0.001
	(0.034)	(0.161)	(0.014)	(0.060)	(0.004)	(0.002)
Time × Treatment variable	−0.037	0.054	0.017	−0.057	0.027	0.012
	(0.052)	(0.260)	(0.022)	(0.091)	(0.007)	(0.031)
Constant	9.595	9.771	5.937	5.802	5.996	6.003
	(1.881)	(2.188)	(0.834)	(0.887)	(0.490)	(0.572)
Number of taxi per min	0.056	0.058	−0.003	−0.004	0.001	0.002
	(0.030)	(0.032)	(0.011)	(0.012)	(0.007)	(0.008)
Number of bus per min	−0.207	−0.216	0.027	0.018	0.027	0.025
	(0.113)	(0.123)	(0.051)	(0.056)	(0.031)	(0.033)
Number ratio taxi/bus per min	−0.862	−0.882	0.112	0.110	0.027	0.021
	(0.481)	(0.507)	(0.148)	(0.153)	(0.096)	(0.099)
Ave. speed before cutoff	7.10	7.10	6.97	6.97	6.63	6.63
Polynomial order	Linear	Quadratic	Linear	Quadratic	Linear	Quadratic
Sample size	96	96	96	96	96	96
R-square	0.356	0.361	0.194	0.220	0.171	0.226
Cutoff	9:30 a.m	9:30 a.m	9:30 a.m	9:30 a.m	9:30 a.m	9:30 a.m

Note Robust standard errors are in parentheses

similar to the results of the foregoing OLS regression, indicating that the results are reliable.

Table 7 Estimated treatment effect of bus partial regression (cutoff: 9:30 a.m.)

Kernel function type	Bandwidth			
	180 min	90 min	60 min	30 min
Rectangular	−0.456	−0.702	−0.624	−0.814
	(0.204)	(0.259)	(0.283)	(0.341)
Triangular	−0.521	−0.659	−0.770	−0.745
	(0.216)	(0.256)	(0.283)	(0.288)
Epanechnikov	−0.485	−0.638	−0.764	−0.739
	(0.213)	(0.263)	(0.294)	(0.302)

Note Robust standard errors are in parentheses

5.2 *Taxi Regression Results*

As mentioned above, although the opening of the EBL is convenient for buses, it may have an adverse effect on social vehicles. Therefore, we analyze the effect of the EBL on taxis (as representative social vehicles).

5.2.1 Graph Analysis of Taxis

As before, prior to performing data regression at the opening time of the EBL, we first examine the discontinuous points in images. Figure 6 shows the variation in the average speed of the taxi section over time, with different fitting times (increasing from 1 to 3 from left to right). These image-fitting results are different from those for buses in the foregoing analysis. Specifically, as the number of fittings increases, the magnitude of the sudden change at the discontinuous point gradually decreases.

The same graph analysis is also performed at the closing time of the EBL. The images under different conditions in Fig. 7 reflect the positive mutation of speed at

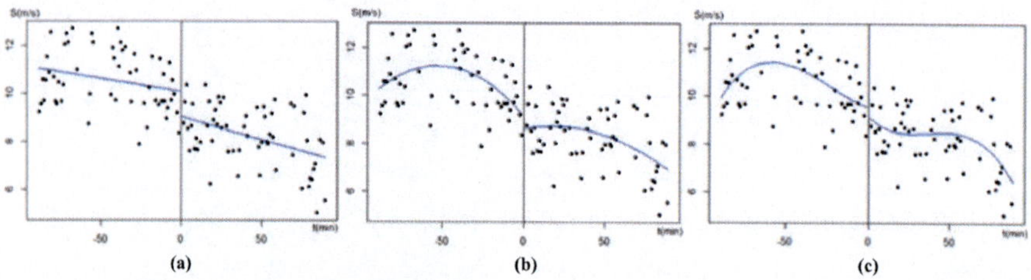

Fig. 6 (**a**) Change of average taxi speed overtime during 6:00–9:00 a.m. (1st order global polynomial). (**b**) Change of average taxi speed overtime during 6:00–9:00 a.m. (2nd order global polynomial). (**c**) Change of average taxi speed overtime during 6:00–9:00 a.m. (3rd order global polynomial)

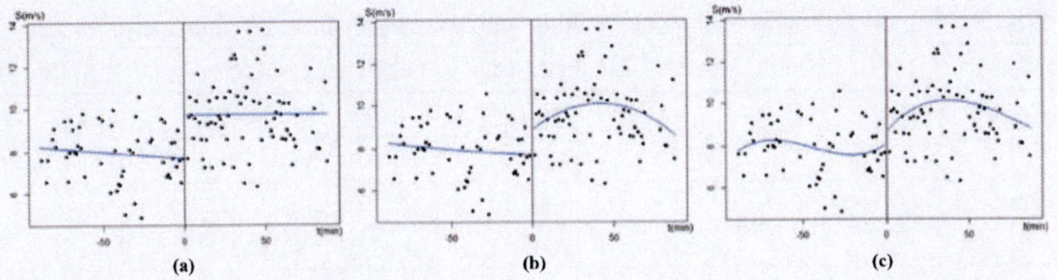

Fig. 7 (**a**) Change of average taxi speed overtime during 8:00–11:00 a.m. (1st order global polynomial). (**b**) Change of average taxi speed overtime during 8:00–11:00 a.m. (2nd order global polynomial). (**c**) Change of average taxi speed overtime during 8:00–11:00 a.m. (3rd order global polynomial)

the discontinuous points, and the mutation value of the linear fitting is significantly greater than that of the quadratic and cubic fittings.

5.2.2 Regression Model Analysis of Taxis

The results of the full review of the EBL opening time period data (see Table 8) also confirm the characteristics reflected by the image, i.e., as the degree of the polynomial increases, the value of the treatment effect decreases. Both OLS(1) and OLS(2) are linear regressions. The former does not use other variables, while the latter uses four other control variables: passenger ratio, number of taxis, number of buses, and ratio of taxis to buses. However, the treatment effect estimates obtained by OLS(1) and OLS(2) are very similar (−1.045 and −0.969, respectively) and their standard errors are relatively small. This shows that in linear regressions, other control variables have a weak influence on the estimated treatment effect, i.e., the average taxi speed. However, when the number of regressions is increased to two, the treatment effect value reduces by nearly half (to −0.493). When the number of regressions is increased to three, the treatment effect value increases to −0.627. However, when the number of regressions is increased by between two and three times, the R^2 value is higher than that for the aforementioned two linear regressions, and the standard errors of the obtained estimates are too large, so the obtained treatment effect estimates are not reliable.

The same regression analysis is carried out at the EBL closing time (9:30 a.m.) The regression results of the overall data, shown in Table 9, show that other control variables have little effect on the linear OLS results. The first two estimates of the treatment effect are very similar (2.039 and 2.161), whereas the results of quadratic and cubic polynomial regression are relatively small (1.249 and 1.571, respectively).

Comprehensive analysis shows that after the EBL opening time, the average taxi speed on the target road section decreases by approximately 10.0%, whereas after the EBL closing time, the average taxi speed increases by approximately 25.9%. This increase after EBL closing is more than twice the decrease after EBL opening, which shows that EBL closing time has a greater effect on the average speed of taxis.

Table 8 Overall regression result of taxi (cutoff: 7:30 a.m.; bandwidth: 180 min)

Model	OLS(1)	OLS(2)	OLS(3)	OLS(4)
Treatment effect	−1.045	−0.969	−0.493	−0.627
	(0.399)	(0.410)	(0.574)	(0.510)
Time	−0.011	−0.007	−0.072	−0.048
	(0.005)	(0.009)	(0.024)	(0.016)
Time × Treatment variable	−0.008	−0.014	0.081	0.050
	(0.008)	(0.010)	(0.033)	(0.021)
Constant	10.085	10.321	9.813	9.998
	(0.287)	(0.934)	(0.920)	(0.900)
Number of taxi with/without passengers	–	0.817	0.337	0.326
	–	(0.665)	(0.662)	(0.654)
Number of taxi per min	–	−0.012	−0.012	−0.013
	–	(0.008)	(0.008)	(0.008)
Number of bus per min	–	0.010	−0.003	−0.001
	–	(0.045)	(0.044)	(0.043)
Number ratio taxi/bus per min	–	−0.581	−0.497	−0.555
	–	(0.370)	(0.360)	(0.355)
Ave. speed before cutoff	10.56	10.56	10.56	10.56
Polynomial order	Linear	Linear	Quadratic	Cubic
Sample size	117	117	117	117
R-square	0.526	0.528	0.560	0.568
Cutoff	7:30 a.m	7:30 a.m	7:30 a.m	7:30 a.m

Note Robust standard errors are in parentheses

It also shows that in the time period around the closing time (9:00–10:00 a.m.), taxis have a greater need to use the outer lane where the EBL is located, probably to pick up and drop off passengers. In contrast, there is no significant reduction in taxis' average speed during the opening period (7:00–8:00 a.m.), which may be because residents do not have a great demand for taxis at this time and the overall traffic operation in this period is fast, such that the effect of taxis on buses is not large. The regression coefficient of the number of taxis with/without passengers in the regression results also confirms this result. In the discontinuity analysis at the opening time, this coefficient is approximately 0.8, whereas, in the discontinuity analysis at the closing time, the coefficient is approximately 2, which shows that the number of passengers has the largest effect on the speed of taxis after the peak period. Therefore, if the EBL is closed so that taxis can freely use the outer lane, i.e. the EBL, the number of taxi passengers can be greatly increased, which increases the average speed of taxis.

Table 9 Overall regression result of taxi (cutoff: 9:30 a.m.; bandwidth: 180 min)

Model	OLS(1)	OLS(2)	OLS(3)	OLS(4)
Treatment effect	2.039	2.161	1.249	1.571
	(0.520)	(0.543)	(0.722)	(0.649)
Time	−0.0002	−0.007	0.049	0.025
	(0.008)	(0.008)	(0.028)	(0.017)
Time × Treatment variable	0.006	0.002	−0.034	−0.023
	(0.011)	(0.011)	(0.040)	(0.025)
Constant	9.701	10.752	9.423	9.610
	(0.345)	(0.889)	(1.078)	(1.044)
Number of taxi with/without passengers	–	1.619	1.836	1.837
	–	(0.771)	(0.785)	(0.786)
Number of taxi per min	–	−0.010	−0.011	−0.011
	–	(0.006)	(0.006)	(0.006)
Number of bus per min	–	−0.094	−0.062	−0.062
	–	(0.046)	(0.048)	(0.048)
Number ratio taxi/bus per min	–	0.086	0.114	0.151
	–	(0.079)	(0.080)	(0.080)
Average speed before the cutoff	7.93	7.93	7.93	7.93
Polynomial order	Linear	Linear	Quadratic	Cubic
Sample size	130	130	130	130
R-square	0.234	0.280	0.344	0.275
Cutoff	9:30 a.m	9:30 a.m	9:30 a.m	9:30 a.m

Note Robust standard errors are in parentheses

5.2.3 Robustness Test of the Taxi Results

To test the robustness of the valuation, we conduct OLS regressions in different data ranges. The regression results of taxis in different bandwidths around the EBL opening time period are listed in Table 10. These show that when the data bandwidth is 60 or 120 min, the treatment effect estimates are largely the same, and the regression coefficient estimates of other control variables are also very similar, indicating that the regression results obtained under these bandwidth conditions are relatively stable. In contrast, when the data bandwidth is 30 min, the treatment effect estimate is significantly reduced, and the regression coefficient of the variable taxi passenger ratio is negative. This result seems to contradict reality because the passenger ratio reflects the ratio of the number of taxis with passengers passing through the road section per unit time compared to the total number of taxis. Generally, the speed of taxis with passengers should be greater than their speed without passengers; thus, the greater the passenger ratio, the greater should be the average speed of taxis on a road section. However, although there is a positive correlation between these variables,

Table 10 Regression result of taxi in different bandwidth (cutoff: 7:30 a.m.)

Bandwidth	30 min	30 min	60 min	60 min	120 min	120 min
Treatment effect	−0.606	0.540	−1.068	−0.493	−0.984	−0.349
	(0.811)	(1.348)	(0.635)	(1.032)	(0.459)	(0.693)
Time	−0.077	−0.270	−0.009	−0.070	−0.03	−0.068
	(0.071)	(0.297)	(0.029)	(0.125)	(0.013)	(0.044)
Time × treatment variable	0.07	0.040	−0.003	0.006	0.019	0.028
	(0.094)	(0.572)	(0.04)	(0.178)	(0.016)	(0.061)
Constant	9.581	8.660	9.291	8.846	9.576	9.456
	(1.563)	(1.927)	(1.437)	(1.680)	(1.043)	(1.070)
Number of taxi with/without passenger	−0.237	−0.052	0.968	1.046	0.806	−0.460
	(1.242)	(1.282)	(1.016)	(1.057)	(0.816)	(0.430)
Number of taxi per min	−0.021	−0.020	−0.014	−0.017	−0.009	−0.014
	(0.015)	(0.016)	(0.014)	(0.015)	(0.009)	(0.010)
Number of bus per min	0.051	0.082	0.053	0.070	0.045	0.045
	(0.072)	(0.082)	(0.07)	(0.075)	(0.051)	(0.052)
Number ratio taxi/bus per min	−0.182	−0.052	−0.248	−0.351	−0.341	−0.460
	(0.512)	(1.282)	(0.507)	(0.569)	(0.387)	(0.430)
Ave. speed before cutoff	9.89	9.89	9.94	9.94	10.43	10.43
Polynomial order	Linear	Quadratic	Linear	Quadratic	Linear	Quadratic
Sample size	27	27	47	47	80	80
R-square	0.276	0.244	0.21	0.178	0.462	0.458
Cutoff	7:30 a.m	7:30 a.m	7:30 a.m	7:30 a.m	7:30 a.m	7:30 a.m

Note Robust standard errors are in parentheses

the regression results do not show an increase in taxi speed when taxis are carrying passengers. This may be because the elected bandwidth is too narrow, such that the amount of data it contains cannot reflect the overall trend change. Indeed, as mentioned above, selecting too small a bandwidth may cause a large deviation in the results obtained by the regression.

The results of partial regression also reflect the same problem, as shown in Table 11. When the bandwidth is 30 min, the treatment effect estimate obtained is smaller than that obtained when using other bandwidths. However, the value obtained by the rectangular kernel function is reasonable (−1.07), and the results under other bandwidths fluctuate between −0.868 and −1.235, with standard errors within an acceptable range.

The same robustness test is performed at the EBL closing time (9:30 a.m.) As shown in Table 12, by narrowing the data range, the treatment effect estimate increases gradually, from 1.353 to 2.218. At this time, parametric regression cannot obtain a stable effect estimate, so the results of non-parametric regression must be compared (as shown in Table 13). Aside from the relatively small estimate in the

Table 11 Estimated treatment effect of taxi partial regression (cutoff: 7:30 a.m.)

Kernel function type	Bandwidth			
	180 min	90 min	60 min	30 min
Rectangular	−1.089	−0.897	−1.235	−1.07
	(0.343)	(0.428)	(0.502)	(0.696)
Triangular	−1.017	−0.868	−1.183	−0.527
	(0.35)	(0.432)	(0.518)	(0.693)
Epanechnikov	−1.048	−0.912	−1.281	−0.518
	(0.348)	(0.426)	(0.515)	(0.675)

Note Robust standard errors are in parentheses

Table 12 Regression result of taxi in different bandwidth (cutoff: 9:30 a.m.)

Bandwidth	30 min	30 min	60 min	60 min	120 min	120 min
Treatment effect	2.218	1.801	1.973	1.537	1.353	0.937
	(0.891)	(1.525)	(0.723)	(1.171)	(0.741)	(1.075)
Time	−0.133	−0.295	−0.004	0.004	0.009	0.051
	(0.101)	(0.438)	(0.036)	(0.155)	(0.019)	(0.081)
Time × treatment variable	0.104	0.587	−0.024	0.038	0.009	−0.020
	(0.112)	(0.509)	(0.047)	(0.201)	(0.024)	(0.097)
Constant	8.182	7.441	7.932	8.230	9.151	9.288
	(1.806)	(1.955)	(1.396)	(1.677)	(1.469)	(1.522)
Number of taxi with/without passenger	0.956	1.121	3.033	3.115	1.805	1.825
	(1.503)	(1.842)	(1.130)	(1.160)	(1.054)	(1.071)
Number of taxi per min	−0.040	−0.052	−0.017	−0.019	−0.007	−0.007
	(0.020)	(0.023)	(0.008)	(0.010)	(0.008)	(0.008)
Number of bus per min	0.056	0.119	−0.066	−0.078	−0.129	−0.122
	(0.109)	(0.122)	(0.063)	(0.069)	(0.066)	(0.068)
Number ratio taxi/bus per min	−0.942	−1.547	0.093	0.091	0.093	0.112
	(1.240)	(1.350)	(0.086)	(0.096)	(0.104)	(0.111)
Ave. speed before cutoff	7.82	7.82	8.00	8.00	7.77	7.77
Polynomial order	Linear	Quadratic	Linear	Quadratic	Linear	Quadratic
Sample size	29	29	49	49	95	95
R-square	0.366	0.354	0.294	0.265	0.353	0.289
Cutoff	9:30 a.m	9:30 a.m	9:30 a.m	9:30 a.m	9:30 a.m	9:30 a.m

Note Robust standard errors are in parentheses

Table 13 Estimated treatment effect of taxi partial regression (cutoff: 9:30 a.m.)

Kernel function type	Bandwidth			
	180 min	90 min	60 min	30 min
Rectangular	2.197	1.119	2.02	2.16
	(0.418)	(0.52)	(0.616)	(0.87)
Triangular	1.82	1.228	1.976	1.836
	(0.39)	(0.512)	(0.648)	(1.044)
Epanechnikov	1.852	1.097	1.958	1.977
	(0.396)	(0.499)	(0.618)	(0.967)

Note Robust standard errors are in parentheses

90-min bandwidth, the estimates are all in the range of 1.8–2.2, which is similar to the foregoing linear regression results. Therefore, the treatment effect at this time can be regarded as 1.8–2.2.

6 Conclusion

As an EBL is a commonly implemented bus-priority policy, it has been widely praised and used by local governments. In response to the call for carbon neutrality and peak emissions of carbon dioxide by a certain year, governments have made great efforts to build bus-serviced cities with clear goals in mind. In this context, effective public transportation policies are crucial, but only empirical methods and simulation methods have previously been used to evaluate the real-life effects of EBLs. This chapter, therefore, uses a new empirical data-based method for evaluating the effect of an EBL, which empirically supplements simulation- and theoretical-based analyzes. The method uses an RD design to quantitatively evaluate the effect of an EBL on the speed of buses and social vehicles. Specifically, the GPS data of taxis and buses in Shenzhen are used as a case study, with Shennan East Road as the target road section. After the opening of the EBL in the study area, the average speed of buses increases by 8.2%, whereas the average speed of taxis decreases by 9.9%. In contrast, after the closing time of the EBL in the study area, the average speed of buses decreases by 10.0%, and the average speed of taxis increases by 25.9%.

These results indicate that the EBL has a significantly positive effect on buses, but these effects vary according to opening hours. In addition, the EBL has a greater effect on buses at its closing time than at its opening time. We suggest that to increase average bus speeds at peak travel times in the target road section, the closing time of the EBL should be postponed appropriately.

The EBL also has a certain effect on taxis, but this is related to the need for taxis to pick up and drop off passengers on the roadside. As this need increases after 9:00 a.m., the speed of taxis increases significantly after the closing time of the EBL. Thus, to allow the passage of buses and ensure the effective implementation of the EPL policy, corresponding restrictions should be imposed on taxis and other social vehicles, such as by establishing a centralized pick-up station. Moreover, signal priority and other supporting priority facilities should be used with the EBL to maintain the smooth and convenient passage of buses in the study area.

References

Angrist, J. D., & Lavy, V. (1999). Using Maimonides' rule to estimate the effect of class size on scholastic achievement. *The Quarterly Journal of Economics, 114*(2), 533–575.

Arasan, V. T., & Vedagiri, P. (2010). Microsimulation study of the effect of EBLs on heterogeneous traffic flow. *Journal of Urban Planning and Development, 136*(1), 50–58.

Black, S. E. (1999). Do better schools matter? Parental valuation of elementary education. *The Quarterly Journal of Economics, 114*(2), 577–599.

Burger, N. E., Kaffine, D. T., & Yu, B. (2014). Did California's hand-held cell phone ban reduce accidents. *Transportation Research Part A: Policy and Practice, 66*, 162–172.

Cox, M. (1975). Reserved bus lanes in Dallas, Texas. *Transportation Engineering Journal of ASCE, 101*(4), 691–705.

Currie, G., Sarvi, M., & Young, B. (2007). A new approach to evaluating on-road public transport priority projects: Balancing the demand for limited road-space. *Transportation, 34*(4), 413–428.

D'Arcangelo, F. M., & Percoco, M. (2015). Housing rent and road pricing in Milan: Evidence from a geographical discontinuity approach. *Transport Policy, 44*, 108–116.

Diakaki, C., Papageorgiou, M., Dinopoulou, V., Papamichail, I., & Garyfalia, M. (2015). State-of-the-art and-practice review of public transport priority strategies. *IET Intelligent Transport Systems, 9*(4), 391–406.

Ding, H., Li, H., & Sze, N. N. (2021). Effects of the abolishment of London western charging zone on traffic flow and vehicle emissions. *International Journal of Sustainable Transportation*, 1–24.

Ebenstein, A., Fan, M., Greenstone, M., He, G., & Zhou, M. (2017). New evidence on the impact of sustained exposure to air pollution on life expectancy from China's Huai River Policy. *Proceedings of the National Academy of Sciences, 114*(39), 10384–10389.

Feather, D., Cracknell, J. A., & Forster, J. A. (1973). Bus priority in greater London. 3. Development of bus lane schemes. *Traffic Engineering & Control,* 14.

Gao, Z., Long, K., Li, C., Wu, W., & Han, L. D. (2019). Bus priority control for dynamic EBL. *CMC-Computers Materials & Continua, 61*(1), 345–361.

Hahn, J., Todd, P., & Van der Klaauw, W. (2001). Identification and estimation of treatment effects with a regression-discontinuity design. *Econometrica, 69*(1), 201–209.

Hymel, K. (2014). Do parking fees affect retail sales? Evidence from Starbucks. *Economics of Transportation, 3*(3), 221–233.

Imbens, G. W., & Lemieux, T. (2008). Regression discontinuity designs: A guide to practice. *Journal of Econometrics, 142*(2), 615–635.

Lee, D. S., & Lemieux, T. (2010). Regression discontinuity designs in economics. *Journal of Economic Literature, 48*(2), 281–355.

Li, S., & Ju, Y. (2009). Evaluation of bus-exclusive lanes. *IEEE Transactions on Intelligent Transportation Systems, 10*(2), 236–245.

Liu, Z., Li, R., Wang, X. C., & Shang, P. (2018). Effects of vehicle restriction policies: Analysis using license plate recognition data in Langfang, China. *Transportation Research Part A: Policy and Practice, 118*, 89–103.

Mo, B., Kong, H., Wang, H., Wang, X. C., & Li, R. (2021). Impact of pricing policy change on on-street parking demand and user satisfaction: A case study in Nanning, China. *Transportation Research Part A: Policy and Practice, 148*, 445–469.

Mora-Garcia, R. T., Cespedes-Lopez, M. F., Perez-Sanchez, J. C., & Perez-Sanchez, R. (2015). The kernel density estimation for the visualization of spatial patterns in urban studies. In *15th International Multidisciplinary Scientific GeoConference SGEM 2015* (pp. 867–874).

Sacks, J., & Ylvisaker, D. (1978). Linear estimation for approximately linear models. *The Annals of Statistics,* 1122–1137.

Shalaby, A. S. (1999). Simulating performance impacts of bus lanes and supporting measures. *Journal of Transportation Engineering, 125*(5), 390–397.

Shin, E. J. (2021). Exploring the causal impact of transit fare exemptions on older adults' travel behavior: Evidence from the Seoul metropolitan area. *Transportation Research Part A: Policy and Practice, 149*, 319–338.

Szarata, M., & Olszewski, P. (2019). Traffic modelling with dynamic bus lane. In *2019 6th International Conference on Models and Technologies for Intelligent Transportation Systems (MT-ITS)* (pp. 1–8). IEEE.

Thistlethwaite, D. L., & Campbell, D. T. (1960). Regression-discontinuity analysis: An alternative to the ex post facto experiment. *Journal of Educational Psychology, 51*(6), 309.

Van der Klaauw, W. (2002). Estimating the effect of financial aid offers on college enrollment: A regression–discontinuity approach. *International Economic Review, 43*(4), 1249–1287.

Viard, V. B., & Fu, S. (2015). The effect of Beijing's driving restrictions on pollution and economic activity. *Journal of Public Economics, 125*, 98–115.

Viegas, J., & Lu, B. (2004). The intermittent bus lane signal setting within an area. *Transportation Research Part C, 12*(6), 453–469.

Yao, J., Shi, F., An, S., & Wang, J. (2015). Evaluation of EBLs in a bi-modal degradable road network. *Transportation Research Part C: Emerging Technologies, 60*, 36–51.

Yu, B., Kong, L., Sun, Y., Yao, B., & Gao, Z. (2015). A bi-level programming for bus lane network design. *Transportation Research Part C: Emerging Technologies, 55*, 310–327.

Zhao, J., & Zhou, X. (2018). Improving the operational efficiency of buses with dynamic use of exclusive bus lane at isolated intersections. IEEE Transactions on Intelligent Transportation Systems, 20(2), 642-653.

Zhao, J., Yu, J., Xia, X., Ye, J., & Yuan, Y. (2019). EBL network design: A perspective from intersection operational dynamics. *Networks and Spatial Economics, 19*(4), 1143–1171.

Zheng, F., Chen, J., Wang, H., Liu, H., & Liu, X. (2020). Developing a dynamic utilisation scheme for EBLs on urban expressways: An enhanced CTM-based approach versus a microsimulation-based approach. *IET Intelligent Transport Systems, 14*(12), 1657–1664.

Chapter 5
Analyzing Spatiotemporal Congestion Pattern on Urban Roads Based on Taxi GPS Data

Abstract With the development of in-vehicle data collection devices, GPS trajectory has become a priority source to identify traffic congestion and understand the operational states of the road network in recent years. This study aims to investigate the relationship between traffic congestion and built environment, including traffic-related factors and land use. Fuzzy C-means clustering was used to conduct an exhaustive study on 24-h congestion pattern of road segments in an urban area. The spatial autoregressive moving average model (SARMA) was then introduced to analyze the output from the clustering analysis to establish the relationship between the built environment and the 24-h congestion pattern. The road segments were classified into four congestion levels. The regression explained 12 traffic-related factors and land use factors' impact on road congestion pattern. The continuous congestion was found to mainly occur in the city center. Factors, such as road type, bus station in the vicinity, ramp nearby, commercial land use have large impacts on congestion formation. In combination with quantitative spatial regression, the proposed Fuzzy C-means clustering approach was employed to develop an overall evaluation process, which could be applied generally to assist the assessment of spatial–temporal levels of road service from the congestion perspective.

Keywords Congestion pattern · Taxi GPS data · Fuzzy C-means clustering · Spatiotemporal regression · Built environment factor

1 Introduction

In an urban road network, the recurrent or non-recurrent congestion of a certain road segment may largely impact the local traffic network efficiency. Consequently, it is important to identify the congested road segments in real-time and implement corresponding traffic mitigation strategies. Fixed facilities, such as inductive loops, traffic surveillance systems and microwave radars are commonly used to collect data for road traffic detection, including traffic speed, traffic volume, density and vehicle classification. However, such facilities are expensive and mostly only serve intersections or freeways. The sparse sensor network makes it difficult to identify the problematic links in real-time. Global Position System (GPS) data enables the simultaneous

S. Zhong and D. Sun, *Logic-Driven Traffic Big Data Analytics*,
https://doi.org/10.1007/978-981-16-8016-8_5

analysis of spatial and temporal travel patterns in a network. While Global Position System (GPS) data collected from vehicles and mobile phones have become increasingly popular, among which taxi GPS data is preferable because of its exhaustive coverage of the road network, high frequency and the acquisition convenience (Ding et al., 2014; Wang et al., 2017). In general, no fixed definitions were proposed for the level of congestion, while travel speed is the most commonly used indicator for traffic congestion assessments (He et al., 2016; Sun et al., 2014a). This study intends to investigate the 24-h congestion pattern of the road network by speed, to classify the road segments by their speed patterns through Fuzzy C-means (FCM) clustering, and to analyze the problematic segments with continuous low speed or unconventional congestion. Spatial models based on the geographical detector, MORAN's Index and spatial regression (SARMA) were developed to analyze the relationship between congestion patterns and the surrounding built environment.

Compared with mobile phone data, floating car data, cargo transport vehicle record and navigation system, taxi GPS trace data is one of the easiest available sources for accurate travel route and travel time records for a wider area with more details. Data mining based on taxi trip can be traced back to the 1970s (Goddard, 1970), which has been applied in a wide range of studies, mainly including activity-based and infrastructure-based fields. The activity-based studies mostly focus on driver behavior, supply–demand pattern, and traffic state analysis, while the infrastructure-based studies mainly focused on lanes channelization (Tang et al., 2015) and signal-timing estimation (Yu & Lu, 2016).

From driver behavior perspective, Zhang et al. (2015) proposed a space–time visualization method to demonstrate taxi daily trajectories by GIS-T to recognize the working time, operating range, and residence location without time division. Qing et al. (2015) compared directly extracted Manhattan GPS data, including travel distance, speed, demand, and supply mismatch of taxi trip, between fair weather and extreme storm, and discovered the reduction in trip distance and supply of services during the extreme storm. Meanwhile, Hwang et al. (2006) used structural equation modeling techniques to improve taxi dispatching service using consumer preference modeling based on questionnaires and GPS data, however, the time variation was not considered. Tang et al. (2016) analyzed drivers' customer searching behaviors by proposing a two-layer model based on GPS data, path size, path distance, and travel time, however, without discussion of geographical factors like land use and traffic-related factors. Yazici et al. (2016) studied New York taxi drivers' decisions on pick-ups or cruising for passengers after the end of trips at the JFK airport using a logistic regression based on temporal and weather factors, among which peak hour was treated as an independent variable. Chen et al. (2014) introduced B-planner for planning bidirectional night bus routes using taxi GPS traces and conducted qualitative analyses using clustering results.

Taxi GPS data also reveal the supply–demand pattern of taxi. Hu et al. (2014) analyzed time of day and day of the week variations in urban taxi drivers' service time and operation frequency using descriptive statistics. However, the authors failed to further construct the relationship between the service and built environment. Qian and Ukkusuri (2015) combined geographically weighted regression (GWR) with

NYC taxi data to establish the relationship between taxi ridership and demographic, land-use and transportation system. However, only the daily ridership was aggregated and analyzed, which couldn't reflect the hourly demand variation. Lu and Li (2014) used taxi GPS data to predict OD distribution, while the statistical methods still treated one-day data as a whole, and the variations were only explained by the time factor without multivariable combination.

Conventional studies based on GPS data regarding traffic state include congestion detection (Montero et al., 2016), link or route travel time, speed and distance measurements (Jiménez-Meza et al., 2013; Tulic et al., 2014), and urban road network accessibility detection (Cui et al., 2016). Most of these studies only focus on peak hours and only use descriptive statistics without in-depth analyses about influencing factors such as time, speed, and distance. Secondary transformation is deficient in state identification. Azimi and Zhang (2010) applied clustering algorithms (K-means, Fuzzy C-means, and CLARA) to sort freeway traffic conditions by traffic flow. However, their result was qualitative, which makes it difficult to apply.

The previous taxi GPS data based researches mainly have three limitations. First, the majority of studies only focused on peak hours or discrete hours, or even aggregated daily data without hourly division, which fail to reflect the hourly volatility and the temporal changes in traffic system. Second, additional analyses generally focused on certain straightforward indices, such as speed, flow rate or travel time, while other indirect or consequential indicators, such as congestion, acquired less attention because of the lack of fixed quantitative definition. Finally, the existing studies failed to utilize independent variables, such as land use, built environment, and traffic-related factors, to illustrate the information extracted from taxi GPS data. To this end, GPS data is used to identify those infrastructure-based factors, as well as land use as mentioned (Pan et al., 2013).

This research aims to explore the speed pattern using a 24-h-based dataset and to reflect the volatility trend by clustering. An analytical framework combining clustering method and spatial regression is proposed to cover the shortage of secondary transformation for congestion and quantitative analysis. Land use variables and traffic-related factors are included in the regression.

The research flow chart is presented in Fig. 1. This study first uses taxi GPS data of Shanghai to classify road segments in an urban area based on their 24-h congestion pattern using Fuzzy C-means clustering (FCM), which allocated objects to clusters by probability. We then set such probability classification as the dependent variables and conduct a spatial regression with the mixed spatial autoregressive moving average model (SARMA) to assess the impacts of environmental factors on speed patterns.

The remainder of the chapter is organized as follows. Section 2 introduces data collection and processing. Section 3 presents FCM clustering of road segments, and the spatial analysis models are proposed in Sect. 4. Finally, conclusion and future research are provided in Sect. 5.

Fig. 1 Research flow chart

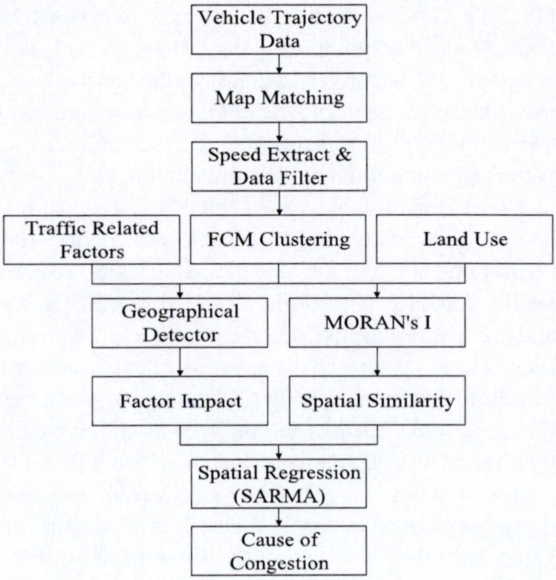

2 Data Preparation

2.1 Data Collection

By 2016, Shanghai has over 58,000 taxis, carrying about 4000,000 passengers each day. In this study, the Taxi FCD data on April 10th, 2015, a sunny Friday with rather heavy traffic was provided by the Qiangshen Company via an online open data competition (http://sodata.io/). The original dataset has 114,633,142 records, with the time interval as 30 s. The original records include fields such as *Taxi ID, Status, Signal Receive Time, Signal Measured Time, Longitude, latitude, speed* and so on (Sun et al., 2014b). Only the occupied taxi trips were kept for speed pattern generation (Cui et al., 2016) because the empty taxies do not reflect real traffic condition resulting from slowdown for passenger pickup, shift change or gas refill.

Focusing on the heavy traffic, only the primary and secondary roads within the Central City (Outer Ring) of Shanghai were considered. Road segment was defined as the link between two main intersections. The segments with length less than 300 m were removed to avoid the excessive influence of intersections. A total of 853 road segments with traffic and built environment data were screened out, representing the generic traffic condition of the road network during a typical workday.

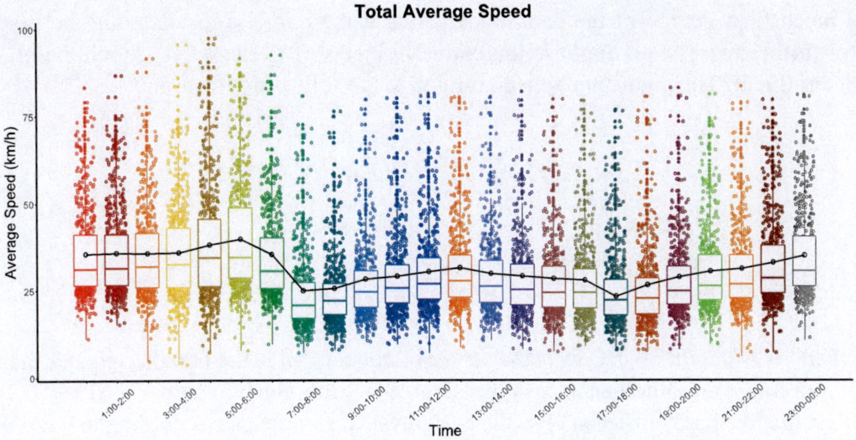

Fig. 2 Hourly average speeds of taxis on the selected road segments during the day of study

2.2 *Data Processing*

Data not matching to the nearest roads within 15 m, staying still for over 5 min, or with speed over 120 km/h and sudden distance deviation over 1 km/min were eliminated. 30 visits per hour were set as the threshold for segments to avoid error propagation. 551 segments were kept for further analysis. We calculated the average speed of each segment by averaging the taxis passed through each hour. Figure 2 presents the boxplot and mean trend of the average speed on all segments by hour. The speed pattern shows obvious valleys during peak hours (7–10 am and 4–5 pm), when the road network suffers from the most severe congestion, as well as the time-of-day congestion degree variations.

3 Clustering of Road Segments

3.1 *Fuzzy C-means Clustering*

24-h speed patterns of 551 selected segments were calculated to reveal the congestion patterns. Clustering analysis, an unsupervised machine learning method, was used to aggregate road segments into groups based on their speed patterns. First, 24-h speed pattern of each segment was expressed with a 24-dimension vector: $y_i = \{S_1, S_2, \ldots, S_{24}\}$, where Si is the average speed of the time interval hour i.

Hard clustering (e.g. K-means) has less flexibility (Azar et al., 2013; Sun & Elefteriadou, 2011, 2012), while the soft clustering, fuzzy C-means (FCM) clustering algorithm (Bezdek et al., 1984; Dunn, 1973), expresses that data points possibly belong to multiple clusters at the same time by membership degrees, which offers a much finer degree of the data model, so that the numeric results could be used for further regression. Suppose there are N objects of C classes (C should be predetermined). The algorithm aims to minimize the following function:

$$J(U, C) = \sum_{i=1}^{N} \sum\nolimits_{j=1}^{C} (u_{ij})^m \|X_i - V_j\|_A^2$$

$$s.t. \sum_{j=1}^{C} u_{ij} = 1; \quad 0 \leq u_{ij} \leq 1 \tag{1}$$

where X_i is the *ith* object, V_j is the center of cluster *j*. *m* is the fuzzifier greater than 1, and a higher value means a higher degree of ambiguity. When *m* is close to 1, it's more like hard clustering. For the best physical significance according to Bezdek (1980), $m = 2$ was adopted in this study. u_{ij} is the grade of membership value of the *ith* object to the *jth* cluster, $\sum\nolimits_{j=1}^{C} u_{ij} = 1$. $\|X_i - V_j\|_A$ is the A-norm on R^n, measuring the similarity of objects to the assigned cluster, according to Bezdek et al. (1984). Equations (2) and (3) illustrate the iterative process to calculate the centroids of clusters and the membership values:

$$V_j = \sum\nolimits_{i=1}^{N} (u_{ij})^m X_i / \sum\nolimits_{i=1}^{N} (u_{ij})^m \tag{2}$$

$$u_{ij} = 1 / \sum\nolimits_{s=1}^{C} \left(\frac{|X_i - V_j|}{|X_i - V_s|} \right)^{\frac{1}{m-1}} \tag{3}$$

The iteration stops when $|U_{p+1} - U_p| < \varepsilon$, where U_p is the membership u_{ij} at the *pth* iteration. Since the membership is within the range of 0 to 1, typically the iteration accuracy ε is set as 0.001 (Bezdek et al., 1984), or the iteration number is fixed at 100 (Schw & Jensen, 2010). ε was chosen as 1e−5 in this research, which is enough to both guarantee the iteration number and accuracy.

As mentioned, the cluster number is predetermined and requires revision by validity indexes to generate a meaningful and explicable result. In principle, an ideal cluster number C could keep a balance between the inter-distance for each pair of centroids, $\|V_i - V_j\| (i \neq j)$, and the intra-distance of clusters, $\|X_i - V_j\| (X_i \subseteq C_j)$. Four validity indexes appropriate for FCM were used to confirm optimal C:

1. **Partition coefficient**: *PC* (Bezdek 1973) emphasizes the intensity of membership, and is expressed by square weighting:

$$PC(C) = \frac{1}{N} \sum_{i=1}^{N} \sum_{j=1}^{C} {u_{ij}}^2 \tag{4}$$

A larger value of *PC* generally indicates a better expression of belongingness.

2. **Fuzzified *PBM***: *PBMF* (Pakhira et al., 2005) indicates the compactness within the same cluster and the segregation between clusters. Such effect is reflected by ratio. The greater, the better.

$$PBMF(C) = \frac{\max_{j,k}\{\|V_j - V_k\|\} \times E_1}{\sum_{i=1}^{N}\sum_{j=1}^{C} u_{ij}{}^m \|X_i - V_j\|} \tag{5}$$

where E_1 is a constant for a fixed sample.

3. **Minimum centroids' distance**: *MCD* (Zhu & Nandi, 2014) is the minimum distance between current cluster centers, aiming to explain the dispersion degree of clusters:

$$MCD(C) = \min_{i \neq j} |V_i - V_j|^2 \tag{6}$$

Generally speaking, *MCD* is monotone decreasing with the increase of *C*, and the suggested *C* should be the point when the recession curve comes to stability.

4. **Fukuyama-Sugeno index**: *FSI* (Fukuyama & Sugeno, 1989) tests both the separation of all objects and the separation of clusters, and the target is to ensure these two separation degrees conforming to each other. The smaller, the better.

$$FSI = \sum_{i=1}^{N}\sum_{j=1}^{C} u_{ij}{}^m \left(|X_i - V_j|^2 - \left| \left(\frac{1}{N} \sum_{k=1}^{N} X_k \right) - V_j \right|^2 \right) \tag{7}$$

These indexes evaluated the cluster number from different points of view and were calculated simultaneously.

3.2 Clustering Result

First, the cluster number was determined through validity indexes. After testing cluster number from 2 to 7, *FSI*, *PBMF,* and *PC* got the best cluster numbers of 5, 3, 5 respectively. For *MCD*, it turns to be stable when the cluster number attained 4 with an abrupt decline in centroid distance. Finally, 4 clusters were chosen because of the physical significance. Cluster 1 has 235 objects, Cluster 2 has 177 objects, Cluster 3 has 107 objects and Cluster 4 has 32 objects.

Table 1 presents the 24-h mean, standard deviation (SD), coefficient of variation (CV) and range for the four clusters.

It can be figured out that Cluster 4 has the largest mean (61.63 km/h, approximate to speed limit of primary roads at 60 km/h) and range, which implies a high level of service on average. However, its *CV* is comparably large, which means a low level of

Table 1 Statistical indexes of each cluster

Statistical index	Cluster 1	Cluster 2	Cluster 3	Cluster 4
MEAN (km/h)	22.59	29.51	41.11	61.63
SD (km/h)	3.38	4.23	6.82	6.27
CV	0.15	0.14	0.17	0.10
RANGE (km/h)	10.93	14.80	24.65	27.42

dispersion. The mean speed of Cluster 3 is 41.11 km/h, which is approximate to the speed limit of secondary road at 40 km/h. The speed MEAN and RANGE experience a progressive increase trend from Clusters 1 to 4, but not for the *SD* and *CV*.

Figure 3 presents the 24-h speed pattern of four clusters resulting from the FCM clustering. The temporal trajectories of road segments are plotted against the primary cluster with the highest membership values. All the scatter points are represented by a gradual change of color and size, shown in the legend. The horizontal stochastic disturbance is added to the points for better visualization, with the black polyline marking the 24-h trend of cluster centers. As presented in Fig. 3a, Cluster 1 with the largest sample size (i.e. 235) is labeled as 'Highly Congested Segments' with the mean speed of 22.58 km/h. Its speed trajectory keeps at a low level with relatively stable trend compared with other clusters. The trend in Cluster 1 implies that the highly congested pattern might be caused by continuous traffic pressure, design flaw, or certain intrinsic attributes of facilities. Cluster 2 (Fig. 3b) is characterized by the comparatively medium speed with 29.51 km/h mean, and could be labeled as 'Normal Speed Segments' because its speed pattern conforms to the typical urban road segment travel speed (Kumar & Vanajakshi, 2013). Cluster 2 has the second largest sample size, 177. Cluster 3 (Fig. 3c) could be regarded as a critical state, which exhibits a higher speed and fluctuation changes during peak hours, thus labeled as 'Unimpeded Segments'. The mean speed, 41.11 km/h, is just about 7 km/h higher than that of Cluster 2, which is a comparatively small value, with the 40 km/h secondary speed limit. The other dominant factor distinguishing Cluster 3 from Cluster 2 is the peak characteristics, and these segments may be on the main commuting corridors and have a tidal phenomenon. Cluster 4 (Fig. 3d) is labeled as 'High-speed Segments', who's mean speed is the highest, 61.63 km/h, with the smallest sample size, 32. Such a continuous non-congestion pattern is rare, closed to the 60 km/h speed limit of the arterial roads in Shanghai. As the segments in Cluster 4 generally are with the highest level of service with almost no congestion, the time-of-day rationality has been proved.

Figure 4 illustrates the spatial distribution of the studied road segments based on their highest membership value. The red line is the Huangpu River, which divides Shanghai into Pudong (right) and Puxi (left). The drop marks the city center. It shows that the segments in Cluster 1 concentrate mainly in Puxi, and these roads connecting the city center with outskirts, carrying the majority of commuting vehicles. This may explain why road segments in Cluster 1 have the worst traffic condition. Segments in Cluster 2 (green line) evenly distribute across the study area. Clusters 3 and 4

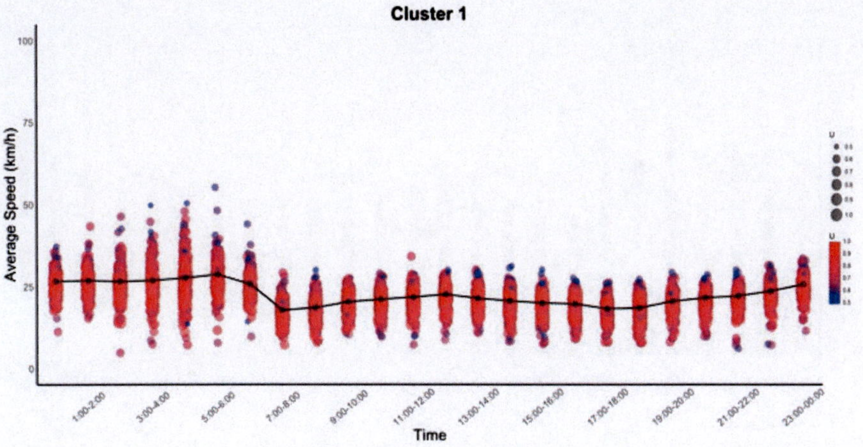

(a) Highly Congested Segments

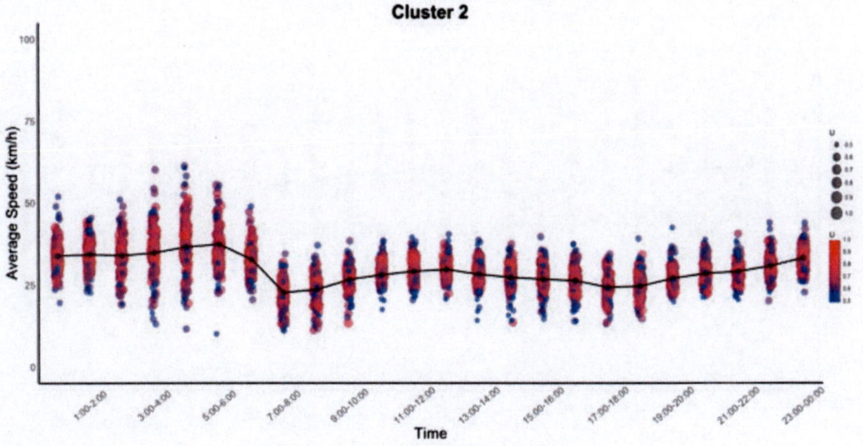

(b) Normal Speed Segments

Fig. 3 Temporal clustering of 24-h speed pattern by segments

distribute in the surrounding parts of the study area, in other words, the roads further away from the city center have a higher probability of high-speed patterns and the probability of congestion is much lower.

4 Spatial Analysis of Road Segments

The quantitative analysis method, SARMA, is introduced to provide further explanation for clustering characteristics and the built environment impacts, while

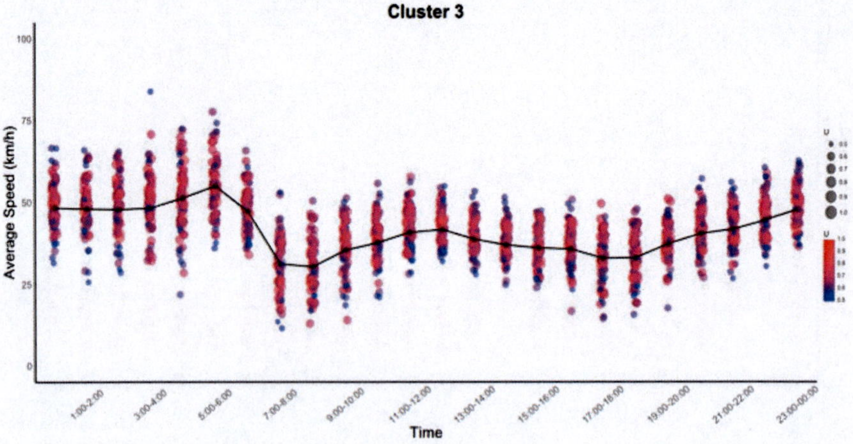

(c) Unimpeded Segments

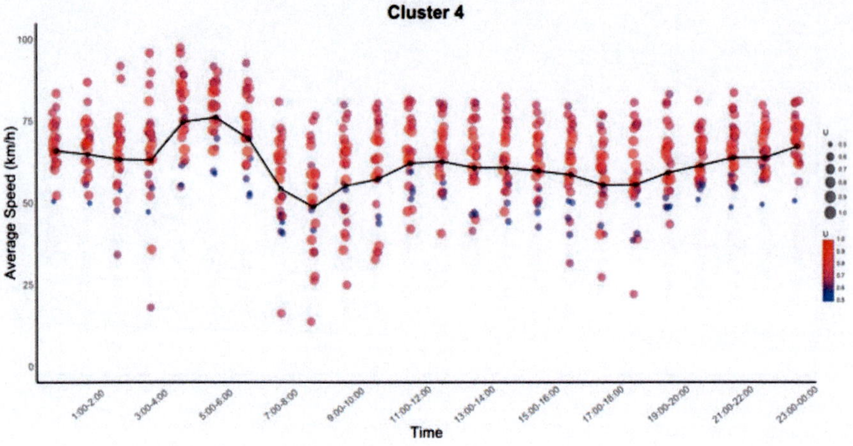

(d) High-speed Segments

Fig. 3 (continued)

geographic detector and MORAN's I were tested beforehand to the applicability of SARMA. The explanatory variables include two categories, traffic-related factors and land use, which are widely used to explain traffic phenomenon but not in combination with taxi GPS data. For example, Zhang et al. (2012) measured the impact of residential density, employment density, land use mix, block size and distance from CBD on vehicle-miles travel. Tian et al. (2015) assessed the relationship between traffic generation and mixed-use development. Briefly, main built environment factors affecting traffic state or travel behavior could be divided into traffic related (Feng et al., 2011;

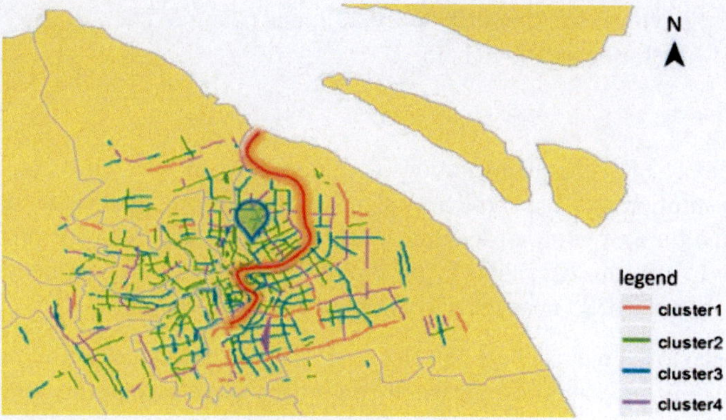

Fig. 4 Clustering result by primary membership

Hahn et al., 2002; Zhang & Levinson, 2017) and land use related (Handy et al., 2005; Nian et al., 2021; Wheaton, 1998) ones.

Based on the previous researches, variables chosen for further analyses are purposed as follows:

1. Traffic-related factors:

F1: Road type, primary or secondary road. In the numeric analysis, 1 for the primary road and 2 for the secondary road. The average speed of the primary road was always higher than that of the secondary road in this study.

F2: Road segment length. Since the segment was defined as the road link between two intersections, the segment length might affect the average speed greatly.

F3: Distance to the nearest ramp. Ramps are bottlenecks for traffic breakdown, and breakdown may affect miles away (Kerner & Klenov, 2006), so distance to the nearest ramp would impact road congestion.

F4: Number of bus stations along the road segment per 100 m. More bus stations might reflect the commuting pressure, bring more frequent lane-changing or heavy vehicle rate (Hahn et al., 2002).

F5: Distance to the nearest metro station. Metro stations carry a large amount of passenger flow and transship of bus or sedan.

F6: The relative location to the urban expressway rings. The urban expressway system in Shanghai has three rings: inner ring, middle ring and outer ring, which typically distinguish the center with the suburb. The regions divided by the three rings are donated by 1, 2, 3, 4 from the inner center to the outside of the outer ring.

F7: Number of parking lots open to the society within 500 m per 100 m. Parking lots are the main destination for private vehicles, which may induce congestion around. Since drivers have to walk to the final destination, parking lots' influencing scope should be the typical walking distance, 0.25 miles (Yang & Diez-Roux, 2012), and the buffer was chosen as 500 m here.

2. Land use

F8: Number of schools/institutions within 500 m per 100 m. Because schools/institutions in China produce a significant traffic pressure on the surrounding roads at the time of going or leaving school due to pickup behavior of parents or school bus (Yu & Liu, 2011), this factor should be considered. The influencing buffer was also chosen as 500 m as the parking lots.

F9: Distance to the nearest hospital. Hospital is the main service institution in urban areas, which is the potential congestion zones because of high traffic demand (Wen et al., 2017).

F10–12: Land use. Land use may have significant interactions with transport (Miller & Evans, 2011; Moeckel, 2016). Land use mainly considers commercial area (**F10**), residential area (**F11**) and transportation area (**F12**), while actors like education and hospital have been involved before. Transportation land cover mainly includes the railway station, airport, transportation hubs, etc. Three factors are expressed by the area proportion within a radius of 500 m:

$$P_i = S_i / \sum\nolimits_{j=1}^{n} S_j \tag{8}$$

where S_i is the target function area, n is the number of land use types, including commercial land, residential land, and transportation land.

Table 2 provides the minimum, mean, and maximum values of the built environment variables. The second column gives the abbreviation of variables, and the last three columns provide a brief description of statistical data according to the aforementioned definitions.

4.1 Geographical Detector

FCM cluster classified the road segments by 24-h speed pattern, however, whether the traffic-related factors and land use have a noticeable impact on congestion remains doubtable. *Geographical detector* (Wang et al., 2010) was introduced to judge the built environment parameters which may be responsible for the road segments clustering. The advantage of using such geographical spatial detectors is that it considers built environment parameters of various units. The power of determinant (*PD*) was introduced to determine whether a spatial factor may be responsible for clustering result.

Assuming there are n objects, and cluster D_i contains n_{Di} objects, $n = \sum_{Di=1}^{4} n_{Di}$. The power of determinant factor R is calculated by:

Table 2 Variables in the segment cluster membership model

Variable	Abbr	Min	Mean	Max
F1: Road type	Rd_type	1	–	2
F2: Road segment length (m)	Rd_len	300	1314	4510
F3: Distance to the nearest ramp (m)	Dist_ramp	8.33	990.9	4304
F4: Number of bus stations along the road segment per 100 m (stations/100 m)	Num_bus	0	0.24	2.47
F5: Distance to the nearest metro station (m)	Dist_metro	8.05	814.9	3213
F6: Relative location to the freeway rings	Ring	0	–	3
F7: Number of parking lot within 500 m	Parking	0	3.15	39
F8: Number of schools/institutions within 500 m per 100 m (schools/100 m)	Num_scho	0	0.46	2.92
F9: Distance to the nearest hospital (m)	Dist_hosp	13.47	979.8	6207
F10: Commercial area proportion (%)	Com_pro	0	5.25	59.12
F11: Residential area proportion (%)	Res_pro	0	29.29	99.41
F12: Transportation area proportion (%)	Trans_pro	0	15.63	100

$$PD_R = 1 - \frac{1}{n\sigma_R^2}\sum_{i=1}^{4} n_{Di}\sigma_{Di,R}^2 \tag{9}$$

where PD_R is the factor *R*'s power of determinant on clustering result, σ_R^2 is the global variance of factor *R* in the study region, and $\sigma_{Di,R}^2$ is the variation of factor *R* in cluster *Di*. Equation (9) interprets the ratio of the n_{Di} weighted variation in single clusters over the global variance. The value range of PD_R is [0, 1], a larger value indicates the factor *R*'s value between clusters is largely distinct, and the determinant power of *R* is stronger. If PD_R equals to 1, factor *R* alone could perfectly classify objects.

Figure 5 presents factors' explanatory power. Bus station factor (0.130) has the highest *PD*, which means more bus stations along the road segment per 100 m is related to the high possibility of congestion, because bus stations of higher density reflect larger commuting volume along the road segments. The secondary factor is road type (0.105), the average speed on the primary road is always higher than that of the secondary road due to speed limit, intersection density, and control strategy (arterial priority). The third highest factor is the distance to the nearest hospital (0.091), which carries a large volume of patients. When the trip is concerned with diseases, people are inclined to take a taxi or private vehicle, causing larger traffic volume. The number of schools within 500 m per 100 m (0.084) and transportation land use (0.071) are also main explanation factors, as schools obviously attract more commuting traffic, and transportation hubs would gather a huge amount of mixed traffic flow to pick up passengers or cargo. Unusual phenomenon comes from the distance to the nearest metro distance with low *PD*, as people using metro line mainly take public transit with less private vehicles. The factors of residential land around,

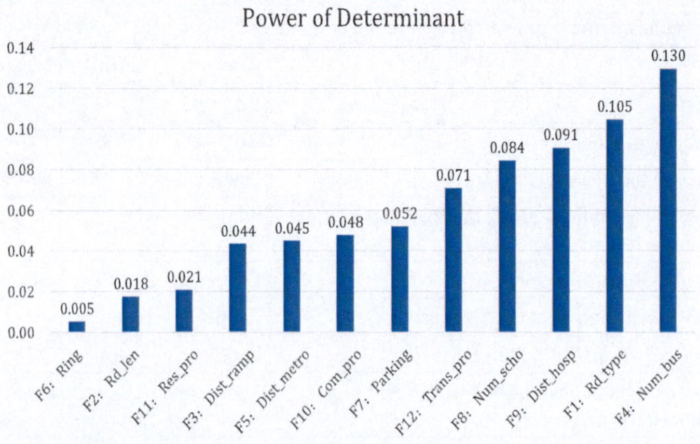

Fig. 5 Power of each determinant in ascending sequence

segment length and location relative to freeway rings are less powerful, meaning less obvious difference between clusters, but they may be significant in a certain cluster. The low power of residential factor is particularly interesting, as our clustering result mainly reflects the overall speed pattern on a 24-h basis while the neighborhoods may only have significant impacts on traffic during peak hours.

By the *geographical detector*, some factors were found affecting the congestion formation in all clusters, which disclose global influence. Others may only have exclusive impacts on certain clusters with unobvious global influence (low *PD*), which may require further investigation for each cluster, respectively.

4.2 *MORAN's I*

GLOBAL MORAN's I (Moran, 1950) measures spatial correlation and tests whether observed objects have similarities with the spatial adjacency objects. For the value of *MORAN's I* ranges, $[-1, 1]$, $I = 0$ means totally spatial independence, $I > 0$ reflects positive correlation and $I < 0$ means negative correlation. The calculation formula is as follows:

$$I = \frac{n\sum_{i=1}^{n}\sum_{j=1}^{n} w_{ij}(X_i - \overline{X})(X_j - \overline{X})}{(\sum_{i=1}^{n}\sum_{j=1}^{n} w_{ij})\sum_{i=1}^{n}(X_i - \overline{X})^2} \quad (i \neq j) \tag{10}$$

where, X_i and X_j are the observed values, which indicate membership of clusters, and $\overline{X}$ is the mean value. W_{ij} is the spatial weight matrix describing the spatial relationship among objects. *MORAN's I* can be calculated cluster by cluster.

Spatial weight matrix plays an important role in the spatial analysis. Binary joint matrix is commonly used to characterize spatial weight matrix (Cliff & Ord, 1982)

Table 3 *MORAN's I* for clusters

Index	Cluster1	Cluster 2	Cluster 3	Cluster 4
MORAN's I	0.246	0.032	0.14	0.271
Z value	5.997	0.816	3.434	6.694
P value	1.01e−09	0.207	0.0	1.09e−11

if two observations directly connect with each other, $W_{ij} = 1$, otherwise $W_{ij} = 0$. However, the binary joint matrix is not suitable for line object, such as road segments. Because of the connectivity of roads and the transmissibility property, road segments would impact each other. The spatial weight matrix used was based on distance decay (Greicius et al., 2003), and the midpoint of the road segment represents the location feature:

$$w_{ij} = \begin{cases} \exp\left[-0.5\left(d_{ij}/b\right)^2\right], & d_{ij} < b \\ 0, & otherwise \end{cases} \tag{11}$$

where $b = 1000$ m, and the matrix is standardized by row, d_{ij} is the distance between midpoints of two road segments.

Z test was used to assess the result of *MORAN's I*, and it could be interpreted by typically *P-value* (Cliff & Ord, 1982):

$$Z = [1 - E(I)]/\sqrt{VAR(I)} \tag{12}$$

where, $E(I)$ and $VAR(I)$ are the mean and variation values of *MORAN's I*, respectively.

MORAN's I evaluated whether objects' memberships to a cluster have aggregation effects spatially, the result is shown in Table 3.

All *MORAN's I* values are larger than 0, and only Cluster 2 fails to pass the 5% level of the significance test. The positive value indicates the bipolar aggregation phenomenon. As in a certain cluster, road segments with higher membership gather together and lower ones as well, indicating a lag effect for neighboring segments, which means neighboring segments have similar possibility and level of congestion. Although *MORAN's* I value is still comparably small, it discloses spatial gathering and explains the effectiveness of FCM clustering.

4.3 Spatial Regression of Road Segments

Results from FCM were used to conduct multiple regression analysis based on continuous membership u_{ij} and environment factors. *Geographical detector* and *MORAN's I* have proved the factors' impact on clustering results and the spatial similarity of the nearby road segments, so in addition to the 12 factors, lagging influence from neighboring segments are also considered. A spatial model involving both spatial autocorrelation and multivariable system is therefore preferred. A spatial lag mode called mixed spatial autoregressive moving average model (*SARMA*) (Anselin et al., 1996) is introduced to consider both dependence and errors with nearer objects having a greater impact. The structure of *SARMA* is shown in Eqs. (13) and (14).

$$y_i = (1 - \rho W)^{-1} X\beta + (1 - \rho W)^{-1} u \tag{13}$$

$$u = (1 - \lambda W)^{-1}\varepsilon, \quad \varepsilon \sim N(0, \sigma^2 I) \tag{14}$$

where y_i is the segment's membership belonging to Cluster i, X is the vector of environmental characteristics; ρ is the spatial autoregressive parameter measuring neighborhood effects, $\rho > 0$ means positive correlation and vice versa; λ is the spatial error coefficient, disclosing and quantifying the inherent similarity or dissimilarity; ε is the random error term; W is the spatial weight matrix mentioned in Sect. 4.2, and β is the coefficient vector.

Before *SARMA* regression, each factor was standardized by Eq. (15) to make the estimated coefficient at a comparable magnitude:

$$y_i' = [y_i - E(y)]/SD(y) \tag{15}$$

where, y_i' is the standardized value, *E*(y) is mean and *SD*(y) is the standard deviation.

The results of the *SARMA* regression are presented in Table 4, in which influential factors and spatial lagging effect have been disclosed. In other words, surrounding location characteristics and neighboring road segments are related to the type of speed pattern on the road segment.

For **Cluster 1 (Highly Congested Segments)**, all significant factors show a continuous traffic pressure. The merging area on a ramp may cause congestion, thus blocking the ground road. Bus stations and parking lots alongside also bring continuous traffic flow. For road segments whose highest membership belonging to Cluster 1, 55.1% of the campuses around are universities or community college, which seldom reflecting the commuting feature of parents pick-up behavior compared with high schools. Moreover, demand for hospitals is generally stupendously high, stimulating private traffic flow and taxis. Noting the coefficient for road type is positive, which indicates that secondary road with lower speed limit causes more impact. An interesting finding is that a higher portion of transportation type land-use lowers the membership degree for Cluster 1, which means less congestion. This is probably due to the fact that the high proportion of transportation-type land use represents transit hub, such as an airport or a railway station, where road network in the vicinity is usually well organized. Small hubs such as highway bus stations or logistic stations are generally accompanied with congested high occupancy traffic.

Cluster 2 (Normal Speed Segments) has a strong trend of commuting traffic demand. Significant factors with commuting phenomenon include schools, metro stations, and commercial land-use effects. However, the influential factors such as

Table 4 *SARMA* models for cluster membership prediction

Factors	Cluster 1		Cluster 2		Cluster 3		Cluster 4	
	Highly congested		Normal speed		Unimpeded		High speed	
(Intercept)	0.3731	***	0.2885	***	0.1332	***	0.0803	***
	(0.0328)		(0.0245)		(0.0138)		(0.0105)	
Rd_type	0.0695	***			−0.0577	***	−0.0181	***
	(0.0113)				(0.0083)		(0.0059)	
Rd_len			0.0229	**				
			(0.0098)					
Dist_ramp	−0.0192	*	−0.0306	***	0.0165	**	0.0395	***
	(0.0114)		(0.0085)		0.0073		(0.0084)	
Num_bus	0.0831	***	0.0186	*	−0.0369	***	−0.0173	***
	(0.0114)		(0.0096)		(0.0085)		(0.0057)	
Dist_metro			−0.0229	**			0.0257	***
			(0.0099)				(0.0080)	
Ring								
Parking	0.0305	**			−0.0160	*		
	(0.0140)				(0.0089)			
Num_scho	0.0373	***	−0.0269	**				
	(0.0139)		(0.0106)					
Dist_hosp	−0.0307	**					0.0227	***
	(0.0140)						0.0082	
Com_pro			2.3256	**				
			(1.1483)					
Res_pro								
Trans_pro	−3.6390	***			0.0239	***		
	(1.2078)				(0.0085)			
Rho	0.1242	**	0.1432	*	0.2978	*	−0.2764	*
	(0.0789)		(0.0767)		(0.0716)		(0.0882)	
Lambda			−0.1979	**	−0.2465	*	0.4858	***
			(0.0915)		(0.0894)		(0.0684)	
LR test	9.9031	***	3.3252	*	9.5079	***	31.497	***
Log likelihood	−37.7427		55.7193		113.7466		296.6868	
AIC	107.49		−79.44		−195.49		−561.37	

Note * indicates $p < 0.1$, ** indicates $p < 0.05$, *** indicates $p < 0.01$. Standard errors are recorded in parentheses bus station, parking lot and hospital in the vicinity come to be less significant, and the overall average speed is improved. Longer road segment generally means fewer signal controls, which is also a critical factor. And ramps nearby will greatly affect the travel speed, which causes more congestion.

Cluster 3 (Unimpeded Segments) has less significant factors, and the impacts are inverted compared with Cluster 1. Road segments in Cluster 3 are mainly primary roads (with negative coefficient), and the disturbing factors, such as ramps, bus stations, and parking lots become fewer or further, while the proportion of transportation land-use turns to be extremely high. These actually relieve the traffic pressure on the road.

For **Cluster 4 (High-Speed Segments)**, primary-membership objects' mean speed is close to 60 km/h speed limit, while the surrounding built environment factors are similar to the ones in suburban, which can also be obtained from Fig. 4. The low density of ramp, bus station, metro station and hospital all prove this conclusion. According to this result, relative location to the expressway rings and the proportion of residential have little impact on the congestion formation. Based on general understanding, residential area is always regarded as the origin of commuting traffic flow and causes congestion. This may be true in peak hours. However, when considering the entire 24-h patterns, the impact of residential district may not be obvious. Furthermore, commercial land, such as CBD only plays a key role in Cluster 2 and contributes little to road clustering analysis. However, it acts as an important threshold between 'Highly Congested' and 'Normal Speed'.

The spatial-lagged dependent variable ρ and spatial error variable λ were chosen and further analyzed. Spatial-lagged dependent variable indicates the contagious or alien of a dependent variable based on positive or negative values. For Clusters 1, 2 and 3, ρ is significantly positive, indicating that the geographical adjacency of road segments have a positive effect on the membership values. This discloses the fact that for segments within 1000 m, their congestion pattern and traffic condition are mainly driven by the surrounding environmental factors. However, ρ value for Cluster 4 is negative, indicating low speed on road segments near a high-speed segment. This may result from disconnectivity of roads or high traffic demand in a special location, and may also indicate the very different traffic situations of elevated expressways and their underneath surface road segments. The significance of λ associated with Clusters 2 and 3 is negative, which implies that unobserved factors nearby impact membership differently. While λ for Cluster 4 is positive, showing unobserved neighboring variables have parallel effects on segment clustering.

The regression result could be implemented to recognize the congested road segments in the urban area of Shanghai. During the process of transportation planning or urban planning, the result could also be applied to assess road network layout and its combination with land use and traffic-related factors. For example, a secondary road segment with a low proportion of transportation and high density of bus station has a higher probability of suffering continuous congestion, such as Cluster 1. With better insights for problematic road segments, some spatial or temporal redesigns, such as setting variable lanes, widening roads, optimizing road function or setting bus transit lane could be conducted.

5 Conclusion and Recommendations

This study uses taxi GPS data to analyze 24-h speed pattern of primary roads and secondary roads. Speed was selected as the main indicator to further disclose the congestion phenomenon of roads. Correlations with 12 built environment factors including traffic-related factors and land use were further investigated using the data from Shanghai, China as a case study.

First, the average speed of road segments per hour was extracted from GPS trajectories. The fuzzy C-means algorithm was applied to cluster the segments with 24-h dimension vector based on average speed to classify roads into 4 different congestion level. *A geographical detector* was then utilized to find key common factors related to congestion patterns. *MORAN's I* was computed based on types of cluster to investigate spatial similarity of adjacent segments and confirmed a spatial lagging effect. Based on previous findings, a spatial regression model was implemented to identify influential environmental factors associated with each cluster and the interaction between neighboring segments.

Compared with previous studies, this study combined clustering method with quantitative spatial analysis for a better explanation. The influencing factors of congestion were explored using spatial regressions, thus to provide a better understanding of the existing congestion level and quick service evaluation based on environmental data and road conditions.

While the results are promising, further studies need to be conducted to improve the performance of the model. First, in this research, the taxi GPS data don't cover the entire scope of Shanghai, confining the study mainly to urban areas. Secondly, only one weekday data were analyzed, while multi-date analysis may have to be carried out in the future. Particularly during the spatial analysis, the proposed models failed to consider the interactions of environmental factors, which may ignore significant impacts. Moreover, certain research for peak hour analysis would also be carried out in the future, since short-period analysis may disclose some important features or distinct phenomenon only appearing in peak duration. Since peak hours generally have more variation and may need additional accuracy, it would also be interesting to have various temporal divisions, which may result in more realistic classifications of congestions. An attempt would be that, based on the general statistics of the current division, further divide the congested hours into half-hour periods considering the ordinary taxi trip would be less than 30 min (or even 15 min), while keeping 1 h for non-peak trips.

References

Anselin, L., Bera, A. K., Florax, R., & Yoon, M. J. (1996). Simple diagnostic tests for spatial dependence. *Regional Science and Urban Economics, 26*(1), 77–104.

Azar, A. T., El-Said, S. A., & Hassanien, A. E. (2013). Fuzzy and hard clustering analysis for thyroid disease. *Computer Methods and Programs in Biomedicine, 111*(1), 1–16.

Azimi, M., & Zhang, Y. (2010). Categorizing freeway flow conditions by using clustering methods. *Transportation Research Record: Journal of the Transportation Research Board, 2173*, 105–114.

Bezdek, J. C. (1973). Cluster validity with fuzzy sets. *Journal of Cybernetics, 3*(3), 58–73.

Bezdek, J. C. (1980). A convergence theorem for the fuzzy ISODATA clustering algorithms. *IEEE Tansactions on Pattern Analysis and Machine Intelligence, 2*(1), 1–8.

Bezdek, J. C., Ehrlich, R., & Full, W. (1984). FCM: The fuzzy c-means clustering algorithm. *Computers and Geosciences, 10*(2–3), 191–203.

Chen, C., Zhang, D., Li, N., & Zhou, Z. H. (2014). B-planner: Planning bidirectional night bus routes using large-scale taxi GPS traces. *IEEE Transactions on Intelligent Transportation Systems, 15*(4), 1451–1465.

Cliff, A. D., & Ord, J. K. (1982). Spatial processes: Models & applications. *Quarterly Review of Biology.*

Cui, J., Liu, F., Janssens, D., An, S., Wets, G., & Cools, M. (2016). Detecting urban road network accessibility problems using taxi GPS data. *Journal of Transport Geography, 51*, 147–157.

Ding, J., Gao, S., Jenelius, E., Rahmani, M., Huang, H., Ma, L., & Ben-Akiva, M. (2014). Routing policy choice set generation in stochastic time-dependent networks: Case studies for Stockholm, Sweden, and Singapore. *Transportation Research Record: Journal of the Transportation Research Board, 2466*, 76–86.

Dunn, J. C. (1973). A fuzzy relative of the ISODATA process and its use in detecting compact well-separated clusters. *Journal of Cybernetics, 3*(3), 32–57.

Feng, H., Li, C., Zhao, N., & Hu, H. (2011). Modeling the impacts of related factors on traffic operation. *Procedia Engineering, 12*, 99–104.

Fukuyama, Y., & Sugeno, M. (1989). A new method of choosing the number of clusters for fuzzy C-means method. Presented at the Proceedings of the 5th Fuzzy System Symposium, Japan.

Goddard, J. B. (1970). Functional regions within the city centre: A study by factor analysis of taxi flows in central London. *Transactions of the Institute of British Geographers, 49*, 161–182.

Greicius, M. D., Krasnow, B., Reiss, A. L., & Menon, V. (2003). Functional connectivity in the resting brain: A network analysis of the default mode hypothesis. *Proceedings of the National Academy of Sciences, 100*(1), 253–258.

Hahn, E., Chatterjee, A., Younger, M. S., Hahn, E., Chatterjee, A., & Younger, M. S. (2002). Macro-level analysis of factors related to areawide highway traffic congestion. *Transportation Research Record: Journal of the Transportation Research Board, 1817*, 11–16.

Handy, S., Cao, X., & Mokhtarian, P. (2005). Correlation or causality between the built environment and travel behavior? Evidence from northern California. *Transportation Research Part D Transport and Environment, 10*(6), 427–444.

He, F., Yan, X., Liu, Y., & Ma, L. (2016). A traffic congestion assessment method for urban road networks based on speed performance index. *Procedia Engineering, 137*, 425–433.

Hu, X., An, S., & Wang, J. (2014). Exploring urban taxi drivers' activity distribution based on GPS data. *Mathematical Problems in Engineering, 2014*(2), 1–13.

Hwang, K., Wu, K., & Jian, R. J. (2006). Modeling consumer preference for global positioning system-based taxi dispatching service: Case study of Taichung City, Taiwan. *Transportation Research Record: Journal of the Transportation Research Board, 1971*, 99–106.

Jiménez-Meza, A., Arámburo-Lizárraga, J., & Fuente, E. D. L. (2013). Framework for estimating travel time, distance, speed, and street segment level of service (los), based on GPS data. *Procedia Technology, 7*(4), 61–70.

Kerner, B. S., & Klenov, S. L. (2006). Probabilistic breakdown phenomenon at on-ramp bottlenecks in three-phase traffic theory: Congestion nucleation in spatially non-homogeneous traffic. *Physics, 1965*(2006), 473–492.

Kumar, V., & Vanajakshi, L. D. (2013). Modewise travel time estimation on urban arterials using transit buses as probes. Presented at the 92nd Annual Meeting of the Transportation Research Board, January 13–17, Washington, D.C.

Lu, Y., & Li, S. (2014). An empirical study of with-in day OD prediction using taxi GPS data in Singapore. *Langmuir the ACS Journal of Surfaces and Colloids, 30*(31), 9567–9576.

Miller, J. S., & Evans, L. D. (2011). Divergence of potential state-level performance measures to assess transportation and land use coordination. *Journal of Transport and Land Use, 4*(3), 81–103.

Moeckel, R. (2016). Constraints in household relocation: Modeling land-use/transport interactions that respect time and monetary budgets. *Journal of Transport and Land Use, 10*(1), 211–228.

Montero, L., Pacheco, M., Barcelo, J., Homoceanu, S., & Casanovas, J. (2016). A case study on cooperative car data for traffic state estimation in an urban network. Presented at the 95th Annual Meeting of the Transportation Research Board, January 10–14, Washington, D.C.

Moran, P. A. (1950). Notes on continuous stochastic phenomena. *Biometrika, 37*(1–2), 17–23.

Nian, G., Sun, J., & Huang, J. (2021). Exploring the Effects of Urban Built Environment on Road Travel Speed Variability with a Spatial Panel Data Model. ISPRS International Journal of Geo-Information, 10(12), 829.

Pakhira, M. K., Bandyopadhyay, S., & Maulik, U. (2005). A study of some fuzzy cluster validity indices, genetic clustering and application to pixel classification. *Fuzzy Sets and Systems, 155*(2), 191–214.

Pan, G., Qi, G., Wu, Z., Zhang, D., & Li, S. (2013). Land-use classification using taxi GPS traces. *IEEE Transactions on Intelligent Transportation Systems, 14*(1), 113–123.

Qian, X., & Ukkusuri, S. V. (2015). Exploring spatial variation of urban taxi ridership using geographically weighted regression. Presented at the 94th Annual Meeting of the Transportation Research Board, January 11–15, Washington, D.C.

Qing, C., Parfenov, S., & Kim, L. J. (2015). Identifying travel patterns during extreme weather using taxi GPS data. Presented at the Transportation Research Board 94th Annual Meeting, January 11–15, Washington, DC.

Schw, M. V., & Jensen, O. N. (2010). A simple and fast method to determine the parameters for fuzzy C-means cluster analysis. *Bioinformatics, 26*(22), 2841–2848.

Sun, D., & Elefteriadou, L. (2011). Lane changing behavior on urban streets: A focus group based study. *Applied Ergonomics: Human Factors in Technology and Society, 42*(5), 682–691.

Sun, D., & Elefteriadou, L. (2012). Lane changing behavior on urban street: An "in-vehicle" field experiment based study. *Computer-Aided Civil and Infrastructure Engineering, 27*(7), 525–542.

Sun, D., Liu, X., Ni, A., & Peng, C. (2014a). Traffic congestion evaluation method for urban arterials: Case study of Changzhou, China. *Transportation Research Record: Journal of the Transportation Research Board, 2461*, 9–15.

Sun, D., Zhang, C., Zhang, L., Chen, F., & Peng, Z. R. (2014b). Urban travel behavior analyses and route prediction based on floating car data. *Transportation Letters, 6*(3), 118–125.

Tang, J., Jiang, H., Li, Z., & Li, M. (2016). A two-layer model for taxi customer searching behaviors using GPS trajectory data. *IEEE Transactions on Intelligent Transportation Systems, 17*, 1–7.

Tang, L., Yang, X., Kan, Z., & Li, Q. (2015). Lane-level road information mining from vehicle GPS trajectories based on Naïve Bayesian classification. *ISPRS International Journal of Geo-information, 4*(4), 2660–2680.

Tian, G., Ewing, R., White, A., Hamidi, S., Walters, J., & Goates, J. P. (2015). Traffic generated by mixed-use developments: Thirteen-region study using consistent measures of built environment. *Journal of Urban Planning and Development, 137*(3), 248–261.

Tulic, M., Bauer, D., & Scherrer, W. (2014). Link and route travel time prediction including the corresponding reliability in an urban network based on taxi floating car data. *Transportation Research Record: Journal of the Transportation Research Board, 2442*, 140–149.

Wang, H., Peng, Z. R., Lu, Q. C., Sun, J., & Bai, C. (2017). Assessing effects of bus service quality on passengers' taxi-hiring behavior. *Transport*. https://doi.org/10.3846/16484142.2016.1275786

Wang, J. F., Li, X. H., Christakos, G., Liao, Y. L., Zhang, T., Gu, X., & Zheng, X. Y. (2010). Geographical detectors-based health risk assessment and its application in the neural tube defects study of the Heshun Region, China. *International Journal of Geographical Information Science, 24*(1), 107–127.

Wen, T. H., Chin, W. C., & Lai, P. C. (2017). Understanding the topological characteristics and flow complexity of urban traffic congestion. *Physica A: Statistical Mechanics and Its Applications, 473*(1), 166–177.

Wheaton, W. C. (1998). Land use and density in cities with congestion. *Journal of Urban Economics, 43*(2), 258–272.

Yang, Y., & Diez-Roux, A. V. (2012). Walking distance by trip purpose and population subgroups. *American Journal of Preventive Medicine, 43*(1), 11–19.

Yazici, M. A., Kamga, C., & Singhal, A. (2016). Modeling taxi drivers' decisions for improving airport ground access: John F. Kennedy airport case. *Transportation Research Part A: Policy and Practice, 91*, 48–60.

Yu, L., & Liu, Y. (2011). Traffic characteristics analysis and suggestions on school bus operation for primary school students in Beijing. *Journal of Transportation Systems Engineering and Information Technology, 11*(5), 193–200.

Yu, J., & Lu, P. (2016). Learning traffic signal phase and timing information from low-sampling rate taxi GPS trajectories. *Knowledge-Based Systems, 110*, 275–292.

Zhang, J., Qiu, P., Duan, Y., Du, M., & Lu, F. (2015). A space-time visualization analysis method for taxi operation in Beijing. *Journal of Visual Languages and Computing, 31*, 1–8.

Zhang, L., Hong, J. H., Nasri, A., & Shen, Q. (2012). How built environment affects travel behavior: A comparative analysis of the connections between land use and vehicle miles traveled in U.S. cities. *Journal of Transport & Land Use, 5*(3), 40–52.

Zhang, L., & Levinson, D. (2017). A model of the rise and fall of roads. *Journal of Transport and Land Use, 10*(2), 1–23. https://doi.org/10.5198/jtlu.2016.887

Zhu, Z., & Nandi, A. K. (2014). Blind digital modulation classification using minimum distance centroid estimator and non-parametric likelihood function. *IEEE Transactions on Wireless Communications, 13*(8), 4483–4494.

Chapter 6
Travel Time Estimation Based on Built Environment Attributes and Low-Frequency Floating Car Data

Abstract This chapter studies the effect of urban built environment attributes on the estimation of road travel time (hereafter denoted "travel time") from low-frequency floating car data without complex global positioning system information, such as speed. In addition, a new method of estimating travel time distribution is developed, which uses the distribution of the number of vehicles on a road, rather than the road's length, as the proportional coefficient of a travel time distribution. To verify the correctness of this novel method, the effect parameters of various built environment attributes on travel time are examined in an example, using the maximum likelihood estimation method. The results show that certain urban built environment attributes around a road will lead to a significant increase in travel time in certain periods. The effect time of schools is from 6:00 a.m. to 7:20 a.m., while that of hospitals and clinics is from 7:00 a.m. to 8:00 a.m.; in addition, a similar travel time increase is caused by intersections in all scenarios. Finally, the likelihood ratio test verifies the reliability of using built environment attribute variables as influencing factors of travel time.

Keywords Urban traffic · Travel time estimation · Floating car data · Built environment · Maximum likelihood estimation · Travel time distribution

1 Introduction

The travel time on a road is one of the most important pieces of information describing the road's operational state and plays an important role in transportation planning, management, and control. In recent years, road travel time estimation (hereafter denoted "travel time") has become a research focus in the field of intelligent transportation systems. Travel time estimation methods can be divided into three categories: complex model-based methods, data analysis methods, and simulation model-based methods (Table 1).

Existing estimation methods based on complex models typically combine the concepts of queuing theory and traffic flow delay theory and consider various traffic parameters in a road network and various traffic factors that cause delays. Hellinga et al. (2008) decomposed total travel time into the travel time of each section on

S. Zhong and D. Sun, *Logic-Driven Traffic Big Data Analytics*,
https://doi.org/10.1007/978-981-16-8016-8_6

Table 1 Summary of previous studies

Research classification	Literature	Research methods
Complex models	Hellinga et al. (2008)	The total travel time is decomposed into the travel time of each section. A likelihood function is defined to calculate the most likely travel time of each road on the route
	Jenelius and Koutsopoulos (2013)	Considering the road traffic conditions, the total travel time is decomposed and the maximum likelihood method is used to estimate the parameters
	Dell' Orco et al. (2016)	A new network traffic assignment model is proposed by using a time-based discrete traffic flow model and meta-heuristic algorithm to improve the authenticity of prediction
	Zhang et al. (2018)	In order to estimate the travel time of Ring Road, an emergency and gentle traffic flow model (UGM) considering viscoelasticity and ramp effect is established, and the vehicles in the traffic flow are divided into emergency and gentle categories
	Cheng et al. (2019)	A travel time estimation model based on random forest is proposed. Seven influence variables were used as candidate variables. The model is verified by the data obtained from VISSIM simulation
	Zhong et al. (2021)	The global regression model and the geographically weighted regression (GWR) model were adopted to explore the main built environment factors affecting the travel time of road
Data analysis method	Hofleitner et al. (2012)	The distribution parameters were estimated by maximum likelihood estimation. The estimated parameters are proportional to the conditional probability product of historical floating car data
	Zheng and Zuylen (2013)	Combined the low-frequency floating car data with the neural network model, the road travel time was estimated
	Shi et al. (2017)	A speed estimation algorithm is developed to estimate the travel time of deterministic links, and a distributed estimation algorithm is proposed to estimate the delay in travel time
	Ma et al. (2017)	A generalized Markov chain method is proposed to estimate the probability distribution of travel time, and the correlation between time and space is considered

(continued)

Table 1 (continued)

Research classification	Literature	Research methods
	Rahmani et al. (2017)	Based on the taxi data in Stockholm, the sensitivity of road travel time estimation is studied, and a fixed-point formula is proposed to solve the problem of path reasoning and travel time estimation at the same time
	Fan et al. (2018)	Based on the data of the expressway electronic toll collection system, a machine learning method embedded in a big data analysis platform is established by using the random forest method and Apache Hadoop to estimate expressway travel time
Simulation model	Liu and Ma (2009)	Using the real-time data acquisition and archiving system developed by the University of Minnesota, a road travel time estimation algorithm based on a traffic simulation model is proposed

a travel route. They developed a likelihood function model to determine the most likely travel time of each section on a travel route and suggested that the parking delay of each section could be allocated in proportion to the possibility of finding a park within each section. On this basis, Jenelius and Koutsopoulos (2013) took the turning characteristics and travel conditions of a section as the explanatory variables for the maximum likelihood estimation of parameters, in which other road travel conditions were also considered. Dell' Orco et al. (2016) proposed a new network traffic assignment model based on a time-based discrete traffic flow model, which improved the authenticity of predictions.

Methods based on data analysis generally use a large number of data and apply linear regression models, Bayesian network models, neural network models, Kalman filter models, and other scientific methods. In addition, estimation methods based on floating car data are often used due to their wide applicability and high reliability. A floating car continuously collects traffic information on a road and can be used as an effective data source to estimate the travel time on the road. For example, Hofleitner et al. (2012) used floating car data to estimate distribution parameters with the maximum likelihood estimation method, such that the estimated parameters were directly proportional to the conditional probability product of the historical floating car data. Zheng and Zuylen (2013) estimated road travel time from low-frequency floating car data combined with a neural network model. In recent years, the methods for estimating travel time have become more diversified. Ma et al. (2017) proposed a generalized Markov chain method to estimate the probability distribution of travel time, which considered the correlation between time and space. Rahmani et al. (2017) studied the sensitivity of travel time estimation and proposed a fixed point formula to simultaneously solve the problems of path reasoning and travel time estimation. Fan et al. (2018) estimated the travel time of an expressway by using a machine learning method to analyze the expressway's electronic toll collection data. In addition to the above two types of methods, traffic simulation is also applied in the estimation of road travel time. Liu and Ma (2009) used a real-time data acquisition and archiving system developed by the University of Minnesota (United States) to develop a travel time estimation algorithm based on a traffic simulation model.

The aforementioned approaches can be used to estimate the travel time on a road and the traffic state on a road, the road parameters, and historical travel time data are more commonly considered in research. However, few researchers have examined the effect of built environment attributes on travel time. These attributes affect travel time by influencing traveler-related aspects, such as travel destination, travel frequency, travel mode, and travel route, which are crucial considerations in urban transportation planning and management. Thus, there is an urgent need for a travel time estimation model that considers the effects of urban built environment factors and thereby generates findings that are relevant for the improvement of real-world road facilities. To this end, Zhong et al. (2021) recently applied a global regression model and a geographically weighted regression (GWR) model to study the main factors affecting travel time from the perspective of an urban built environment. However, although they effectively solved the spatial heterogeneity problem regarding the effect of urban built environment on travel time, they did not propose an estimation method for the distribution of travel time. In addition, studies on estimating travel time based on floating car data often collected a variety of information, such as vehicle speed and direction angle. As a result, an effective way to use floating car data to estimate travel time when the above information is missing has yet to be developed.

As such, this chapter establishes the relationship between urban built environment attributes and travel time based on the location data of low-frequency floating car reports, with only longitude and latitude coordinate information being required (i.e., vehicle speed and direction angle data are not required). This novel method takes the surrounding facilities (e.g., schools, hospitals, and downstream intersections) as the variables that directly affect travel time. Moreover, in contrast to the conventional method of estimating travel time based on global positioning system (GPS) coordinates and time data, this novel method does not need to speculate and match a vehicle travel path in the study area. A statistical modeling method is also used to simplify the calculation process, which has universal applicability to different traffic conditions. An entire road is divided into several small segments, and the distribution of travel time on these segments is studied. The distribution of the number of floating cars on a segment is used instead of distance as the travel time distribution coefficient, which improves the accuracy of the travel time estimation results. Finally, a case study of Jinshan Street in Dandong, China, is used to verify the novel method and analyze the effect of urban built environments on travel time.

2 Description of a Floating Car System

A basic understanding of a floating car system enables the processing and interpretation of data. Thus, this section introduces the composition and advantages of a floating car system.

2.1 Composition of a Floating Car System

A floating car system consists of an information acquisition system, a data processing system, and an information release system. The information acquisition system consists of vehicles equipped with positioning equipment; the data processing system comprises an information collection and processing system; and the information release system establishes mutual communication between data processing results and user terminals.

The information acquisition system uses automatic positioning technology to collect real-time vehicle data, such as position coordinates, speed, position time, direction angle, and load condition. Different floating car systems contain different types of data. The methods and case studies introduced in this chapter using floating car data that do not include speed and direction angle data, see Table 2 for details.

Table 2 Example of the floating car data of Dandong City used in this chapter

Group billing number	Car numbers	SIM card number	X-coordinate	Y-coordinate	Load condition	Positioning time
9000075344	Liao F95403	13470496XXX	44.7123	143.8302	Empty car	2011-06-11 00:00:07
89000168349	Liao F93783	13942571XXX	44.7824	144.5141	Heavy car	2011-06-11 00:00:40
89000168349	Liao F92871	15842588XXX	44.7779	144.4272	Heavy car	2011-06-11 00:00:48
89000168349	Liao F91232	15842599XXX	44.7706	144.3597	Empty car	2011-06-11 00:00:52

2.2 Advantages of Floating Car Systems

In contrast to fixed, coil-based car detectors, floating car systems can be constructed without damaging a road or interrupting traffic. Moreover, a floating car system has low capital construction costs and maintenance costs. In addition, there is never a problem of insufficient coverage (which can occur if there are insufficient fixed detectors), because a vehicle positioning and information transceiver device operates inside a floating vehicle traversing a road network, and this device always maintains contact with the data center. Thus, a floating car continuously collects data and records road traffic conditions and other data as it travels: data is collected wherever there are floating cars, and their high mobility means that high coverage can be obtained without much equipment. Overall, the advantages of floating cars are that they continuously collect traffic data, have high spatiotemporal coverage, cause no damage to road infrastructure, do not interrupt traffic, and have low equipment investment and maintenance costs. Therefore, floating car technology is highly suited to the collection of urban traffic information.

3 Methodology

3.1 Relationship Between the Number of Reports Sent by a Floating Car and Travel Time

The more congested a road, the longer the travel time, and thus the more likely a floating car is to send a report. The event of a floating car sending a report on a road is taken as a random variable, and the entire road is divided into several segments; this establishes the relationship between the number of floating cars sending a report on each segment and the travel time on a segment.

It is known that the time interval of the floating car to send the report is fixed, so the possibility of each floating car sending the report at any time is the same. Suppose that the frequency of the report sent by the floating car at any time is ε, which can be expressed as:

$$\varepsilon = \frac{1}{T} \tag{1}$$

where T is the time interval between two reports sent by the floating car (s), ε is the frequency of reports sent for floating cars (s^{-1}).

The whole road is divided into K segments, It is easy to know that in any small segment x_k, the possibility ρ_x of the floating car reporting its position in the segment x_k is directly proportional to the travel time $t(x_k)$ (s) of the floating car in the segment, as shown in Eq. (2).

$$\rho_x = \varepsilon t(x_k) = \frac{t(x_k)}{T}, \quad \text{where } t(x_k) < T \tag{2}$$

If the floating car stays at a certain segment for more than u reporting periods, namely $t(x_k) > uT$, where $u \in N_+, u = \left[\frac{t(x_k)-T}{T}\right]$, then the minimum number of times to send the report is u, and the possibility ρ_{ux} of sending report for $u+1$ times is:

$$\rho_{ux} = \varepsilon(t(x_k) - uT) = \frac{t(x_k) - uT}{T} \tag{3}$$

It is assumed that the traffic state of a segment remains unchanged during the study period, which means that the travel time of vehicles on each segment is the same. Therefore, each segment of a road that a floating car traverses is regarded as a random event. Assuming that there is no difference in the running state of a floating car during the period when the traffic state is constant, multiple floating cars traversing segments can be regarded as independent repeated tests that obey a Bernoulli distribution. While $t(x_k) < T$, among the total number of vehicles m passing through each segment, the probability p_x that the number of vehicles sending the report is equal to n_x is given as follows:

$$p_x = C_m^{n_x} \rho_x^{n_x} (1 - \rho_x)^{m-n_x} = C_m^{n_x} \left(\frac{t(x_k)}{T}\right)^{n_x} \left(1 - \left(\frac{t(x_k)}{T}\right)\right)^{m-n_x} \tag{4}$$

When $t(x_k) > uT, u \in N_+$, that is, when the dwell time of the floating car in the segment is longer than u report sending cycles, the probability p_{ux} that the number of vehicles sending the report in the segment is equal to n_x can be expressed as follows:

$$\begin{aligned} p_{ux} &= C_m^{n_x-mu} \rho_{ux}^{n_x-mu} (1 - \rho_{ux})^{m-n_x+mu} \\ &= C_m^{n_x-mu} \left(\frac{t(x_k) - uT}{T}\right)^{n_x-mu} \left(1 - \left(\frac{t(x_k) - uT}{T}\right)\right)^{m-n_x+mu} \end{aligned} \tag{5}$$

where $n_x - mu$ is the number of vehicles that send the report again in the segment after removing the number of times the report was sent in the previous u cycles, and $0 < n_x - mu < m$, namely $mu < n_x < m(u+1)$. It is reasonable to assume that in each segment, the number of reports sent by vehicle differs at most one time, considering that the data of low-frequency floating cars are used.

In this section, the relationship between the number of floating cars sending reports in a certain segment and the travel time in that segment is established considering that the possibility of floating cars sending reports on a given segment is proportional to the travel time in that segment.

3.2 Relationship Between Travel Time and the Built Environment Attributes

An entire road is divided into several segments according to the length of the road. In previous studies, the travel time of each segment was assumed to be affected by the attributes of the segment, such as its length and distance from a downstream intersection. Thus, the influence of the built environment attributes (facilities) on the travel time of a segment is considered in this chapter, and is expressed as the distance from the segment to the built environment facilities.

A linear structure is used to express the impact of J explanatory variables (including road attributes, intersections, built environment attributes) on the travel time $t(x_k)$ of a certain segment. It can be expressed as:

$$t(x_k) = \sum_{j=1}^{J} \alpha_j A_j \tag{6}$$

where A_j is the value of the jth explanatory variable (m), including the length of the segment, the distance between the segment and the downstream intersection or the built environmental facilities. α_j is the influence degree of the jth explanatory variable on the segment travel time (s/m), which is the parameter to be estimated. The specific relationship is shown in Fig. 1.

Equation (6) establishes the linear combination relationship between the travel time of the road segment and the built environment attributes. Then the travel time estimation of each segment is transformed into a maximum likelihood estimation problem for the influence degree parameter, as follows:

$$\max \prod_x p_x = \prod_x C_m^{n_x} \rho_x^{n_x} (1 - \rho_x)^{m - n_x}$$

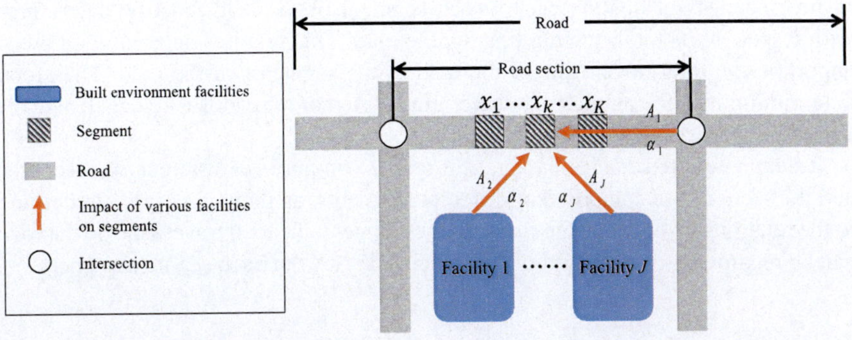

Fig. 1 Relationship between segment travel time and built environment attributes

$$= \prod_x C_m^{n_x} \left(\frac{t(x_k)}{T} \right)^{nx} \left(1 - \left(\frac{t(x_k)}{T} \right) \right)^{m-n_x}$$

$$= \prod_x C_m^{n_x} \left(\frac{\sum_{j=1}^{J} \alpha_j A_j}{T} \right)^{nx} \left(1 - \left(\frac{\sum_{j=1}^{J} \alpha_j A_j}{T} \right) \right)^{m-n_x} \quad (7)$$

$$\max \prod_x p_{ux} = \prod_x C_m^{n_x - mu} \rho_{ux}^{n_x - mu} (1 - \rho_{ux})^{m-n_x+mu}$$

$$= \prod_x C_m^{n_x - mu} \left(\frac{t(x_k) - uT}{T} \right)^{n_x - mu} \left(1 - \left(\frac{t(x_k) - uT}{T} \right) \right)^{m-n_x+mu}$$

$$= \prod_x C_m^{n_x - mu} \left(\frac{\sum_{j=1}^{J} \alpha_j A_j - uT}{T} \right)^{n_x - mu} \left(1 - \left(\frac{\sum_{j=1}^{J} \alpha_j A_j - uT}{T} \right) \right)^{m-n_x+mu} \quad (8)$$

The estimated results are the values of each parameter, so the segment travel time can be calculated by considering various built environment attributes, according to Eq. (6). Then, the travel time under the influence of these built environment attributes can be calculated according to the relationship between the road and the segment.

3.3 Distribution of Travel Time

Between the two reports, the trajectory of the floating car is likely to pass a part of certain segments. Thus, an important problem when using low-frequency floating car data to estimate travel time is to determine the relationship between the travel time and the known interval between floating cars sending reports. Thus, the proportion of time spent by a floating car on each segment of the road must be determined with respect to its total running time on the road. This requires determining how to apportion the total travel time of the road to the segments of the road. Therefore, determining how to calculate the travel time of an entire road is a key challenge for determining travel time distribution.

It should be noted that in this study, the road is composed of different road sections, and the sections are composed of different segments, as shown in Fig. 1. According to the relationship between the road and the segment, the total travel time on the road can be regarded as the integral of the travel time $t(x_k)$ of each segment, namely:

$$t_s = \int_0^S t(x_k) dx \quad (9)$$

where S is the total length of the road (m), t_s is the total travel time of the road (s). Therefore, the travel time of a road section is the integral of the travel time of each segment in this road section, and it can be expressed as:

$$t_i = \int_{l_i}^{l_{i+1}} t(x_k)dx \tag{10}$$

where l_i is the distance from the road section i to the starting point of the road (m).

The expected number of the floating cars is equal to the product of the probability $p(x_k)$ that the floating car reports its position in the segment and the number of tests m (i.e., the total number of vehicles passing through the segment), namely:

$$E(x_k) = mp(x_k) \tag{11}$$

The number of vehicles n_x that are observed to send reports on a segment is an unbiased estimation of the expected value, and the travel time of a floating car on the segment is directly proportional to the possibility of a floating car sending reports on the segment. Therefore, it can be considered that the travel time of a floating car on the segment is directly proportional to the number of times a floating car sends reports on the segment, namely $t(x_k) \propto p(x_k) \propto E(x_k) \propto n_x$. Then, the total number of vehicles sending reports in a certain time interval is counted, and the ratio of the travel time on each segment to the total travel time of the entire road is equal to the ratio of the number of vehicles sending reports on the segment to the total number of vehicles sending reports on the entire road.

$$Q_1 = \frac{t_1}{ts} = \frac{\int_{l_1}^{l_2} t(x_k)dx}{\int_0^S t(x_k)dx} = \frac{\int_{l_1}^{l_2} n(x_k)dx}{\int_0^S n(x_k)dx} \tag{12}$$

where Q_1 is the ratio of the travel time of the first road section to the total travel time of the road. t_1 is the travel time of the first road section (s). l_1, l_2 is the distance from the first and second road sections to the starting point of the road (m). $n(x_k)$ is the number of times of vehicles sending reports in segment x_k (count).

The above approach is also followed when studying the travel time distribution of different roads, as it serves as an independent repeated test of vehicles passing any location with two or more road segments in the same traffic state. The ratio of the travel times of two road segments is obtained as the ratio of the total number of reports sent by vehicles traversing the two road segments at the same time.

$$\frac{t_{s1}}{t_{s2}} = \frac{\int_0^{S_1} n(x_k)dx}{\int_0^{S_2} n(x_k)dx} \tag{13}$$

where t_{s1} and t_{s2} represent the travel time of two road sections (s). S_1 and S_2 represent the length of two road sections (m). Then the ratio of travel time between all road sections can be obtained, and the problem of travel time distribution between road sections can be solved.

4 Case Study

18.68 million pieces of data generated by 603 taxis in Dandong, China, were used to estimate the corresponding parameter values of the variables that affect the travel time of a segment in different time periods. The likelihood-ratio test shows that the built environment facilities have a significant effect on the travel time on the segment. Comparison with the value given by Baidu Maps also proves the rationality of the estimation result.

4.1 Study Area

The study area is located on the road from the First Company of Dandong Public Transportation Corporation to Dandong Environmental Science Research Institute, Jinshan Street, Zhenxing District, Dandong (see Fig. 2). Dandong is located in the southeast of Liaoning province and is a coastal and riverside border city in the center

Fig. 2 Study area: Jinshan Street, Dandong City, China

of northeast Asia. Jinshan Street is the main road in Dandong and has a total length of 2.7 km. There are 10 intersections along the road, most of which are roundabouts and non-signalized intersections, which means that the traffic conditions on the road are complex. There are also many schools on the road (e.g., Jingye High School, Korean Middle School, and Fuchunjie Primary School), medical facilities (e.g., women's and children's hospitals), gas stations, and other convenience facilities. Thus, Jinshan Street is an excellent choice as a representative research object to examine the effect of built environment attributes on travel time.

4.2 Correction of Floating Car Data

Floating car data must be processed before it can be used for research. First, the encrypted data are affine transformed onto the target road network. Affine transformation linearly transforms data from one set of two-dimensional coordinates to another. Thus, it maintains the linearity of data (i.e., a straight line remains a straight line after transformation) and the parallelism of data such that the relative position relationship of graphic elements remains unchanged (i.e., a parallel line remains a parallel line, and the order of the points on the line remains unchanged).

Second, ArcGIS software, a family of client software, server software, and online geographic information system (GIS) services developed and maintained by Esri, is used to correct the coordinates in this research (Scott & Janikas, 2010). The relationship between the display position of a floating car report location and the actual position of the floating car in the road network is used as the transformation matrix input, and the "calculate geometry" tool is used to change its coordinate attributes to those of the transformed coordinate in its attributes table. The coordinate comparison before and after transformation is shown in Table 3. The corrected data points are displayed in ArcGIS as shown in Fig. 3.

Third, outliers are eliminated, such that only the floating cars on the road network are used to estimate the road condition. If a floating car only stops outside the road network, as shown in the point far away from the road network in Fig. 3, it will not affect the road network operation state, nor will it reflect the road network condition. Therefore, the data of this car is eliminated; this is achieved by setting a maximum matching distance threshold in the map matching process.

Table 3 Coordinate comparison before and after spatial correction

X-coordinate before correction	Y-coordinate before correction	X-coordinate after correction	Y-coordinate after correction
143.5712	44.69388	124.1432045	39.87715703
144.4186	44.77984	124.3828909	40.1144931
143.58	44.69447	124.1448548	39.87964019
144.4793	44.78142	124.3873226	40.13174936

Fig. 3 Distribution of report points in the road network after spatial correction

4.3 Parameter Selection and Estimation

The case study road section is divided into 16 segments denoted ID1–ID16, and the length of each segment is set as an influencing factor. Five types of built environment facilities are selected as the factors affecting travel time (intersections, schools, hospitals, clinics, and gas stations), and the distance from each segment to one of these facilities represents a variable. Each variable is regarded as a decreasing function of distance to reflect the fact that the closer a vehicle is to a built environment facility, the more its travel time is affected. When the distance between a segment and a facility is sufficiently large, the effect of the facility can be ignored; here, this distance is considered to be $\geq$1 km. The value of the distance variable of each segment within 1 km is taken as $1 - D_s/1000$, where D_s is the distance from the built environment facility to the study segment, and the value of the distance variable of each segment beyond 1 km is set to 0. Intersections are used to select the distance from a segment to a downstream intersection, and each segment has only one downstream intersection; thus, the number of intersection variables in any segment should be less than or equal to 1.

To observe changes in parameter values, the time interval is divided into 10 min, i.e., a group of estimated values is obtained every 10 min. Considering that little data are obtained from floating cars from 6:00 a.m. to 6:30 a.m., the three time intervals within this period are combined into a single 30-min time interval. Substituting the value of each variable A_j into Eqs. (7) and (8), the parameter results obtained are shown in Table 4.

4.4 Analysis of Results

As shown in Table 4, the parameter values of all variables are positive, which verifies that the built environment variables are positively correlated with travel time in the

Table 4 Estimated travel time parameters

Time period	α_j (s/m)									
	ID1	ID2	ID3	ID4	ID5	ID6	ID7	ID8	ID9	ID10	ID11
6:00–6:30	0.000	0.177	0.035	0.087	0.151	0.127	0.054	0.105	0.237	0.169	0.052
6:30–6:40	0.000	0.259	0.096	0.100	0.192	0.171	0.088	0.124	0.145	0.140	0.126
6:40–6:50	0.050	0.257	0.122	0.120	0.161	0.080	0.088	0.115	0.213	0.207	0.058
6:50–7:00	0.000	0.214	0.126	0.145	0.217	0.106	0.095	0.136	0.271	0.050	0.042
7:00–7:10	0.000	0.201	0.127	0.135	0.181	0.159	0.073	0.127	0.268	0.174	0.141
7:10–7:20	0.000	0.178	0.085	0.116	0.211	0.168	0.123	0.143	0.058	0.205	0.129
7:20–7:30	0.044	1.349	0.143	0.141	0.275	0.077	0.126	0.159	0.151	0.174	0.144
7:30–7:40	0.000	0.277	0.133	0.087	0.247	0.030	0.102	0.187	0.000	0.140	0.104
7:40–7:50	0.000	0.321	0.147	0.119	0.269	0.541	0.087	0.169	0.000	0.248	0.133
7:50–8:00	0.000	0.325	0.104	0.105	0.283	0.151	0.077	0.155	0.160	0.154	0.151
Time period	α_j (s/m)										
	ID12	ID13	ID14	ID15	ID16	Intersection	School		Hospital	Clinic	Gas station
6:00–6:30	0.036	0.215	0.067	0.108	0.135	0.047	0.041		0.038	0.064	0.064
6:30–6:40	0.045	0.182	0.090	0.109	0.125	0.059	0.015		0.056	0.052	0.020
6:40–6:50	0.061	0.121	0.097	0.102	0.164	0.055	0.006		0.071	0.038	0.013

(continued)

Table 4 (continued)

6:50–7:00	0.092	0.186	0.130	0.153	0.250	0.009	0.024	0.004	0.040	0.031
7:00–7:10	0.107	0.188	0.137	0.182	0.193	0.000	0.006	0.067	0.075	0.071
7:10–7:20	0.094	0.219	0.141	0.155	0.251	0.029	0.017	0.112	0.059	0.018
7:20–7:30	0.125	0.253	0.102	0.166	0.143	0.040	0.000	0.101	0.053	0.000
7:30–7:40	0.132	0.104	0.117	0.129	0.260	0.015	0.002	0.118	0.116	0.003
7:40–7:50	0.133	0.105	0.118	0.159	0.160	0.014	0.001	0.116	0.090	0.040
7:50–8:00	0.107	0.167	0.132	0.147	0.219	0.000	0.000	0.176	0.101	0.074

study area. In addition, the parameter value corresponding to the distance variable to the intersection does not change markedly with time, indicating that the delay at the intersection during the morning peak is relatively constant. As only one intersection within the study area is a signalized intersection, this case study does not distinguish between signalized and non-signalized intersections. The implementation of non-signalized control at most intersections also indicates that the traffic conditions of these intersections are good. In addition, the parameter values corresponding to the school variable in the time period of 6:00–7:20 a.m. are larger than those in other time periods, which reflects the effect on the travel time of vehicle parking around schools (due to students going to school) and the associated increase in traffic.

It can also be seen from the estimated parameter value corresponding to the distance variable to the hospital that this estimated value increases in certain time periods. In addition, these time periods lag the period in which estimated values for the parameters corresponding to the distance variable to the school increase, especially after 7:00 a.m. This is because some hospitals workers start at 7:30–8:00 a.m., and because people go to the hospital for medical treatment later in the day than students go to school. Furthermore, the estimated values of parameters corresponding to the distance variable to hospitals are larger than those for other variables. This may be because the front entrance of a hospital faces toward the main road of the city, while that of schools (especially primary schools and kindergartens) generally faces toward a branch road, for safety and to facilitate a teaching-friendly environment. In addition, people entering and leaving a hospital prefer to take private cars and taxis, and thus there is a greater delay caused by vehicle parking outside hospitals than outside schools.

4.5 Test of Parameter Values

The likelihood ratio test is used to evaluate the rationality of the built environment explanatory variables. The likelihood ratio test formula is $R_{\text{LR}} = -2[\ln(L_1) - \ln(L_2)]$, where R_{LR} is the likelihood ratio test value, L_1 is the maximum likelihood value with the built environment explanatory variables added, and L_2 is the maximum likelihood value without the built environment variables. The obtained likelihood ratio is:

$$\lambda = \frac{L_1}{L_2} \tag{14}$$

For the likelihood ratio of multiple parameters, $R_{\text{LR}} = -2\ln(\lambda)$ obeys χ^2 distribution.

In addition, to test whether the difference between the likelihood values of the two models is significant, the degree of freedom is also considered. In the commonly used likelihood ratio test, the degree of freedom is equal to the number of model parameters added to the complex model. In this study, it refers to the number of the built environment explanatory variables; thus, the degree of freedom is 5.

Then, according to the critical value table of the χ^2 distribution, we can judge whether the effect of a built environment variable is significant. The values of the maximum likelihood function with and without a built environment variable are compared by taking the opposite number of logarithms, as shown in Table 5.

Table 5 shows the minimum likelihood ratio $R_{LR} = -2[\ln(L_1) - \ln(L_2)] = 30$. It can be determined that the χ^2 value with a degree of freedom of 5 and $\alpha = 0.05$ is 11.071, so R_{LR} is greater than the critical value, indicating that the travel time estimation method based on built environment variables is more effective than a method not based on these variables. This verifies the effect of the built environment attributes on travel time and confirms the rationality of taking the built environment attributes as the interpretation variable.

4.6 *Calculation of Travel Time*

Using the obtained parameters, the travel time from the First Company of Dandong Public Transportation Corporation to Dandong Institute of Environmental Sciences along Jinshan Street is calculated using Eqs. (6)–(8). The results are shown in Table 6 and Fig. 4. Overall, the travel time on Jinshan Street increases from 6:00 a.m. to 8:00 a.m., which is consistent with the actual morning-peak phenomenon; It began to fall back near 7:30 when the travel speed of vehicles on the road gradually increased.

A comparison of the estimated time in Table 5 with the travel time of "total length of approximately 2.8 km, approximately 5 min" measured on Baidu Maps shows that the times are consistent. That is, from 6:00 a.m. the travel time on the road gradually increases, which is consistent with the actual situation; this further verifies the rationality of considering the effect of built environment attributes on travel time.

5 Conclusion

In this chapter, selected built environment attributes are taken as the explanatory variables of travel time. Then, based on the position data reported by floating cars, the relationship between these built environmental attributes and travel time is constructed using statistical methods, without using speed data or travel path estimates of the floating cars in the research area. Jinshan Street in Dandong, China, is used as a case study to verify the effect of the built environment attributes on travel time. The results show that the built environment attributes have a significant effect on the travel time in the study area. The effect of schools mainly occurs at 6:00–7:20 a.m., while that of hospitals and clinics is concentrated at 7:00–8:00 a.m. However, the increase in travel time caused by intersections is relatively stable over time.

Table 5 Likelihood ratio test of the built environment explanatory variables

Time period	6:00–6:30	6:30–6:40	6:40–6:50	6:50–7:00	7:00–7:10	7:10–7:20	7:20–7:30	7:30–7:40	7:40–7:50	7:50–8:00
$\ln(L1)$	2704	1554	1784	2071	1723	1710	1658	2644	2658	2691
$\ln(L2)$	2761	1572	1799	2091	1744	1744	1673	2660	2677	3436
RLR	114	36	30	40	42	68	30	32	38	1490

Table 6 Variation of travel time with time

Time period	Travel speed (m/s)	Travel time (s)	total distance (m)
6:00–6:30	35.77	277.11	2753.63
6:30–6:40	34.14	290.37	2753.63
6:40–6:50	35.25	281.23	2753.63
6:50–7:00	27.40	361.78	2753.63
7:00–7:10	26.04	380.63	2753.63
7:10–7:20	26.98	367.37	2753.63
7:20–7:30	20.78	477.04	2753.63
7:30–7:40	20.83	476.02	2753.63
7:40–7:50	21.09	469.93	2753.63
7:50–8:00	24.67	401.81	2753.63

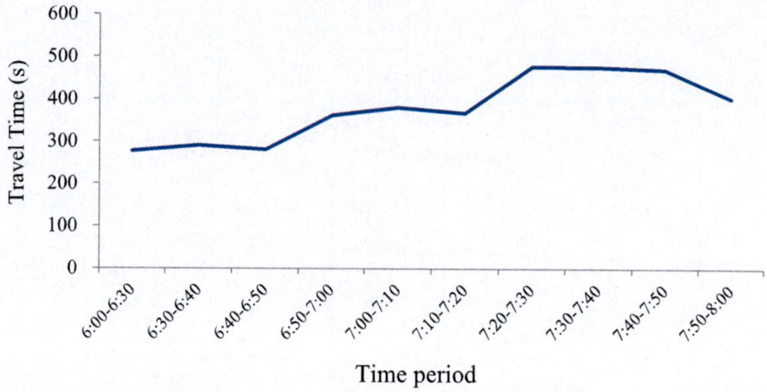

Fig. 4 Changes of travel time

In this case study, the increase in travel time in peak periods is primarily due to the frequent access and parking of vehicles at schools and medical treatment facilities and for refueling. Thus, more standardized traffic management measures should be implemented to smooth vehicle access to and parking near schools, hospitals, and other facilities, to ensure normal traffic conditions exist for these vehicles.

Furthermore, this chapter proposes a method to estimate the travel time distribution on roads. This method can be used to estimate the travel time distribution on roads after establishing a historical database of travel times. Furthermore, as this method uses the distribution of the number of floating cars on a road instead of distance as the distribution coefficient of travel time, the accuracy of travel time estimation is improved.

References

Cheng, J., Li, G., & Chen, X. (2019). Developing a travel time estimation method of freeway based on floating car using random forests. *Journal of Advanced Transportation,* 1–13.

Dell' Orco, M., Marinelli, M., & Silgu, M. A. (2016) Bee colony optimization for innovative travel time estimation, based on a mesoscopic traffic assignment model. *Transportation Research Part C: Emerging Technologies, 66*(1), 48–60.

Fan, S. K. S., Su, C. J., Nien, H. T., et al. (2018). Using machine learning and big data approaches to predict travel time based on historical and real-time data from Taiwan electronic toll collection. *Soft Computing, 22*(17), 5707–5718.

Hellinga, B., Izadpanah, P., Takada, H., et al. (2008). Decomposing travel times measured by probe435 based traffic monitoring systems to individual road segments. Transportation Research Part C: 436 Emerging Technologies, *16*(6), 768–782. 437

Hofleitner, A., Herring, R., & Bayen, A. (2012). Arterial travel time forecast with streaming data: A hybrid approach of flow modeling and machine learning. Transportation Research Part B: Methodological, 46(9), 1097–1122.

Jenelius, E., & Koutsopoulos, H. N. (2013). Travel time estimation for urban road networks using low frequency probe vehicle data. *Transportation Research Part B: Methodological, 53*, 64–81.

Liu, H. X., & Ma, W. T. (2009). A virtual vehicle probe model for time-dependent travel time estimation on signalized arterials. *Transportation Research Part C: Emerging Technologies, 17*(1), 11–26.

Ma, Z. L., Koutsopoulos, H. N., Ferreira, L., et al. (2017). Estimation of trip travel time distribution using a generalized Markov chain approach. *Transportation Research Part C: Emerging Technologies, 74*, 1–21.

Mil, S., & Piantanakulchai, M. (2018). Modified Bayesian data fusion model for travel time estimation considering spurious data and traffic conditions. *Applied Soft Computing, 72*, 65–78.

Rahmani, M., Jenelius, E., & Koutsopoulos, H. N. (2015). Non-parametric estimation of route travel time distributions from low-frequency floating car data. *Transportation Research Part C: Emerging Technologies, 58*(SEP. PT. B), 343–362.

Rahmani, M., Koutsopoulos, H. N., & Jenelius, E. (2017). Travel time estimation from sparse floating car data with consistent path inference: A fixed-point approach. *Transportation Research Part C: Emerging Technologies, 85*, 628–643.

Scott, L. M., & Janikas, M. V. (2010). Spatial statistics in ArcGIS. In *Handbook of applied spatial analysis* (pp. 27–41). Springer.

Shi, C., Chen, B. Y., & Li, Q. (2017). Estimation of travel time distributions in urban road networks using low-frequency floating car data. *ISPRS International Journal of Geo-information, 6*(8), 253.

Tang, K., Chen, S. Y., Liu, Z. Y., et al. (2018). A tensor-based Bayesian probabilistic model for citywide personalized travel time estimation. *Transportation Research Part C Emerging Technologies, 33*(3), 387–397.

Zhang, Y., Smirnova, M. N., Bogdanova, A. I., Zhu, Z., et al. (2018). Travel time estimation by urgent-gentle class traffic flow model. *Transportation Research Part B: Methodological, 113*, 121–142.

Zheng, F. F., & Zuylen, H. (2013). Urban link travel time estimation based on sparse probe vehicle data. *Transportation Research Part C: Emerging Technologies, 31*(1), 145–157.

Zhong, S. P., Wang, Z., Wang, Q. Z., et al. (2021). Exploring the spatially heterogeneous effects of urban built environment on road travel time variability. *Journal of Transportation Engineering Part A: Systems, 147*(1), 4020142.

Chapter 7
Exploring the Spatially Heterogeneous Effects of Urban Built Environment on Road Travel Time Variability

Abstract Most studies of road travel time estimation have been based on traffic flow theory or data-driven methods and generally neglect the influence of urban built environment on road travel time. A global regression model and a geographically weighted regression model were thus established to analyse the spatial heterogeneity of the effects of urban built environment on road travel time. The estimated results of the global regression model indicate that the occupancy rate of taxis, the distance from the nearest intersection, and the speed limit show positive correlations with a road's travel speed, whilst the number of bus stops and the distance from the nearest school show negative associations with the travel speed of the road. Furthermore, based on the results of the geographically weighted regression model, the spatially varying relationships between urban built environment and road travel time can be established, thus providing important information for decision-makers to reduce road travel time by adjusting the attributes of urban built environment.

Keywords Urban built environment · Road travel time variability · Spatial heterogeneity · GWR model · Floating car data

Notation

The following symbols are used in this study

b_i	bandwidth of road segment i;
D_p	nearest intersection type to the road segment, which is a regulating variable;
d_{ij}	distance between road segment i and road segment j;
m	number of independent variables;
m'	number of independent variables that are statistically significant in the global regression model;
n	number of regulating variables;
$\mathbf{S}$	vector of dependent variables;
S	average speed of a road, which is a dependent variable;
s_i	average speed of road segment i;

S. Zhong and D. Sun, *Logic-Driven Traffic Big Data Analytics*,
https://doi.org/10.1007/978-981-16-8016-8_7

$\mathbf{W}(i)$	weight matrix corresponding to road segment i;
$\mathbf{W}_j(i)$	spatial weight of road segment j adjacent to road segment i;
$\mathbf{X}$	matrix of independent variables with $m' + 1$ columns and a column of 1s for the intercept;
x_k	attribute of the built environment, which is an independent variable;
x_{ik}	kth independent variable associated with road segment i;
β_o	constant;
β_{i0}	constant of road segment i;
β_{ik}	regression coefficient corresponding to x_{ik};
β_k	regression coefficient corresponding to x_k;
$\hat{\boldsymbol{\beta}}(i)$	matrix of regression parameters at road segment i;
ε	random error;
ε_i	random error of road segment i;
η_p	regression coefficient corresponding to D_p; and
λ_{kp}	coefficient of the impact of the interaction between independent variables and regulating variables.

1 Introduction

In recent years, with increasing awareness of the negative effects of travel time on transportation network efficiency, road travel time estimation has attracted increasing attention in the field of intelligent transportation systems (Jenelius & Koutsopoulos, 2013; Karimpour et al., 2019; Li et al., 2006; Liu & Ma, 2009; Liu et al., 2005; Pirc et al., 2016; Rahmani et al., 2017; Tang et al., 2018; Yu et al., 2018; Zheng & Zuylen, 2013). Road travel time refers to the total time required for a vehicle to pass along a designated route (from point to point), including running time on the road, stop time, queuing delay time, and intersection delay time. Excessive road travel time not only increases monetary costs for travellers but also greatly reduces the operating efficiency of the whole road network, thereby further aggravating the jamming of the urban 'lifeline'. Most studies of road travel time estimation have been based on data-driven methods or traffic flow theory (Karimpour et al., 2019; Mori et al., 2015; Rahmani et al., 2015). Methods based on traffic flow theory can be further divided into traffic delay models (Dell'Orco et al., 2016; Jiang et al., 2016; van Lint, 2004), queuing theory (Liu et al., 2005) and simulation models (Liu & Ma, 2009). These methods focus on transportation-related factors and neglect other key influences on road travel time.

An understanding of the key factors that influence road travel time is an essential basis for research into road travel time reduction. Urban land is spatially separated into zones of diverse use, which gives rise to different types of transportation demand. The separation of workplaces and residences results in commuting travel demand, and the spatial distance between residences and shopping areas causes service travel demand. In addition, features of urban built environment, as variables that characterize the spatial attributes of various urban regions, have an important impact on inhabitants' travel behaviour (Fan & Khattak, 2008; Liu et al., 2020; Sun et al., 2018; Wang et al., 2011). This impact mainly manifests in two aspects: first, the layout of various types of land use in urban spaces, such as places of work, residences, and shopping areas, determines travel destinations and travel frequency; and second, features of urban built environment, such as street design, public transportation availability, and destination accessibility, may also influence the residents' choice of travel mode and travel routes (Cervero & Kockelman, 1997). From this perspective, urban built environment has an impact on residents' travel behaviour at the source; it influences traffic flow distribution on the road network and ultimately affects road travel time. Therefore, establishing a relationship between urban built environment and road travel time from the perspective of urban land use and transportation integration provides a feasible direction for exploration of the key factors that influence road travel time.

Most studies have focused on the influence of urban built environment on residents' travel behaviour (Cao et al., 2009; Cervero & Kockelman, 1997; Ding et al., 2014; Ewing & Cervero, 2001). Further study is required to uncover the inherent relationship between urban built environment and road travel time and identify the key factors that affect the latter from the perspective of the former. Moreover, the complexity of urban built environment makes its impact on road travel time spatially heterogeneous (i.e., its impact varies in different areas). The study of the spatially heterogeneous effects of urban built environment on road travel time has important planning and practical implications; ignoring such spatial heterogeneity may result in the loss of a significant amount of detailed information about variables, leading to a biased result. Unlike global regression analysis, which can only determine the average relationships among factors, spatial heterogeneity analysis can help decision-makers identify the local relationship between urban built environment and road travel time, which is crucial to determine the causes of traffic delays in local areas.

This study aims to fill the research gap in the spatial heterogeneity analysis of the impact of urban built environment on road travel time. A global regression model and a geographically weighted regression (GWR) model are used to quantitatively address this issue based on three sources of data obtained in Shenzhen City: floating car (taxi) global positioning system (GPS) data, land use data, and road network data. To the best of the authors' knowledge, this is a pioneering study to explore the key factors that influence road travel time while taking urban built environment into consideration and the study to analyse the spatial heterogeneity of the impact of urban built environment on road travel time.

The rest of this study is organized as follows. The second section reviews the literature. The third and fourth sections introduce the case study and present the global regression model and the GWR model. The estimated results and the practical implications of the two models are provided in the fifth and sixth sections. The seventh section draws the conclusions of the study.

2 Literature Review

2.1 *Urban Built Environment and Travel Behaviour*

The urban built environment refers to man-made environments that provide the required space for human activities, specifically including interactive spatial environments characterized by multiple factors such as land use, transportation infrastructure, and urban design (Handy et al., 2002). In recent decades, scholars have paid increasingly close attention to the relationship between urban built environment (also known as urban form or neighbourhood type) and travel behaviour, motivated by the sprawl of car-oriented cities and increasingly serious problems (such as traffic jams, environmental pollution, and infrastructure waste) that are caused by reliance on private cars. Cervero and Kockelman (1997) and Ewing and Cervero (2001) found that travel behaviour was closely related to the attributes of urban built environment, which ultimately influence road travel time by affecting citizens' travel behaviour, such as travel destination, mode, frequency, and route. Studies have found that urban built environment has an important impact on car use (Ding et al., 2014; Maat & Timmermans, 2009), the running mileage of cars (Cervero & Jin, 2010), and the effects of the implementation of travel demand-management policies (Zhong & Bushell, 2017; Zhong et al., 2015). However, no studies have focused on the influence of urban built environment on road travel time.

In terms of research methods, global regression models are the most common choice for investigating the relationship between urban built environment and travel behaviour. However, because such models explore the average impact of urban built environment on residents' travel behaviour across the entire spatial area under study, they can only solve spatially homogeneous problems. In reality, although samples from the same region may have certain similarities in travel behaviour because of the common attributes of urban built environment, they may differ from those of other regions due to spatial heterogeneity. Neglecting such heterogeneity in research can often result in the loss of much detailed information about the relationships among variables, and the conclusions obtained will be biased.

2.2 *Research on Spatial Heterogeneity*

The term *spatial heterogeneity* is widely used to describe the inhomogeneity and complexity of spatial distributions in ecological processes and patterns. According to Anselin (1988), spatial heterogeneity refers to the fact that the characteristics of objects and phenomena in certain regions differ from those in other regions. In specific models, these differences are presented by variables, parameters, and error terms associated with different regions. To identify spatial heterogeneity, the GWR model proposed by Fotheringham et al. (2002), which uses local regression analysis and variable parameters based on local smoothing polynomials, has been widely used in many research fields, such as ecogeography, economics, medicine, and transportation (Chiou et al., 2015; Elldér, 2014; Feuillet et al., 2015; Wang & Khattak, 2011).

It should be noted that the traditional fixed-bandwidth GWR model has two drawbacks: first, instability occurs if the coefficients of the model suffer from collinearity of explanatory variables (Páez et al., 2011; Wheeler & Tiefelsdorf, 2005); second, inflexibility occurs if the model assumes the same degree of spatial variation across each set of coefficients (Murakami et al., 2018). Regarding the first limitation (instability), Murakami et al. (2018) found that compared with the fixed-bandwidth GWR model, the adaptive-bandwidth GWR model effectively alleviated the instability of model results caused by changes in spatial scale. In addition, when the sample size is large, the GWR model is robust to high levels of multicollinearity (Oshan & Fotheringham, 2018). However, for the second limitation (inflexibility), Murakami et al. (2018) found that for small-scale spatial variations, even the adaptive-bandwidth GWR model is vulnerable to reduced estimation accuracy.

The impacts of urban built environment on road travel time differ between regions, evidencing spatial heterogeneity. A quantitative description of this heterogeneity can contribute to a better understanding of the spatial impact of urban built environment on road travel time. In recent years, scholars have used the GWR model to solve complicated spatially heterogeneous problems in transportation research, such as estimation of average daily traffic on a non-highway road (Zhao & Park, 2004), discovery of the relationship between transportation accessibility and land value (Du & Mulley, 2006), exploration of the effects of travel information on travel decisions (Wang & Khattak, 2011), study the spatial variations in the relationships between the number of crashes and explanatory variables (Ariannezhad et al., 2020; Pirdavani et al., 2014), identification of the key factors that influence the utilisation rates of public transportation (Chiou et al., 2015; Ma et al., 2018), evaluation of the sensitivity of on-street parking occupancy to price changes (Pu et al., 2017), and examination of the effect of transportation accessibility on regional economic resilience (Chacon-Hurtado et al., 2020). Although the GWR model has been validated as an effective approach to study spatial heterogeneity in the transportation field, few works have focused on the relationship between urban built environment and road travel time.

2.3 Research on Road Travel Time

Most studies of road travel time estimation have been based on traffic flow theory or data-driven methods (Karimpour et al., 2019; Mori et al., 2015; Rahmani et al., 2015). Traffic flow theory is used to estimate road travel time through traffic delay models, queuing theory, and simulation models. These methods consider various traffic parameters in the road network and factors that cause traffic delay (Dell'Orco et al., 2016; Hellinga et al., 2008; Liu & Ma, 2009; Vanajakshi, 2005). For example, van Lint (2004) used such a method to predict the traffic conditions in future time intervals and estimated the travel time from the predicted traffic conditions and variables. Liu et al. (2005) used a traffic delay model and queuing theory, which are traditional traffic flow theories, to predict road travel time. In contrast, unlike traffic flow theory, data-driven methods process large amounts of data using informatics methods such as linear regression, Bayesian networks, artificial neural networks, Kalman filtering, fuzzy algorithms, and machine learning or deep learning and thus avoid the difficulties of relying on strict mathematical derivations (Lee et al., 2006; Ma et al., 2017; Rahmani et al., 2015; Tang et al., 2018; Xu et al., 2020; Zheng & Zuylen, 2013).

Elsewhere, other researchers have combined traffic flow theory with data-driven methods to estimate road travel time (Hofleitner et al., 2012). However, whether these studies are based on traffic flow theory, data-driven methods or combined methods, they have mostly paid attention to road parameters and historical travel time data, while usually neglecting the factors of urban built environment that influence road travel time. It is important to clarify the relationship between the attributes of urban built environment and road travel time, as the latter can be minimized by adjusting the former.

This study aims to contribute to the existing literature in the following aspects.

1. Identify the key factors of urban built environment that influence road travel time using a global regression model with interaction terms.
2. Incorporate the impact of the occupancy rate of taxis when using taxi GPS data to estimate road travel time.
3. Analyse the spatial heterogeneity of the relationships between urban built environment and road travel time using a GWR model.
4. Provide important information for transportation planning and practical decision-making to reduce road travel time by adjusting the attributes of urban built environment.

3 Case Study

3.1 Selected Study Region

The target route is situated in Nanshan District, Shenzhen, starting from the intersection of Industrial 8th Road and Houhai Road and ending at the intersection of Qiaocheng East Road and Baishi Road. This route passes through 17 intersections with a total length of approximately 10 km. The study route and the intersection names are presented in Fig. 1, which is modified from Google Maps, where C1–C17 refer to intersections 1–17.

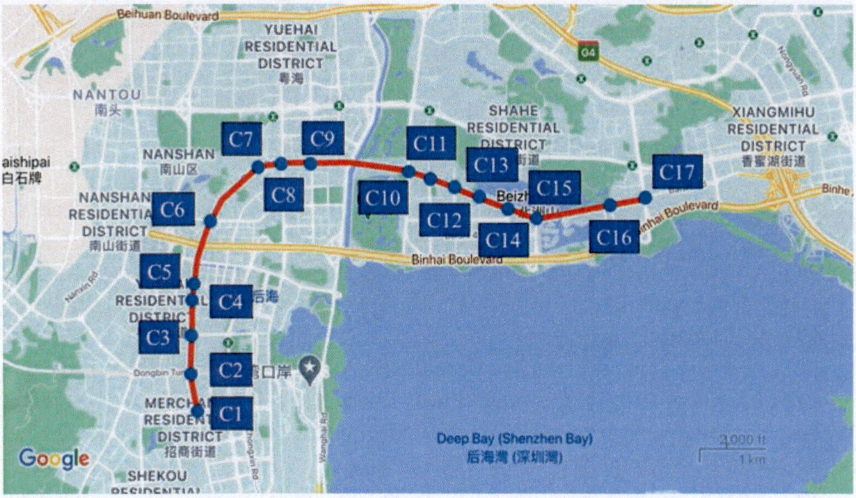

Fig. 1 Study route (© Google)

3.2 Data Collection

Taxi GPS data, land use data, and road network data of Shenzhen were collected. Depending on road operational conditions, the taxi GPS data were uploaded at a frequency of approximately three or four times per minute. Along the study route (see Fig. 1), 8546 records of taxi GPS data, including licence number, time, latitude and longitude, driving direction, speed, and occupancy state of the taxi, were extracted from 1302 taxis in a 2-h period from 7:30 to 9:30 am on 3 days (i.e., Tuesday 10 June to Thursday 12 June 2014). The land use data geographically represent different types of land use in Shenzhen, consisting of the spatial positions and attributes of supermarkets, schools, hospitals, hotels, and other buildings. The road network data comprise geographic information of the Shenzhen road network and data on Shenzhen from Baidu Maps.

After collecting the required data, ArcGIS was used to divide the research route into 397 road segments of 25 m each. Each segment was treated as an independent sample. The average speed of each road segment and the occupancy rate of taxis were then obtained by screening and matching the taxi GPS data. Next, ArcGIS was used to analyse the land use data within a radius of 500 m around the study route to obtain the numbers of buildings of each type and public facilities around each road segment. Preliminarily, three different ranges were tested for this buffer around the road segments: 200, 500, and 1000 m. Considering the requirements for statistical significance and the interpretability of the model, 500 m was chosen. Finally, to elucidate the interactive effects of the intersection types and urban built environment on road travel time, dummy variables were added into the model. These dummy variables were used to classify the intersection types in three respects: the number of entrance lanes, presence or absence of left-turn lanes, and whether the left-turn lanes are independent.

3.2.1 Average Speed and Occupancy Rate of Taxis

For a fixed road distance, the road travel time varies inversely with the average travel speed. Therefore, the road travel time was represented by the average travel speed of a road segment and calculated by screening, correcting, and matching the taxi GPS data (Axhausen et al., 2003). The use of taxi GPS data to estimate the travel time of the road segments is subject to some sources of error, which may lead to over- or underestimated values. Specifically, when the occupancy rate of taxis on a road segment is high, the travel time of the segment is usually overestimated because occupied taxis usually travel faster than normal private vehicles. Conversely, when the occupancy rate of taxis on a road segment is relatively low, the travel time of the segment is, in general, underestimated, as unoccupied taxis usually travel more slowly than normal private vehicles. To calculate and control the impact of the occupancy rate of taxis on road travel time, the occupancy rate of taxis was added as an independent variable. Note that if data were available, the use of a broader database (i.e., one that contains all floating car data) would also alleviate this biased estimation problem.

3.2.2 Measurements of Urban Built Environment

For this study, 14 measurements of urban built environment were selected: the number of office buildings, banks, hotels, pharmacies, car parks, supermarkets, restaurants, bus stops, and schools; the distances (Euclidean distance between a road segment and the centroid of the objective) from the nearest school, from the nearest intersection (intersection on the study route) and from the nearest bus stop; the speed limit within a range of 500 m around the study route; and the geometric design attributes of the roads (number of lanes in one direction). Each road segment has the same length (25 m); thus, the number of facilities reflects the density of facilities.

3.2.3 Classification of Intersection Types

When studying the impact of other variables on road travel time, the type of intersection must be considered because different intersection types often have different impacts. The 17 intersections on the study route were divided into four types according to the number of entrance lanes, presence or absence of left-turn lanes, and whether the left-turn lanes are independent. The resulting classification and a schematic diagram of each class of intersection are shown in Table 1 and Fig. 2,

Table 1 Classification of intersection types

Intersection types	Features	Classification result
Type 1	The number of entrance lanes does not exceed four, and there are independent left-turn lanes	C2, C5, C7, C8, C9, C11, C12
Type 2	The number of entrance lanes does not exceed four, and there are left-turn lanes but they are not independent	C3, C16
Type 3	The number of entrance lanes does not exceed four, and there is no left-turn lane	C13, C14, C15
Type 4	The number of entrance lanes exceeds four, and there are independent left-turn lanes	C1, C4, C6, C10, C17

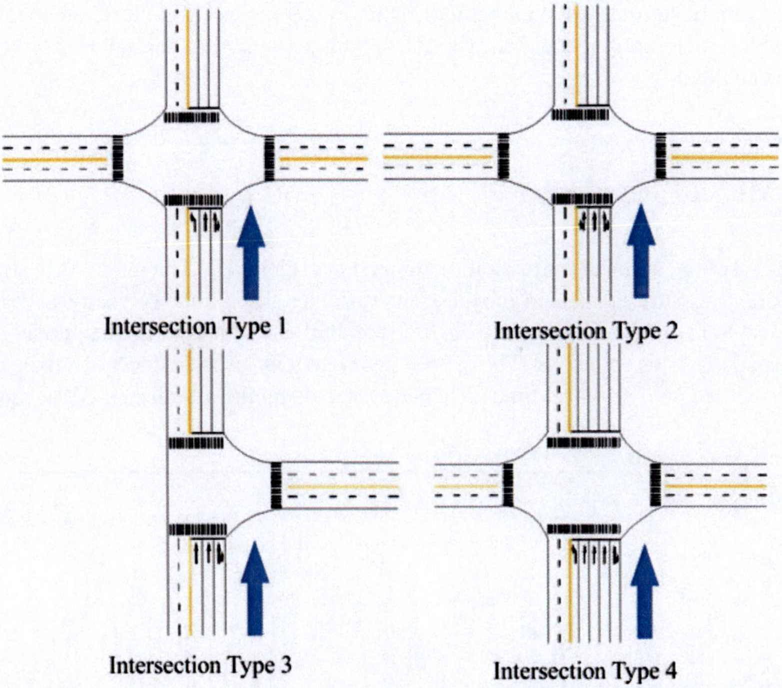

Fig. 2 Schematic diagram of four intersection types

respectively. An independent left-turn means that the intersection has an independent zone in which the left-turning traffic can wait. For example, in Fig. 2, intersection Types 1 and 4 have independent left-turn lanes.

3.3 Descriptive Statistics of Variables

In this study, the dependent variable is the average travel speed of a road segment. Figure 3 presents the distribution of average travel speed on each of the 397 road segments. The average travel speed is greatly affected by delays at intersections, so the average travel speeds at all of the intersections (with a maximum value of 22.72 km/h) are clearly lower than that of the research route as a whole (34.32 km/h).

In the global regression model with interaction terms, the attributes of urban built environment and the occupancy rate of taxis were set as the independent variables. The descriptive statistics of the independent variables are provided in Table 2. In addition, the interaction between the intersection types and the independent variables was considered. Dummy variables were introduced to specifically quantify the impact of intersection types on road travel time, and intersection Type 4 was chosen as a reference item. In summary, the global regression model included 59 initial independent variables, including independent variables, dummy variables, and interaction terms between them.

4 Model Formulation

In this section, a global regression model is first established. The model is designed not only to extract the main independent variables that influence road travel time but also to reveal whether the effects of urban built environment on road travel time are spatially heterogeneous. The spatial heterogeneity of the effects of urban built environment on road travel time is then analysed using the established GWR model.

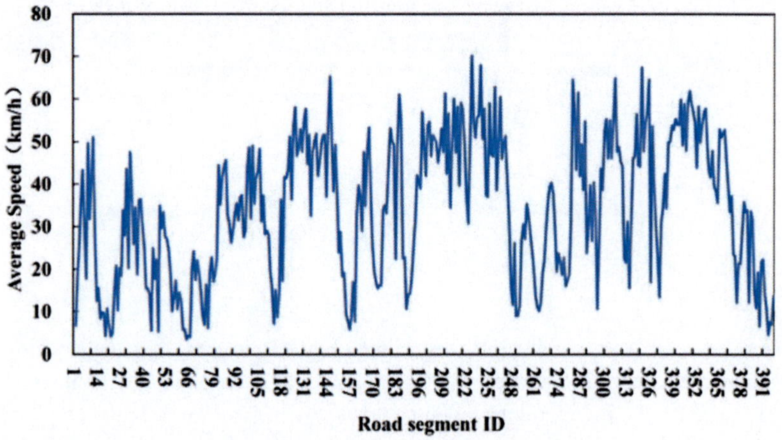

Fig. 3 Distribution of the average travel speed on each road segment

Table 2 Descriptive statistics of the independent variables

Variable	Unit	Minimum value	Maximum value	Average value	Standard deviation
Number of office buildings	count	0	8	1.68	2.41
Number of banks	count	0	33	7.21	8.25
Number of hotels	count	0	8	1.20	1.90
Number of pharmacies	count	0	21	4.25	5.52
Number of car parks	count	0	20	4.31	4.94
Number of supermarkets	count	0	126	22.39	31.41
Number of restaurants	count	0	75	16.26	19.45
Number of bus stops	count	0	38	13.67	8.89
Number of schools	count	0	19	4.35	5.27
Distance from the nearest school	m	8	995	350.43	282.27
Distance from the nearest intersection	m	0	725	202.08	164.08
Distance from the nearest bus stop	m	0	575	195.03	140.02
Speed limit	km/h	50	60	54.56	4.99
Number of lanes in one direction	count	3	4	3.14	0.34
Occupancy rate of taxis	–	0	1	0.57	0.19

4.1 Global Regression Model

The impact of the attributes of urban built environment on the average travel speed of a road segment is affected by the type of the nearest intersection. Therefore, in the global regression model, the type of the nearest intersection to the road segment is set as the regulating variable to analyse the interaction between them. As the road travel time is equal to the length of the road divided by the average travel speed, the average travel speed is adopted as the dependent variable instead of the road travel time. Specifically, in the global regression model, the average travel speed of a road segment is set as a dependent variable, the attributes of urban built environment and the occupancy rate of taxis are set as independent variables, and the dummy variables of the nearest intersection type to the road segment are set as regulating variables. For the interaction under the framework of a regulating variable, the effect of the independent variables on the dependent variable depends on the value of the regulating variable. The structure of the regression model is formulated as:

$$s = \beta_o + \sum_{k=1}^{m} \beta_k x_k + \sum_{p=1}^{n} \eta_p D_p + \sum_{k=1}^{m} \sum_{p=1}^{n} \lambda_{kp} x_k D_p + \varepsilon, \tag{1}$$

where s is the average travel speed of a road; β_o is a constant; x_k represents an attribute of the built environment, which is an independent variable (see Table 2); β_k represents the regression coefficient corresponding to x_k; m is the number of independent variables; D_p is the nearest intersection type to the road segment, which is a regulating variable; η_p represents the regression coefficient corresponding to D_p; n is the number of regulating variables; λ_{kp} is a coefficient of the impact of the interaction between independent variables and regulating variables; and ε is a random error.

The global regression model can be estimated based on the ordinary least-squares method, which minimizes the square of the distance between the observed values and the estimated values, namely, the residual sum of squares. The results of the global regression model reveal whether an interaction lies between the independent variables and the regulating variables.

4.2 GWR Model

In the global regression model, the spatial heterogeneity of variables cannot be captured, as all of the differentiated spatial units are represented by mean values. This can lead to a great loss of detailed information, causing the obtained result to be biased. Unlike the global regression model, in the GWR model, the estimated parameters of each unit vary with location. This allows the structural relationships among variables in various locations and the spatial heterogeneity to be easily established. The spatial locations of variables are embedded in the regression parameters, which can be estimated using the locally weighted least-squares method. The structure of the GWR model is formulated as follows:

$$s_i = \beta_{i0} + \sum_{k=1}^{m'} \beta_{ik} x_{ik} + \varepsilon_i, \tag{2}$$

where s_i means the average speed of road segment i; β_{i0} is a constant of road segment i; x_{ik} represents the kth independent variable associated with road segment i; β_{ik} is the regression coefficient corresponding to x_{ik}; m' is the number of independent variables that are statistically significant in the global regression model; and ε_i is the random error of road segment i.

When estimating the parameters of road segment i using only its own set of data, the result is unbiased, but the standard deviation of the result is large, and the accuracy is poor. To remedy this problem, more samples around road segment i must be added. A consequent problem is that if adjacent samples of road segment i are added, some extent of deviation is inevitable, and this deviation increases with the number of samples. To balance the deviation of the estimated parameters in the GWR model against the standard deviation, it is necessary to select appropriate adjacent samples.

To determine which sample road segments adjacent to road segment i should be used for estimation of model parameters, a spatial weight matrix $\mathbf{W}(i)$ is introduced to give a spatial weight value to each adjacent segment of road segment i. This establishes a function of the distance between the regression segments and the other sample segments through a spatial weight function. In practice, the spatial distribution of sample points tends to be uneven. Some areas are dense in sample points, while others are sparse. The study route includes both straight and curved road segments. Because the density of curved road segments is higher than that of straight road segments, if the bandwidth is fixed in the weight function, more sample road segments will be selected in the areas with high sample density, whilst fewer will be selected in the areas with low sample density. In response to this problem, the adaptive-bandwidth method is adopted to assign different optimal bandwidth values to regions with different sample densities. In the case of adaptive bandwidth, the number of road segments included in the bandwidth is kept constant, making the bi-square weight function more suitable for the Gaussian weight function. The expression of the bi-square weight function is as follows:

$$\mathbf{W}_j(i) = \begin{cases} \left[1 - d_{ij}^2/b_i^2\right]^2, & d_{ij} \le b_i \\ 0, & d_{ij} > b_i \end{cases}, \tag{3}$$

where $\mathbf{W}_j(i)$ represents the spatial weight of road segment j adjacent to road segment i; d_{ij} is the distance between road segments i and j; and b_i is the bandwidth of road segment i. Then, the regression parameter matrix $\hat{\boldsymbol{\beta}}(i)$ at road segment i is estimated as follows:

$$\hat{\boldsymbol{\beta}}(i) = \left[\mathbf{X}^T\mathbf{W}(i)\mathbf{X}\right]^{-1}\mathbf{X}^T\mathbf{W}(i)\mathbf{S}, \tag{4}$$

where $\mathbf{X}$ is the matrix of independent variables with $m' + 1$ columns and a column of 1s for the intercept; $\mathbf{S}$ is the vector of dependent variables; and $\mathbf{W}(i)$ is the weight matrix corresponding to road segment i, which is a diagonal matrix. Lloyd (2010) found that the results of GWR models were more sensitive to the bandwidth of a specific weight function than to the type of weight function. Here, the bandwidth was selected by the minimisation of Akaike information criterion (AIC) (Fotheringham et al., 2002).

5 Results and Analysis

The average travel speed of the road segments was estimated using the global regression model and the GWR model. It is worth noting that the local estimated coefficients may not be consistent with the global estimated results, because some estimated global parameters may take the opposite sign at some locations. The results are presented in the following two sections, together with figures depicting the spatial variations.

5.1 Results and Analysis of the Global Regression Model

The global regression model was estimated using SPSS Statistics 17.0, and the results are shown in Table 3. Considering the large number of independent variables, only the statistically significant independent variables are listed in Table 3. The F-value of the model is 13.678, and the corresponding number of degrees of freedom is 61. Therefore, the null hypothesis is rejected; that is, at least one coefficient of independent variables is significantly different from 0. R_a^2 is 0.47, indicating that the independent variables in the model can explain 47% of the variation in the average speed of the road segments.

It can be seen from Table 3 that the distance from the nearest intersection, the speed limit, and the occupancy rate of taxis show positive correlations with the average speed of the road segments, whilst the corresponding correlations with the number of bus stops and the distance from the nearest school are negative. These results are intuitively reasonable. For example, the higher the speed limit and the further from the nearest intersection, the higher the average speed of the road segment. In addition, Table 3 also suggests that intersection Type 1 shows a positive correlation with the average speed of the road segments, which indicates that left-turn lanes are conducive to improving the speed of vehicles as they pass through an intersection.

Inspection of the interaction terms verifies that interaction does exist between the intersection type and urban built environment and between the intersection type and the occupancy rate of taxis. Another noteworthy finding is that factors that have a significant influence on the average speed of the road segments vary with location. This indicates that the influence of urban built environment on the average speed of the road segments is not uniform across the entire research route, as the type of the nearest intersection varies for different road segments, which leads to spatial heterogeneity. However, the global regression model only summarises the overall pattern, and the result is the average value obtained on the entire route. Therefore, the GWR model is necessary to reveal the underlying spatial heterogeneity.

Table 3 Estimated results of the global regression model and the GWR model

Models	Global regression model		GWR model				
Variable	Coefficient	*p*-value	Min	1st Quantile	Median	3rd Quantile	Max
Constant	−82.94	0.001	−122.11	−18.28	18.54	31.59	76.73
Number of bus stops (a)	−1.80	0.000	−0.52	−0.08	0.32	0.55	1.05
Distance from the nearest school (b)	−0.03	0.032	−0.03	−0.01	0.01	0.03	0.04
Distance from the nearest intersection (c)	0.04	0.001	0.04	0.05	0.07	0.07	0.08
Speed limit (d)	1.93	0.000	−0.91	−0.23	−0.08	0.79	2.33
Occupancy rate of taxis (e)	28.22	0.000	−4.88	5.91	11.92	16.31	22.00
Dummy							
Intersection type 1 (D1)	323.85	0.000	–	–	–	–	–
Interaction term							
a × D1	2.48	0.000	–	–	–	–	–
a × D2	2.87	0.007	–	–	–	–	–
a × D3	2.52	0.004	–	–	–	–	–
b × D1	0.03	0.039	–	–	–	–	–
c × D1	0.05	0.000	–	–	–	–	–
c × D3	0.03	0.051	–	–	–	–	–
d × D1	−4.88	0.000	–	–	–	–	–
d × D2	−3.75	0.019	–	–	–	–	–
e × D1	−19.12	0.010	–	–	–	–	–
R^2	0.48		0.59				
R_a^2	0.47		0.56				
AIC	3113.91		3050.34				

5.2 *Estimated Results and Analysis of the GWR Model*

Spatial heterogeneity refers to the differences between objects and phenomena in a certain geographic position from those in other regions (Anselin, 1988; Fotheringham et al., 2002). Analysis of the interaction terms in the global regression model verifies that the effects of urban built environment on road travel time are spatially heterogeneous. Therefore, the GWR model is used to study this spatial heterogeneity and the reasons for its existence.

5.2.1 Model Estimation

The independent variables used in the GWR model were determined according to the result of the global regression model. In addition, the adaptive-bandwidth method was used, wherein the optimum bandwidth was determined by minimising the AIC value. The impact of each road segment's spatial location on its regression coefficient was described by a bi-square weight function. The MGWR V2.2.1 software package was used to estimate the GWR model (Oshan et al., 2019). The results are the corresponding regression coefficients of the 397 road segments. Due to space limitations, only the minimum value, first quantile value, median, third quantile value, and maximum value of the regression coefficients for each independent variable are shown in Table 3.

Table 3 shows that the same independent variable has different effects on the average speed of various road segments. Specifically, some independent variables show positive correlations with the average speed on some road segments, but negative correlations with the average speed on other road segments. Therefore, it is necessary to analyse the spatial influence of each independent variable on the average speed of the road segments.

5.2.2 Variable Analysis

The following analysis focuses on five factors: the number of bus stops, the distance from the nearest school, the distance from the nearest intersection, the occupancy rate of taxis, and the speed limit.

- Number of bus stops

Figure 4 shows the spatial distribution of the regression coefficients and the *p*-values of the number of bus stops on the study route. A noteworthy finding is that the number of bus stops has a positive correlation with the average speed of the road segments between intersections 5 and 6, between intersections 7 and 9, and between intersections 16 and 17. Although the total length of the three segments is less than 40% of the entire study route, 60% of the bus stops are located within 500 m of these segments. In all three segments, the average speed is sensitive to the number

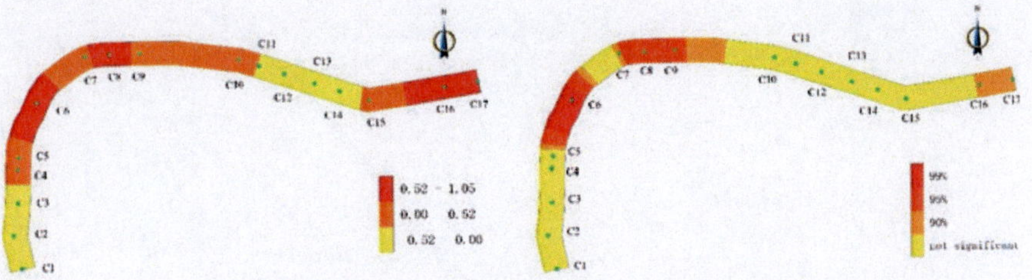

(a) (b)

Fig. 4 **a** Spatial distribution of the regression coefficients of the number of bus stops, **b** spatial distribution of the *p*-values of the number of bus stops

of bus stops; specifically, the greater the number of bus stops, the shorter the road travel time. This relationship has two causes. First, the taxi GPS data were collected during the hours of exclusive bus lane service (7:30–9:30 am). Therefore, despite a large number of bus stops on these road segments, the bus stops do not have a negative impact on the speed of private vehicles due to the exclusive bus lanes and the harbour-shaped bus stops, which segregate buses from other vehicles. Second, the more bus stops there are, the higher the probability of commuters travelling by bus (Sun & Guan, 2016) and the lower the probability of commuters travelling by other modes (including taxi). Correspondingly, the possibility of taxis decelerating to pick up passengers is lower. Therefore, the average speed of the road segments is higher when represented by the travel speed of taxis.

- Distance from the nearest school

Figure 5 presents the spatial distribution of the regression coefficients and the *p*-values of the distance from the nearest school on the study route. It can be seen that the distance from the nearest school has a positive correlation with the average speed of the road segments between intersections 3 and 7. That is, on these segments, the shorter the distance from the nearest school, the longer the road travel time. Figure 6a shows that the building density on these road segments is high. Most vehicles that

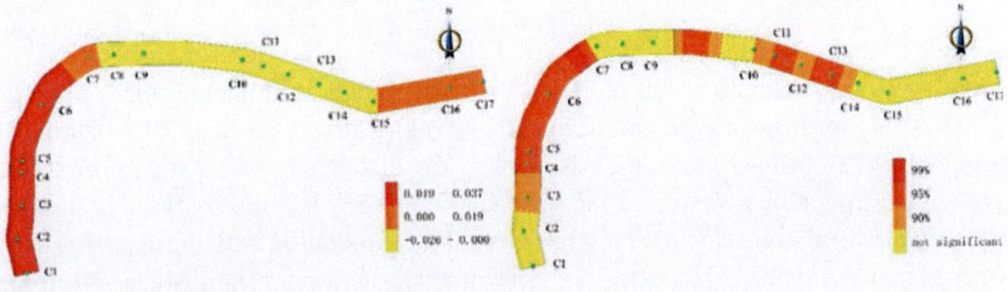

(a) (b)

Fig. 5 **a** Spatial distribution of the regression coefficients of the distance from the nearest school, **b** spatial distribution of the *p*-values of the distance from the nearest school

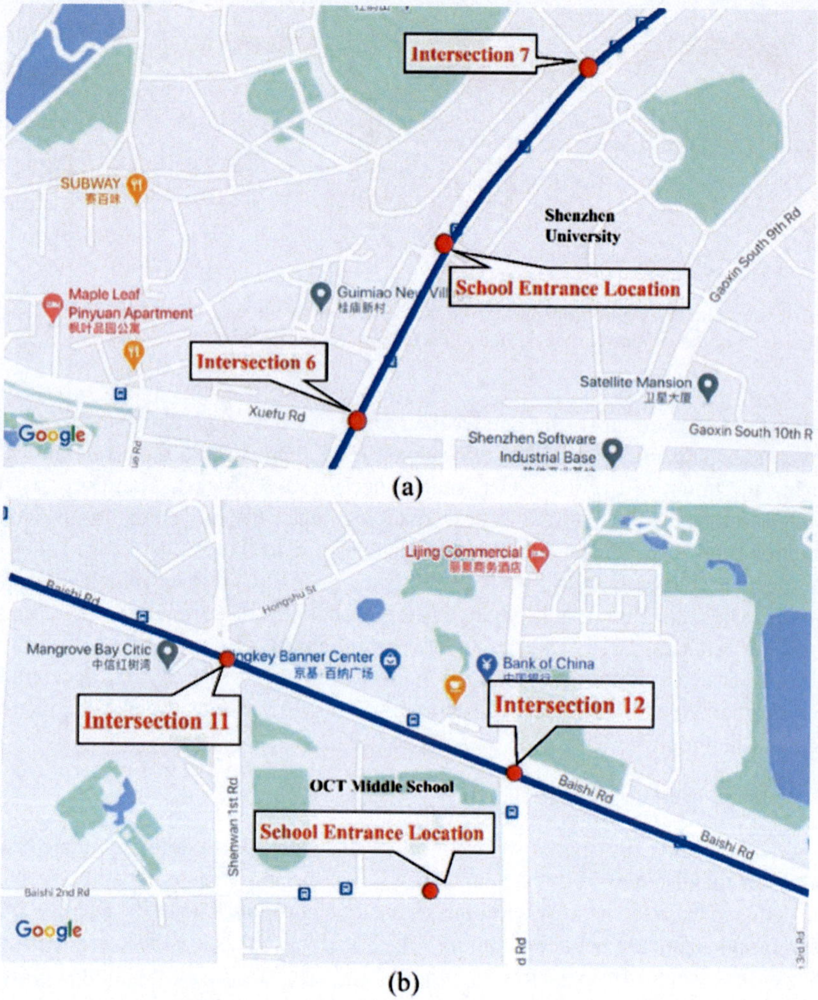

Fig. 6 **a** Road segments between intersections 6 and 7 on the study route, **b** road segments between intersections 11 and 12 on the study route (modified from Google Maps)

travel to nearby schools go through these road segments, and the entrances of some schools also face these segments. Thus, these segments are particularly sensitive to the presence of schools, such that the shorter the distance to the nearest school, the greater the impact. However, for the road segments between intersections 11 and 14, the distance from the nearest school has a negative association with the average speed of the road segments. That is, on these segments, the longer the distance from the nearest school, the lower the average speed. This is because these road segments are close to the rear side of the school, whilst the school entrance is on a road parallel to these segments (see Fig. 6b). This means that the road segments with a shorter direct

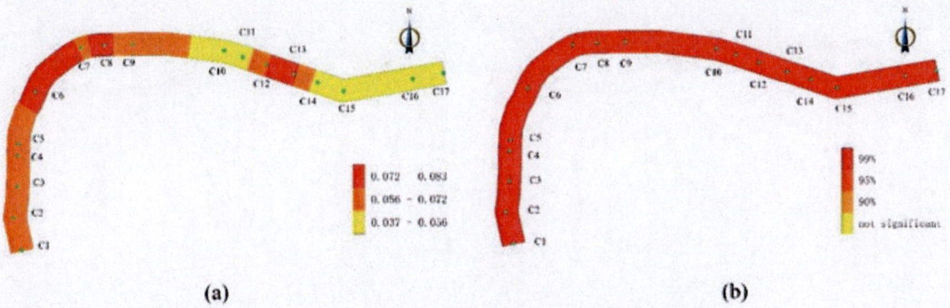

Fig. 7 **a** Spatial distribution of the regression coefficients of the distance from the nearest intersection, **b** spatial distribution of the *p*-values of the distance from the nearest intersection

Euclidean distance from the school are in fact further from the school entrance. In contrast, the road segments with a longer direct distance from the school are closer to the school entrance in terms of the actual road network. Consequently, these road segments are more likely to be affected by the pedestrian flow and the traffic flow in front of the school.

Overall, the impact of the distance from the nearest school on the average speed of the road segments is statistically significant between intersections 3 and 7 and between intersections 11 and 14. The school density is higher on these road segments than on others, and different locations of the main entrances of the nearby schools have different effects on the average speed of the segments. These results confirm that schools have an important impact on the travel time on surrounding roads, and the shorter the distance from the school entrance to the road, the greater the impact. Therefore, to reduce road travel time, schools should be built as far as possible from main urban roads. If schools are already located near the main roads, their impact can be reduced by relocating the school entrances.

- Distance from the nearest intersection

Figure 7 illustrates the spatial distribution of the regression coefficients and the *p*-values of the distance from the nearest intersection on the study route. The figure shows that the distance from the nearest intersection has a positive correlation with the average speed on the whole study route. That is, the shorter the distance from the nearest intersection, the lower the average speed of the road segments, and the longer the road travel time. By comparing road segments with respect to the nearest intersection type, it is found that if the nearest intersection is Type 1 or Type 4, the regression coefficient is relatively large, whilst if the nearest intersection is Type 2 or Type 3, the regression coefficient is relatively small. This indicates that the presence of left-turn lanes has a great impact on the average speed of a road segment. Holding all else equal, the average speed of a segment whose nearest intersection has an independent left-turn lane is higher.

- Occupancy rate of taxis

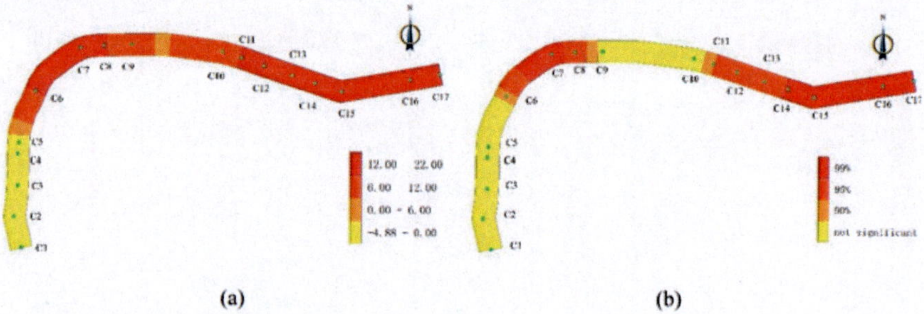

Fig. 8 **a** Spatial distribution of the regression coefficients of the occupancy rate of taxis, **b** spatial distribution of the *p*-values of the occupancy rate of taxis

Figure 8 shows the spatial distribution of the regression coefficients and the *p*-values of the occupancy rate of taxis on the study route. The occupancy rate has a positive correlation with the average speed of the road segments between intersections 6 and 8 and between intersections 11 and 17. That is, the average speed of these road segments increases as the taxis' occupancy rate increases. This indicates two things. First, the frequency of passengers boarding and alighting from taxis is high on these road segments; thus, the average speed of these segments is highly dependent on the occupancy rate of taxis. Therefore, to reduce the interference of taxis with private vehicles, temporary taxi stands could be designed and constructed on these road segments to segregate taxi drivers and passengers from other vehicles. Second, the occupancy rate of taxis has a great impact on road travel time, because taxis may deliberately slow down and stop to pick up passengers at any time when vacant. Therefore, when estimating road travel time based on taxi GPS data, neglecting the status (i.e., occupied or empty) of taxis will lead to biased estimations.

- Speed limit

The target route of this study is approximately 10 km long, with the speed limit ranging from 50 to 60 km/h on different sections, which may have a direct impact on the travel time of each section. Figure 9 depicts the spatial distribution of the regression coefficients and the *p*-values of the speed limit on the study route. The speed limit only impacts on the average speed of the road sections between intersections 7 and 10, where the speed limit has a positive correlation with the average speed of these road sections. That is, the higher the speed limit, the higher the average speed of the road sections, and the shorter the road travel time. Generally, speed limits are set in accordance with the road conditions. Road sections capable of supporting a higher travel speed tend to have a higher speed limit. Here, the fact that the speed limit is only significant between intersections 7 and 10 indicates that that part of the road is in good condition, thus drivers are more likely to exceed the speed limit there (road sections 158-250, see Fig. 3). This finding suggests that the transportation department should actively monitor speeding behaviour to ensure the safety of

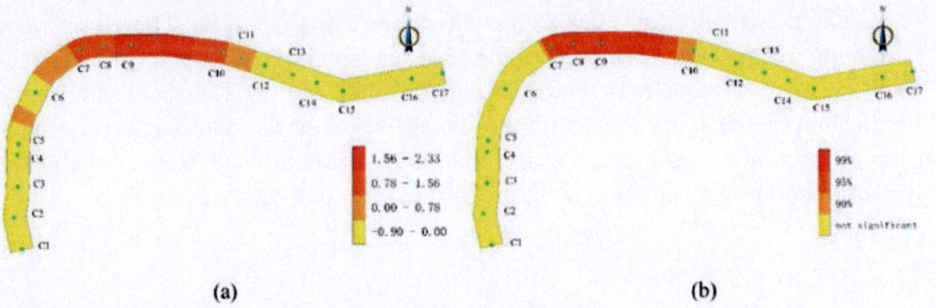

Fig. 9 **a** Spatial distribution of the regression coefficients of the rate of the speed limit, **b** spatial distribution of the *p*-values of the rate of the speed limit

the road network. Another action would be to set more reasonable speed limits by considering the actual conditions of each road section.

6 Method Application and Policy Implications

The method proposed in this study can be applied in two practical aspects. First, the method can assist transportation planners and managers to better understand the relationship between urban built environment and road travel time. The traditional data-driven methods (referring to those that use quantitative data, as opposed to those driven by models or theories) cannot analyse the internal mechanism of the impact of various factors on road travel time. Overcoming this limitation of the traditional data-driven methods, this study established a general relationship between urban built environment and road travel time, which can capture the internal factors that influence road travel time. Second, the proposed method and the obtained results can help decision-makers develop targeted urban planning and management strategies. By adjusting the attributes of urban built environment (such as the type of bus stops, the location of school entrances, and the type of intersections), road travel time can be reduced.

By analysing the spatial impact of urban built environment on the average speed of the road segments, this study provides the following important policy implications for transportation planning and management: (1) The coordination of exclusive bus lanes and harbour-shaped bus stops can reduce the negative effects of bus stops on road travel time, which corroborates the findings of Jia et al. (2009) and Arasan and Vedagiri (2010); (2) The impact of a school on road travel time depends mainly on the location of the school entrance, and the negative effects can be reduced by relocating the entrance, which is consistent with previous findings (e.g., Mackie, 2010); (3) Along the whole study route, the distance from the nearest intersection has a positive correlation with the average speed of the road segments. When the nearest intersection to the road segment has an independent left-turn lane, the corresponding regression coefficient is larger and the average speed of the segment is higher; (4)

The occupancy rate of taxis has a great effect on road travel time. Therefore, when estimating the average travel time of the road segments based on taxi GPS data, the taxis' occupancy status must be taken into account; and (5) The speed limit has a significant positive correlation with the average speed of the road sections on some parts of the route, so the urban transportation department is advised to establish a system to monitor speeding in these areas to ensure road safety.

7 Conclusions

Thus far, most studies on road travel time estimation have been based on traffic flow theory or data-driven methods. These approaches take close account of traffic status and historical travel time data but largely fail to capture the key factors of urban built environment that influence road travel time. In view of this, this study presented pioneering research that identified and examined the key factors that influence road travel time by using a global regression model and a GWR model from the perspective of urban built environment.

The results of the global regression model show that the distance from the nearest intersection, the occupancy rate of taxis, and the speed limit have positive correlations with the average speeds of the road segments, whilst the corresponding correlations with the number of bus stops and the distance from the nearest school are negative. In addition, there is an interaction between the intersection type and the features of urban built environment and between the intersection type and the occupancy rate of taxis. Moreover, factors that significantly affect the average speed of the road (or, equivalently, the road travel time) depend on the type of intersection nearest to the road segment because of spatial heterogeneity. The GWR model results showed that the number of bus stops, the distance from the nearest school, the distance from the nearest intersection, the occupancy rate of taxis, and the speed limit can have either positive or negative associations with the average speeds of the road segments. These factors only have significant effects on parts of the study route.

Last, but not least, this study addressed two drawbacks of the traditional fixed-bandwidth GWR model: instability and inflexibility. By using the adaptive-bandwidth GWR model and a large sample of taxi GPS data, the impact of collinearity on the model results was effectively alleviated, ensuring the stability of the estimated results. For the second limitation (inflexibility), Murakami et al. (2018) found that small-scale spatial variations affect the estimation accuracy of the adaptive-bandwidth GWR model. Therefore, given the small spatial scale, caution should be exercised when using the results of this study. One possible future research direction is to use the flexible adaptive-bandwidth GWR model to solve this problem. Compared with the non-flexible adaptive-bandwidth GWR model, the flexible variant can not only ensure the stability of the estimated results (e.g., in the presence of collinearity), but also control the varying scales of different sets of coefficients. It is worth noting, however, that compared with the traditional GWR model, the computational efficiency of the flexible GWR model will be reduced, especially for large data sets.

References

Anselin, L. (1988). *Spatial econometrics: Methods and models*. Kluwer Academic Publishers.

Arasan, V. T., & Vedagiri, P. (2010). Microsimulation study of the effect of exclusive bus lanes on heterogeneous traffic flow. *Journal of Urban Planning and Development, 136*(1), 50–58.

Ariannezhad, A., Karimpour, A., & Wu, Y. J. (2020). Incorporating mode choices into safety analysis at the macroscopic level. *Journal of Transportation Engineering, Part A: Systems, 146*(4), 04020022.

Axhausen, K. W., Schönfelder, S., Wolf, J., et al. (2003). 80 weeks of GPS traces: Approaches to enriching the trip information. *Transportation Research Record, 1870*, 46–54.

Cao, X., Mokhtarian, P. L., & Handy, S. L. (2009). Examining the impacts of residential self-selection on travel behaviour: A focus on empirical findings. *Transport Reviews, 29*(3), 359–395.

Cervero, R., & Jin, M. (2010). Effects of built environments on vehicle miles traveled: Evidence from 370 US urbanized areas. *Environment and Planning A: Economy and Space, 42*(2), 400–418.

Cervero, R., & Kockelman, K. (1997). Travel demand and the 3Ds: Density, diversity, and design. *Transportation Research Part D: Transport and Environment, 2*(3), 199–219.

Chacon-Hurtado, D., Kumar, I., Gkritza, K., et al. (2020). The role of transportation accessibility in regional economic resilience. *Journal of Transport Geography, 84*, 102695.

Chiou, Y. C., Jou, R. C., & Yang, C. H. (2015). Factors affecting public transportation usage rate: Geographically weighted regression. *Transportation Research Part A: Policy and Practice, 78*, 161–177.

Dell'Orco, M., Marinelli, M., & Silgu, M. A. (2016). Bee colony optimization for innovative travel time estimation, based on a mesoscopic traffic assignment model. *Transportation Research Part C: Emerging Technologies, 66*(1), 48–60.

Ding, C., Lin, Y., & Liu, C. (2014). Exploring the influence of built environment on tour-based commuter mode choice: A cross-classified multilevel modeling approach. *Transportation Research Part D: Transport and Environment, 32*, 230–238.

Du, H., & Mulley, C. (2006). Relationship between transport accessibility and land value: Local model approach with geographically weighted regression. *Transportation Research Record: Journal of the Transportation Research Board, 1977*, 197–205.

Elldér, E. (2014). Residential location and daily travel distances: The influence of trip purpose. *Journal of Transport Geography, 34*, 121–130.

Ewing, R., & Cervero, R. (2001). Travel and the built environment: A synthesis. *Transportation Research Record, 1780*(1), 87–114.

Fan, Y., & Khattak, A. J. (2008). Urban form, individual spatial footprints, and travel: Examination of space-use behavior. *Transportation Research Record, 2082*(1), 98–106.

Feuillet, T., Charreire, H., Menai, M., et al. (2015). Spatial heterogeneity of the relationships between environmental characteristics and active commuting: Towards a locally varying social ecological model. *International Journal of Health Geographics, 14*(1), 12.

Fotheringham, A. S., Brunsdon, C., & Charlton, M. (2002). *Geographically weighted regression: The analysis of spatially varying relationships*. Wiley.

Handy, S. L., Boarnet, M. G., Ewing, R., et al. (2002). How the built environment affects physical activity: Views from urban planning. *American Journal of Preventive Medicine, 23*(2), 64–73.

Hellinga, B., Izadpanah, P., Takada, H., et al. (2008). Decomposing travel times measured by probe-based traffic monitoring systems to individual road segments. *Transportation Research Part C: Emerging Technologies, 16*(6), 768–782.

Hofleitner, A., Herring, R., & Bayen, A. (2012). Arterial travel time forecast with streaming data: A hybrid approach of flow modeling and machine learning. *Transportation Research Part B: Methodological, 46*(9), 1097–1122.

Jenelius, E., & Koutsopoulos, H. N. (2013). Travel time estimation for urban road networks using low frequency probe vehicle data. *Transportation Research Part B: Methodological, 53*(4), 64–81.

Jia, B., Li, X., Jiang, R., & Gao, Z. (2009). The influence of bus stop on the dynamics of traffic flow. *Acta Physica Sinica, 58*(10), 6845–6851.

Jiang, Y., Szeto, W. Y., Long, J., & Han, K. (2016). Multi-class dynamic traffic assignment with physical queues: Intersection-movement-based formulation and paradox. *Transportmetrica A: Transport Science, 12*(10), 878–908.

Karimpour, A., Ariannezhad, A., & Wu, Y. J. (2019). Hybrid data-driven approach for truck travel time imputation. *IET Intelligent Transport Systems, 13*(10), 1518–1524.

Lee, S. H., Viswanathan, M., & Yang, Y. K. (2006) A hybrid soft computing approach to link travel speed estimation. In *International Conference on Fuzzy Systems and Knowledge Discovery* (pp. 794–802). Springer.

Li, R., Rose, G., & Sarvi, M. (2006). Evaluation of speed-based travel time estimation models. *Journal of Transportation Engineering, 132*(7), 540–547.

Liu, H., van Zuylen, H. J., van Lint, H., et al. (2005). Prediction of urban travel times with intersection delays. In *Proceedings of 2005 IEEE Intelligent Transportation Systems* (pp. 402–407).

Liu, H. X., & Ma, W. (2009). A virtual vehicle probe model for time-dependent travel time estimation on signalized arterials. *Transportation Research Part C: Emerging Technologies, 17*(1), 11–26.

Liu, Q., Ding, C., & Chen, P. (2020). A panel analysis of the effect of the urban environment on the spatiotemporal pattern of taxi demand. *Travel Behaviour and Society, 18*, 29–36.

Lloyd, C. D. (2010). *Local models for spatial analysis*. CRC Press.

Ma, X., Zhang, J., Ding, C., & Wang, Y. (2018). A geographically and temporally weighted regression model to explore the spatiotemporal influence of built environment on transit ridership. *Computers, Environment and Urban Systems, 70*, 113–124.

Ma, Z., Koutsopoulos, H. N., Ferreira, L., et al. (2017). Estimation of trip travel time distribution using a generalized Markov chain approach. *Transportation Research Part C: Emerging Technologies, 74*, 1–21.

Maat, K., & Timmermans, H. J. P. (2009). Influence of the residential and work environment on car use in dual-earner households. *Transportation Research Part A: Policy and Practice, 43*(7), 654–664.

Mackie, H. (2010) *Improving school travel systems*. NZ Transport Agency.

Mori, U., Mendiburu, A., Álvarez, M., et al. (2015). A review of travel time estimation and forecasting for advanced traveller information systems. *Transportmetrica A: Transport Science, 11*(2), 119–157.

Murakami, D., Lu, B., Harris, P., et al. (2018). The importance of scale in spatially varying coefficient modeling. *Annals of the American Association of Geographers,* 1–21.

Oshan, T. M., & Fotheringham, A. S. (2018). A comparison of spatially varying regression coefficient estimates using geographically weighted and spatial-filter-based techniques. *Geographical Analysis, 50*, 53–75.

Oshan, T. M., Li, Z., Kang, W., Wolf, L., & Fotheringham, A. S. (2019). MGWR: A python implementation of multiscale geographically weighted regression for investigating process spatial heterogeneity and scale. *ISPRS International Journal of Geo-information, 8*(6), 269.

Páez, A., Farber, S., & Wheeler, D. (2011). A simulation-based study of geographically weighted regression as a method for investigating spatially varying relationships. *Environment and Planning A: Economy and Space, 43*, 2992–3010.

Pirc, J., Turk, G., & Žura, M. (2016). Highway travel time estimation using multiple data sources. *IET Intelligent Transport Systems, 10*(10), 649–657.

Pirdavani, A., Bellemans, T., Brijs, T., & Wets, G. (2014). Application of geographically weighted regression technique in spatial analysis of fatal and injury crashes. *Journal of Transportation Engineering, 140*(8), 04014032.

Pu, Z., Li, Z., Ash, J., et al. (2017). Evaluation of spatial heterogeneity in the sensitivity of on-street parking occupancy to price change. *Transportation Research Part C: Emerging Technologies, 77*, 67–79.

Rahmani, M., Jenelius, E., & Koutsopoulos, H. N. (2015). Non-parametric estimation of route travel time distributions from low-frequency floating car data. *Transportation Research Part C: Emerging Technologies, 58*, 343–362.

Rahmani, M., Koutsopoulos, H. N., & Jenelius, E. (2017). Travel time estimation from sparse floating car data with consistent path inference: A fixed point approach. *Transportation Research Part C: Emerging Technologies, 85*, 628–643.

Sun, D. J., & Guan, S. (2016). Measuring vulnerability of urban metro network from line operation perspective. *Transportation Research Part A: Policy and Practice, 94*, 348–359.

Sun, D. J., Zhang, K., & Shen, S. (2018). Analyzing spatiotemporal traffic line source emissions based on massive didi online car-hailing service data. *Transportation Research Part D: Transport and Environment, 62*, 699–714.

Tang, K., Chen, S., Liu, Z., & Khattak, A. J. (2018). A tensor-based Bayesian probabilistic model for citywide personalized travel time estimation. *Transportation Research Part C: Emerging Technologies, 90*, 260–280.

van Lint, J. W. C. (2004). *Reliable travel time prediction for freeways: Bridging artificial neural networks and traffic flow theory* (Doctoral Dissertation), Civil Engineering & Geosciences, TU Delft, Netherlands.

Vanajakshi, L. D. (2005) *Estimation and prediction of travel time from loop detector data for intelligent transportation systems applications* (Doctoral Dissertation), Texas A & M University.

Wang, D., Chai, Y., & Li, F. (2011). Built environment diversities and activity–travel behaviour variations in Beijing, China. *Journal of Transport Geography, 19*(6), 1173–1186.

Wang, X., & Khattak, A. (2011). Role of travel information in supporting travel decision adaption: Exploring spatial patterns. *Transportmetrica A: Transport Science, 9*(4), 316–334.

Wheeler, D., & Tiefelsdorf, M. (2005). Multicollinearity and correlation among local regression coefficients in geographically weighted regression. *Journal of Geographical Systems, 7*, 161–187.

Xu, D., Wei, C., Peng, P., Xuan, Q., & Guo, H. (2020). GE-GAN: A novel deep learning framework for road traffic state estimation. *Transportation Research Part C: Emerging Technologies, 117*, 102635.

Yu, B., Wang, H., Shan, W., et al. (2018). Prediction of bus travel time using random forests based on bear neighbors. *Computer-Aided Civil and Infrastructure Engineering, 33*, 333–350.

Zhao, F., & Park, N. (2004). Using geographically weighted regression models to estimate annual average daily traffic. *Transportation Research Record: Journal of the Transportation Research Board, 1879*, 99–107.

Zheng, F., & Zuylen, H. V. (2013). Urban link travel time estimation based on sparse probe vehicle data. *Transportation Research Part C: Emerging Technologies, 31*, 145–157.

Zhong, S., & Bushell, M. (2017). Impact of the built environment on the vehicle emission effects of road pricing policies: A simulation case study. *Transportation Research Part A: Policy and Practice, 103*, 235–249.

Zhong, S., Wang, S., Jiang, Y., et al. (2015). Distinguishing the land use effects of road pricing based on the urban form attributes. *Transportation Research Part A: Policy and Practice, 74*, 44–58.

Chapter 8
Taxi Driver Speeding: Who, When, Where and How? A Comparative Study Between Shanghai and New York

Abstract This study proposes a Driver-Road-Environment Identification (DREI) method to investigate the determinant factors of taxi speeding violations. Driving style characteristics, together with road and environment variables were obtained based on the GPS data and auxiliary spatio-temporal data in Shanghai and New York City (NYC). The daily working hours of taxi drivers in Shanghai (18.6 h) was far more than NYC (8.5 h). The average occupancy speed of taxi drivers in Shanghai (21.3 km/h) was similar to that of NYC (20.3 km/h). Speeders in both cities had shorter working hours and longer daily driving distance than the ordinary taxi drivers, while their daily income was similar. Speeding drivers routinely took long distance trips (>10 km) and they preferred to choose relative faster routes rather than the shortest ones. Length of segments (1.0–1.5 km) and good traffic condition were associated with high amount of speeding rate while CBD area and secondary road were associated with low amount of speeding rate. Moreover, many speeding violations were identified occurring between 4:00 AM and 7:00 AM in both Shanghai and NYC and the worst period was between 5:00 AM and 6:00 AM in both cities. Findings of this study may assist to stipulate relevant laws and regulations such as stronger early morning, long segments supervision, shift-rule regulation and working hour restriction to mitigate the risk of potential crashes.

Keywords Taxi driver · Speeding · DREI method · GPS data · Comparative analysis

1 Introduction

Speeding is one of the most significant contributors to traffic accidents in many countries, such as China (Traffic Management Bureau, 2012) and the United States (National Center for Statistics and Analysis, 2013; Royal, 2003; Schroeder et al., 2013). Many literatures have studied the factors affecting driver's speeding behavior on both highways and urban arterials (Harré et al., 1996; Giles, 2004; Lefeve, 1956; Oppenlander, 1966; Rakauskas et al., 2007; Richard et al., 2013; Williams et al., 2006; Zhang et al., 2014). However, limited studies have been conducted on the issue of taxi speeding. Compared to ordinary drivers, taxi drivers are more likely to

S. Zhong and D. Sun, *Logic-Driven Traffic Big Data Analytics*,
https://doi.org/10.1007/978-981-16-8016-8_8

commit risk behaviors (Burns and Wilde, 1995; Mayhew, 2000; Tseng, 2013; Yeh et al., 2015). Between 1996 and 2000 in New South Wales, Australia, 7923 taxi drivers were involved in crashes, nearly 10% (n = 750) were killed or injured (Lam, 2004). In South Africa, a study showed that 33.8% of taxi drivers had been involved in a car accident (Peltzer & Renner, 2003) whereas another study in Hanoi reported that 276 of 1214 taxi drivers (22.7%) were involved in at least one crash (La et al., 2013). Studying taxi speeding is important because of the time-on-the-road of taxi drivers were considerable (Dalziel & Job, 1997) and they have more opportunities to be involved in a speeding violation (Tseng, 2013).

Only a few studies have focused on the taxi drivers' speeding problem. In an observational study, Burns and Wilde (1995) found that taxi drivers with sensation seeking personality were more frequently convicted of a speeding violation. Newnam et al. (2014) utilized questionnaires and found that age and educational level had no significant relationship with taxi drivers' speeding behaviour in Ethiopia. Using the data from a national survey in Taiwan, Tseng (2013) explored the determinant personal factors of taxi drivers' speeding violations and found that age, job experience, operation styles, daily driving distance, driving late at night, and monthly off-duty days were significantly associated with the speeding violations. According to the same survey source in Taiwan, Yeh et al. (2015) analysed the factors of female taxi drivers' speeding offenses and found that female taxi drivers' speeding offenses were significantly related to age, education level, and mileage driven while job experience, business operating style, and vehicle engine size are not significantly affecting the percentage of having at least one speeding ticket. Shi et al. (2014) designed questionnaires to explore the factors affecting aberrant behaviour of taxi drivers in Beijing. They found that economic pressure, ownership of taxi, and the complaint system would affect taxi drivers' aberrant behaviours, including speeding. However, these taxi driver speeding studies neglected some important factors. For ordinary drivers, previous studies have demonstrated that speeding behaviours are also triggered by exogenous pressure, such as road attribution (Giles, 2004; Zhang et al., 2014), traffic condition (Zhang et al., 2014), vehicle parameters (Giles, 2004; Williams et al., 2006; Zhang et al., 2014), time of day (NHTSA, 2013c; Oppenlander, 1966; Zhang et al., 2014), light condition factor (Lefeve, 1956; Zhang et al., 2014) and location factor (Giles, 2004; Rakauskas et al., 2007). However, whether these situational factors lead to speeding violations remain unclear for taxi drivers.

In summary, with respect to taxi speeding, driving style characteristics and situational factors have not been adequately studied. The major reason was the scarcity of reliable taxi driver speeding-related data. Data required to explore the factors associated with taxi speeding included demographic characteristics, road attributions, vehicle parameters and other environmental variables. Most of the previous studies (La et al., 2013; Shi et al., 2014; Tseng, 2013; Yeh et al., 2015) have used survey data to investigate the driver-related variables. The conventional methods were easy to manage and were useful to identify the demographic characteristics. However, the results may have self-reported biases and are insufficient to explore situational factors. The present study explored these determinant factors that lead to taxi speeding using FCD methods. Some studies have demonstrated the advantages of implementing FCD methods into taxi traffic violation research. Compared to the survey studies, the FCD method is capable of capturing spatial–temporal information of taxi speeding behaviour. The sample size of more than 10,000 GPS equipped taxis can improve the accuracy of the study. The three objectives of this study are:

1. To identify the driving style characteristics of taxi drivers in Shanghai and NYC using taxi GPS data and make a comparative analysis
2. To explore the influence of different driving style characteristics on the frequency of speeding (who and how?)
3. To explore the influence of driving style characteristics, road attributes and environmental factors on the speeding rate (when, where and how?).

2 Methodology

2.1 Overview of the DREI Method

This section provides the Driver-Road-Environment Identification (DREI) method to explore the relationships among driving style characteristics, roadway attributes and environment factors associated with speeding for taxi drivers. The method consists of three steps: (a) filtering the data and matching the data on the map; (b) identifying driving style characteristics and comparing the basic characteristics between speeders and ordinary taxi drivers in two cities; (c) adding the situational data and evaluating the significance and strength of driving style characteristics and other situational variables on speeding rate.

2.2 Data

The GPS data in Shanghai was collected in 2015, with 13,475 taxis for one month (April 1–30, 2015). The data on April 10 (Friday) was used to identify the characteristics of taxi driving style and speeding behaviour. The GPS data of NYC were collected in 2013. The data on April 12 (Friday) was adopted to identify the speeding behaviour.

2.3 Measure and Variables

2.3.1 Comparisons of Taxi Speeding in Shanghai and NYC: Who Is Speeding? and Who Speeds the Most?

The speed choice of a driver was fairly consistent over time (Haglund & Åberg, 2000; Ahie et al. 2015) and was highly influenced by their usual speeds (Ahlin, 1979). In this chapter, we defined the taxi drivers who frequently travel faster than other taxi drivers as "speeder." These drivers would be more likely to get a better position or speed advantage whenever they have a chance. As a control group, ordinary taxi drivers were randomly selected from the entire taxi driver sample.

Previous studies already demonstrated that daily working hours, daily driving distance, driving late at night, weekly off days (Tseng, 2013; Wali et al. 2017) would significantly be affecting taxi drivers' speeding behaviours. However, the business operating style of taxi driver was a controversial factor affecting taxi speeding (Tseng, 2013; Yeh et al., 2015). In this study, the business operating style was explored using long-distance trip ratio, route preference, time occupancy rate factors, and daily profit of drivers. Many taxi drivers in Shanghai prefer long-distance trips, such as from/to an airport, because they can earn the long distance surcharge when the trip is longer than 10 km (no such surcharge in NYC). It is possible that these drivers may have a higher probability to commit a speeding violation in a long-distance trip. Besides, it is doubtful that drivers with different route preference would have different speeding behaviour. The previous study (Liu et al., 2010) already showed that some taxi drivers may choose the shortest route from the origin to the destination under occupancy status while others tend to choose the relatively faster path to deliver passengers. In order to examine the effect of route preference, the route directness index (RDI) is chosen to reflects the circuity of the route and the calculation formula to calculate the RDI can refer to (Cui et al., 2016). Considering that taxi drivers with different operational performance have different working experience and working pressure, it is possible that drivers with various operation performance commit different speeding violations. Therefore, the operation performance of taxi drivers was investigated through daily profit (DP) and time occupancy rate (TOR). The TOR is calculated as the passenger occupancy time divided by the total working time.

The present study is to explore the difference of these driving style characteristics between speeders and ordinary taxi drivers in Shanghai and New York to answer the question "who is speeding in Shanghai and NYC". Then, for the question of "who speeds the most", a stepwise regression was employed to evaluate the relative significance and strength of the influential characteristics variables on speeding violations, so as to eliminate the effect of multi-collinearity among variables. Likewise, the results of Shanghai and NYC were compared to figure out the common and different characteristics variables influencing taxi driver speeding offense.

2.3.2 Road-Level Analysis in Shanghai: When, Where and How?

Previous studies explored the determinant factors leading to speeding behaviour and demonstrated that time of day (Oppenlander, 1966; Zhang et al., 2014), traffic condition (Zhang et al., 2014), location (Rakauskas et al., 2007), road type and road length (Giles, 2004; Zhang et al., 2014) would significantly influence drivers' speeding behaviour. The assumption of the present study is that these variables are also associated with taxi drivers' speeding behaviour. Thus, these situational explanatory variables, were included in this study. Considering that the NYC dataset did not include the detailed taxi trajectories on the road, we did this analysis only using data from Shanghai. The area coloured in yellow in Fig. 1 was chosen as our study area, which contained the outer ring of expressways and two main airports in Shanghai. Totally 265 arterial roads and 421 secondary roads segments with the length longer than 300 m were included.

By matching the GPS traces on the road, spatio-temporal situational features can be identified, such as time of day, road type, road length, speed limits of the road segments and regional information. The traffic condition of each road segment can be roughly estimated by the average speed and speed variation of other taxis driving on that road segment in a period of time. In this study, the hourly average speed of each road segment was considered as an independent sample, which (686 * 24) were clustered into five groups by their average speed and standard deviation features using K-means clustering method, as shown in Fig. 2. The proportion of road segments in each cluster is presented in Fig. 3, which changes for each single hour.

A generalized linear model (GLM) was employed to identify how the characteristics and situational factors leading to the speeding rate. The GLM is a systematic extension of linear model for non-normal data. In this study, the dependent variable y of GLM represented the speeding rare. The explanatory variables were based on previous studies and the assumption of this study included daily working hours, daily driving distance, daily profit, long distance trip rate, driving late at night, route directness index, time of day, road speed category, time occupancy rate, road length, location index and road type.

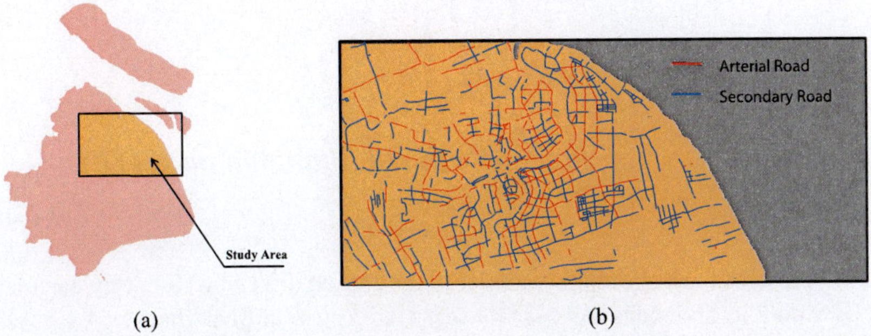

Fig. 1 Study area in Shanghai, **a** an overview, **b** arterials and secondary roads within the area

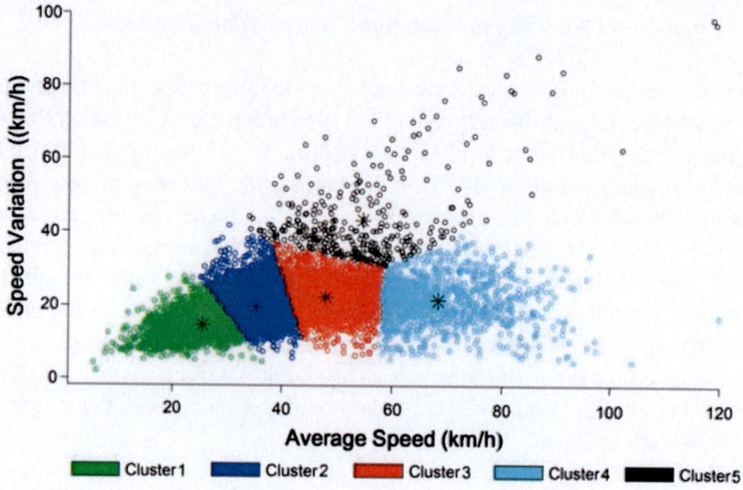

Fig. 2 Clustering result of each road segments groups

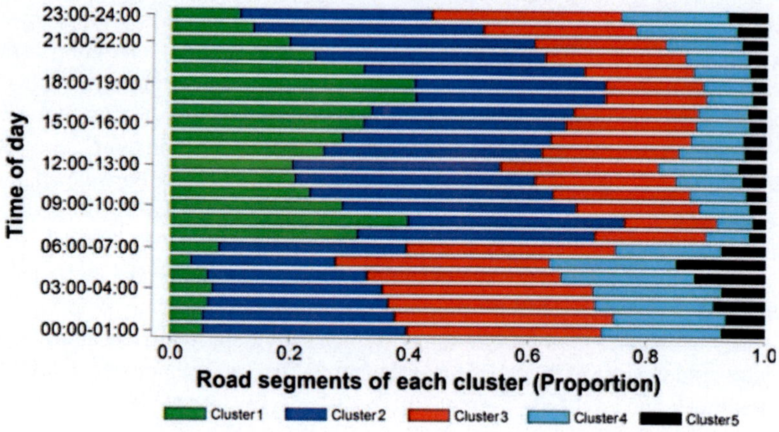

Fig. 3 Proportion of each cluster group during each hour

3 Result Analysis

3.1 Comparisons of Taxi Speeding in Shanghai and NYC: Who Is Speeding?

Table 1 presents the basic characteristics of the chosen drivers, which were calculated by GPS data in Shanghai and NYC. Most taxi drivers in Shanghai choose to work in a 24-h shift and then rest for an entire day after work while most taxi drivers in NYC choose to operate by two shifts per day, and seldom have a whole day rest.

Table 1 Basic characteristics of ordinary taxi drivers and speeders in Shanghai and NYC

Characteristics	Shanghai		NYC	
	Ordinary taxi driver ($N = 2385$)	Speeder ($N = 2281$)	Ordinary taxi driver ($N = 4908$)	Speeder ($N = 4556$)
Daily working hours, h	18.6	15.9	8.5	7.9
Driving late at night (%)	71.4	61.8	39.3	31.7
Income, USD	153.3	152.4	335.5	336.6
Daily driving distance, km	322.6	353.5	146.2[b]	161.8[b]
Daily occupancy driving distance, km	226.8	241.3	101.2	112.0
Weekly off, days	3.5[a]	3.5[a]	0.9	0.7
Weekly working hours, h	65.1	55.7	51.9	49.8
Average occupancy speed, km/h	21.3	25.9	20.3	23.3
Time occupancy rate	0.59	0.57	0.58	0.58
Long distance rate	0.43	0.56	0.34	0.51
Route directness index	1.28	1.36	1.38	1.41

[a] Setting to 3.5, as each Shanghai taxi generally has two alternative drivers
[b] Estimated by setting the same rate of average occupancy speed and idle speed as Shanghai (1.63)

Although the weekly working hours between Shanghai and NYC were similar, the daily working hours in Shanghai (18.6 h) was far more than NYC (8.5 h) because of different shift rules. The average occupancy speed of taxi drivers in Shanghai (21.3 km/h) was similar to that of NYC (20.3 km/h). In both cities, compared to ordinary taxi drivers, speeders had shorter working hours but higher driving distance. Meanwhile, the income of both ordinary taxi drivers and speeder groups was similar. Moreover, in both cities, speeders routinely take long distance trips and choose relatively fast routes rather than the shortest ones.

Figures 4 and 5 present the distribution of speeds relative to the posted speed (PS) for each hour in Shanghai and NYC, respectively. The category of each column was based on the speeding penalty classification in Shanghai and NYC (e.g., PS + 20% represented driving at speeds between posted speed and 20% above the posted speed because drivers at that interval would largely be issued with a speeding ticket).

The detailed speeding penalty rules were shown in Table 2. All drives that were 20% or more below the posted speed in Shanghai and 5 mph or more below the posted speed in NYC were removed. These driving records may not at the "free-flow time", defined as periods of time that drivers have opportunities to exceed the speed limit (NHTSA, 2013c). It was found that the speeding feature was time-varying,

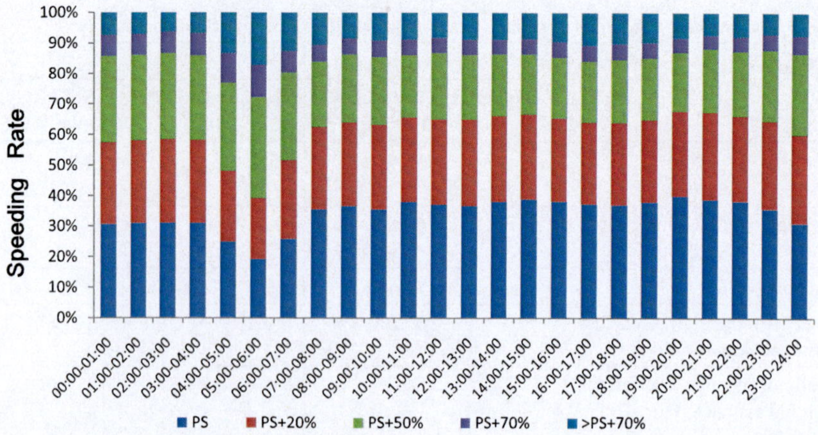

Fig. 4 Percentages of the different speeding rate for each hour in Shanghai

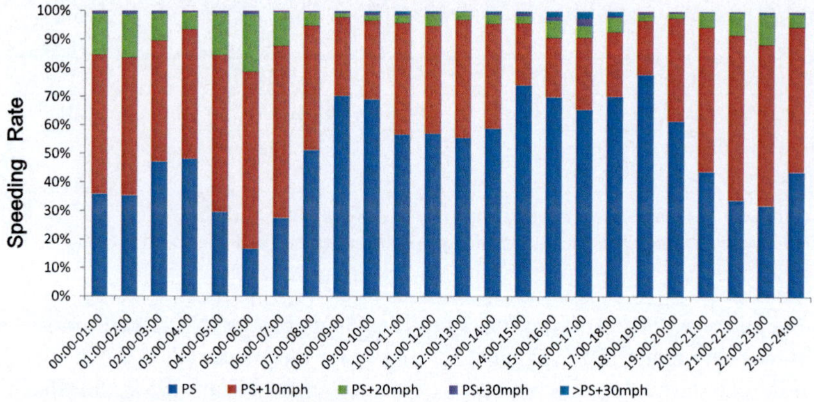

Fig. 5 Percentages of the different speeding rate for each hour in NYC

drivers were more likely to commit speeding violations between 4:00 AM and 7:00 AM and they commit speeding violations most between 5:00 AM and 6:00 AM in both Shanghai and NYC.

3.2 Analysis of Driver Characteristics on Speeding Frequency: Who Speeds the Most?

The stepwise regression was used to explore the significance and strength of the driving style characteristics leading to speeding violations. The results of Shanghai and NYC are shown in Tables 3 and 4. Among the explanatory variables, daily

Table 2 Speed limits and penalty in Shanghai and NYC

City	Speed limits	Speeding level	Penalty points and fine
Shanghai	Arterial roads: 60 km/h Secondary roads: 40–50 km/h	<20%	3 points and 50–100 RMB
		20–50%	6 points and 100–150 RMB
		50–70%	12 points and 300–500 RMB
		>70%	12 points and 500–1000 RMB
NYC	Before Nov.2014: 30 mph Since Nov. 2014: 25 mph	<10 mph (40%)	3 points and 90–150 USD
		10–20 mph (40–80%)	4 points and 180–300 USD
		20–30 mph (80–120%)	6 points and 180–300 USD
		More than 30 mph(>120%)	8–11 points and 360–600 USD

driving distance, daily profit, and long-distance trip rate are relative important factors significantly influencing taxi speeding.

In Shanghai, drivers with long daily working hour and long driving distance generally had more speeding violations because of the long duration exposure on the road. However, drivers with high daily working hour (>10 h) were less likely to have speeding violations in NYC ($B = -0.05, p = 0.003$). Drivers who always drove late at night were more likely to speed in both Shanghai ($B = 16.431$, $p = 0.001$) and NYC ($B = 0.041, p = 0.001$). Moreover, drivers who took long distance trips in both Shanghai (long distance rate > 0.4) and NYC (long distance rate > 0.5) were more likely to commit speeding ($p < 0.05$). The results also showed that the variable of time occupancy rate significantly influenced the speeding violation in NYC ($p = 0.001$) but not in Shanghai.

3.3 GLM Analysis of Determinant Factors: When, Where, and How?

In this section, the GLM analysis was employed to further explore the determinant factors of taxi drivers' committing the speeding violations. The dependent variable y represented the amount of speed travelling over speed limits. The explanatory variables included the driver style characteristics factors and situational factors. The results of GLM analysis are provided in Table 5. The condition number (kappa = 12.09) is less than 100, which indicates that the multi-collinearity effect barely exists in the analysis.

Table 3 Stepwise regression for estimating speeding violations in Shanghai

Characteristics parameters	Ordinary taxi drivers ($N = 2385$)			
	B	*S.E*	*t*	*Sig*
(Intercept)	113.970	11.221	10.157	0.000
Daily working hours				
<16	0[a]	–	–	–
16–17	25.695	6.746	3.809	0.000
18–19	34.895	7.674	4.547	0.000
20–21	36.200	9.183	3.942	0.000
>22	62.954	11.618	5.418	0.000
Daily driving distance				
<250	0[a]	–	–	–
250–299	46.134	9.329	4.945	0.000
300–349	56.203	9.761	5.758	0.000
350–399	61.169	10.222	5.984	0.000
>400	122.111	11.133	10.968	0.000
Daily profit, RMB				
<700	0[a]	–	–	–
700–899	59.704	7.672	7.782	0.000
900–1099	87.671	9.126	9.607	0.000
>1100	140.617	10.592	13.276	0.000
Long distance rate				
<0.3	0[a]	–	–	–
0.3–0.39	11.049	8.342	1.324	0.185
0.4–0.49	16.197	7.831	2.068	0.039
0.5–0.59	18.395	7.854	2.342	0.019
>0.6	34.373	7.180	4.788	0.000
Driving late at night				
No	0[a]	–	–	–
Yes	16.431	4.764	3.449	0.001
Route directness index				
<1.15	0[a]	–	–	–
1.15–1.24	7.941	9.287	0.855	0.393
1.25–1.34	19.970	8.649	2.309	0.021
>1.35	39.509	8.525	4.635	0.000
Adjusted R^2 0.684 **Kappa** 12.70				

Dependent variable y: speeding frequency per day

[a] Set to be 0

Table 4 Stepwise regression for estimating speeding violations in NYC

Characteristics parameters	Ordinary taxi drivers ($N = 4908$)			
	B	*S.E*	*t*	*Sig*
(Intercept)	0.033	0.016	2.066	0.039
Daily working hours				
<8	0[a]	–	–	–
8–10	−0.009	0.014	−0.603	0.546
>10	−0.054	0.018	−.931	0.003
Daily occupancy driving distance				
<100	0[a]	–	–	–
100–119	0.034	0.015	2.364	0.018
>120	0.121	0.019	6.242	0.000
Time occupancy rate				
<0.55	0[a]	–	–	–
0.55–0.65	−0.077	0.012	−6.309	0.000
>0.65	−0.130	0.015	−8.699	0.000
Daily profit, USD				
<300	0[a]	–	–	–
300–399	0.085	0.015	5.682	0.000
>400	0.116	0.023	5.064	0.000
Long distance rate				
<0.3	0[a]	–	–	–
0.3–0.5	0.048	0.012	4.121	0.000
>0.5	0.148	0.013	10.960	0.000
Driving late at night				
No	0[a]	–	–	–
Yes	0.041	0.012	3.389	0.001
Adjusted R^2 0.366 **Kappa** 14.68				

Dependent variable y is a dummy variable with y = 1 if have speeding records or y = 0 otherwise
[a] Set to be 0

It is indicated that drivers with long working hours (e.g., longer than 22 h) did not speed at a high rate of the speed limit ($B = -0.056$, $p = 0.000$), and those with short working hours (e.g., less than 16 h) were more likely to drive at high amount of speed over speed limits. The results also showed that drivers with long daily driving distance (longer than 400 km) were associated with high speeding rate. Although drivers with high-income exceeded the speed limits more often, the result of GLM analysis indicated that these drivers did not drive at high amount of speed over speed limits. Drivers who routinely took long distance trips (LDR > 0.6) were more likely to drive at high speeding rate ($B = 0.009$, $p = 0.000$). The result of GLM also indicated

Table 5 GLM analysis for estimating speeding rate in Shanghai

Variables	*B*	*S.E*	*t*	*Sig*	Variables	*B*	*S.E*	*t*	*Sig*
Intercept	0.206	0.003	80.446	0.000	*Route directness index*				
Daily working hours, h					<1.15	0[a]	–	–	–
<16	0[a]	–	–	–	1.15–1.24	0.007	0.001	7.729	0.000
16–17	−0.032	0.001	−27.241	0.000	1.25–1.34	0.004	0.001	4.6	0.000
18–19	−0.041	0.001	−34.192	0.000	>1.35	0.008	0.001	7.998	0.000
20–21	−0.048	0.001	−34.52	0.000	*Road speed category*				
>22	−0.056	0.002	−32.407	0.000	Group 1	0[a]	–	–	–
Daily driving distance, km					Group 2	0.072	0.001	67.958	0.000
<250	0[a]	–	–	–	Group 3	0.17	0.001	148.454	0.000
250–299	0.009	0.001	7.078	0.000	Group 4	0.254	0.001	204.927	0.000
300–349	0.025	0.001	18.218	0.000	Group 5	0.503	0.001	389.661	0.000
350–399	0.04	0.002	25.437	0.000	*Time occupancy rate*				
>400	0.061	0.002	35.459	0.000	<0.50	0[a]	–	–	–
Daily profit, RMB					0.50–0.60	0.002	0.001	2.583	0.010
<700	0[a]	–	–	–	>0.60	0.009	0.001	9.365	0.000
700–899	−0.012	0.001	−8.007	0.000	*Road length km*				
900–1099	−0.014	0.002	−8.012	0.000	<0.50	0[a]	–	–	–
>1100	−0.012	0.002	−6.061	0.000	0.50–0.99	−0.011	0.002	−4.848	0.000
Long distance trip rate					1.00–1.49	0.042	0.002	19.595	0.000
<0.30	0[a]	–	–	–	1.50–1.99	0.016	0.002	7.131	0.000
0.30–0.39	0.003	0.001	2.914	0.004	>2.00	0.009	0.002	4.556	0.000
0.40–0.49	0.018	0.001	19.841	0.000	*Location index*				
0.50–0.59	0.029	0.001	29.192	0.000	CBD	0[a]	–	–	–
>0.60	0.068	0.001	69.125	0.000	Urban area	0.039	0.001	57.917	0.000
Driving late at night					Suburban area	0.006	0.001	4.614	0.000
No	0[a]	–	–	–	*Road type*				
Yes	0.022	0.001	15.841	0.000	Arterial road	0[a]	–	–	–
Time of day					Secondary road	−0.291	0.001	−434.905	0.000
Daytime	0[a]	–	–	–	**AIC**	−36539			
Night	0.028	0.001	44.978	0.000	**Kappa**	12.09			

Dependent variable y: speeding rate

[a] Set to be 0

that driving late at night and high route directness index also associated with high amount of speeding rate.

The speeding rate of taxi drivers was low when the roads were crowded with low average speed and variation. The speeding rate was high when the traffic was light, and the highest speeding rate occurred on the roads with highest speed variation (Group 5, $B = 0.503$, $p = 0.000$) rather than the roads with highest average speed (Group 4, $B = 0.254$, $p = 0.000$). The temporal data indicated that drivers generally had a higher speeding rate at night ($B = 0.028$, $p = 0.000$). The result of time occupancy rate, another time-varying indicator, indicated that drivers were more likely to commit speeding violations during the period with high taxi demand (time occupancy rate > 0.6, $B = 0.009$, $p = 0.000$).

The road segments which length was 1.0–1.5 km would contribute to the highest speeding rate ($B = 0.042, p = 0.000$). As to location factor, the highest speeding rate was more likely to occur in urban area ($B = 0.039$, $p = 0.000$) and suburban area ($B = 0.006$, $p = 0.000$) than the CBD area ($B = 0.000$). In addition, drivers were less likely to drive at a high speeding rate on secondary roads compared with arterial roads ($B = -0.291, p = 0.000$).

4 Discussion

Taxi drivers in both Shanghai and NYC had high workload and little rest time. The weekly working hours of ordinary taxi drivers in Shanghai and NYC were 65.1 h and 51.9 h, respectively. This finding was consistent with previous studies that the time-on-the-road of taxi driver was often considerable (Dalziel & Job, 1997; Tseng, 2013; Yeh et al., 2015). Daily working hour was another determinant factor for taxi speeding violations in Shanghai and the drivers with higher working hours were more likely to commit speeding violation. However, this finding was not in line with taxi speeding violations in NYC. The reason may be that taxi drivers in Shanghai needed to work much harder than NYC taxi drivers during one shift. The working hours of most taxi drivers in Shanghai were longer than 18 h and over 70% Shanghai taxi drivers worked driving late at night (00:00–06:00) while the average working hours of NYC taxi drivers were 8.5 h per day and less than 40% of them drove late at night or drove early in the morning. Thus, under extreme high workload, taxi drivers in Shanghai were more likely to get angry and exceed the speed limit. In both Shanghai and NYC, the income of speeders and ordinary taxi drivers was similar while speeders tended to have less working hours and more break time than ordinary taxi drivers. It is possible that if taxi drivers desired to get off work earlier and keep the income, they would choose to deliver the passengers to the destination as fast as possible. If they did not earn sufficient income to support their family, they needed to improve their earning by increase working hours, which was consistent with findings in Beijing (Shi et al., 2014) and in Hanoi (La et al., 2013).

Many previous studies pointed out that situational factors are the important predictor of ordinary driver speeding violations (Giles, 2004; Oppenlander, 1966;

Rakauskas et al., 2007; Williams et al., 2006; Zhang et al., 2014). The present study also indicates that situational factors are determining factors for taxi drivers' speeding violations. The speeding rate at night (18:00–06:00) was significantly higher than the daytime (06:00–18:00) and taxi drivers in both Shanghai and NYC were more likely to commit speeding violations between 4:00 AM and 7:00 AM and the worst time period was during 5:00 AM to 6:00 AM. This finding was in line with the previous study (Tseng, 2013; Zhang et al., 2014) which indicated that speeding violations were more likely to take place at night. One of the possible reasons is that traffic condition at night was much better than the daytime. The second reason is that driving at night is less likely to be caught by the police (Tseng, 2013).

It was also found that the speeding rate on the arterial roads was significantly higher than on the secondary roads, which is consistent with extant research that arterial roads in China were the least safe roads (Zhang et al. 2017). The length of the road segment was also an important factor affecting taxi drivers' speeding rate. Taxi drivers were more likely to commit speeding violations at a higher speeding rate on the road segments with 1.0–1.5 km in length. It is possible that without frequently being disrupted by crossings or traffic signal, drivers were able to accelerate to a high travel speed in long road segments. This is consistent with Greibe (2003) that long segments were associated with high crash frequency. In addition, a high speeding rate is more likely to take place outside the inner ring expressways in Shanghai. As indicated in Rakauskas et al. (2007), drivers in such areas may have lower perceptions of the risks and perceive lower value in government-sponsored traffic safety interventions than the central area.

References

Ahie, L. M., Charlton, S. G., & Starkey, N. J. (2015). The role of preference in speed choice. *Transportation Research Part F: Traffic Psychology and Behavior, 30*, 66–73.

Ahlin, F. J. (1979). *An investigation into the consistency of drivers' speed choice* (Ph.D. thesis), Department of Civil Engineering, University of Toronto.

Burns, P. C., & Wilde, G. J. (1995). Risk taking in male taxi drivers: Relationships among personality, observational data and driver records. *Personality and Individual Differences, 18*(2), 267–278.

Cui, J., Liu, F., Hu, J., Janssens, D., Wets, G., & Cools, M. (2016). Identifying mismatch between urban travel demand and transport network services using GPS data: A case study in the fast growing Chinese city of Harbin. *Neurocomputing, 181*, 4–18.

Dalziel, J. R., & Job, R. F. S. (1997). Motor vehicle accidents, fatigue and optimism bias in taxi drivers. *Accident Analysis and Prevention, 29*(4), 489–494.

Giles, M. J. (2004). Driver speed compliance in Western Australia: A multivariate analysis. *Transport Policy, 11*(3), 227–235.

Governor's Traffic Safety Committee. (2014). Speeding and speed limits. Available at: http://www.safeny.ny.gov/spee-ndx.htm. Accessed August 7, 2021.

Greibe, P. (2003). Accident prediction models for urban roads. *Accident Analysis and Prevention, 35*(2), 273–285.

Haglund, M., & Åberg, L. (2000). Speed choice in relation to speed limit and influences from other drivers. *Transportation Research Part F: Traffic Psychology and Behavior, 3*(1), 39–51.

Harré, N., Field, J., & Kirkwood, B. (1996). Gender differences and areas of common concern in the driving behaviors and attitudes of adolescents. *Journal of Safety Research, 27*(3), 163–173.

La, Q. N., Lee, A. H., Meuleners, L. B., & Duong, D. V. (2013). Prevalence and factors associated with road traffic crash among taxi drivers in Hanoi, Vietnam. *Accident Analysis and Prevention, 50*(4), 451–455.

Lam, L. T. (2004). Environmental factors associated with crash-related mortality and injury among taxi drivers in New South Wales, Australia. *Accident Analysis and Prevention, 36*(5), 905–908.

Lefeve, B. A. (1956). Relation of accidents to speed habits and other driver characteristics. *Highway Research Board Bulletin, 120*, 6–30.

Liu, L., Andris, C., & Ratti, C. (2010). Uncovering cabdrivers' behavior patterns from their digital traces. *Computers, Environment and Urban Systems, 34*(6), 541–548.

Mayhew, C. (2000). *Violent assaults on taxi drivers: incidence patterns and risk factors.* Australian Institute of Criminology, Report No. 178.

National Center for Statistics and Analysis. (2013). *Speeding: Traffic safety facts 2011 data.* NHTSA, US Department of Transportation, DOT HS 811 751.

National Research Council (US). Committee for Guidance on Setting, & Enforcing Speed Limits. (1998). *Managing speed: Review of current practice for setting and enforcing speed limits.* National Academy Press.

Newnam, S., Mamo, W. G., & Tulu, G. S. (2014). Exploring differences in driving behavior across age and years of education of taxi drivers in Addis Ababa, Ethiopia. *Safety Science, 68*(10), 1–5.

Oppenlander, J. C. (1966). *Variables influencing spot speed characteristics.* Highway Research Board Special Report, No. 89.

Peltzer, K., & Renner, W. (2003). Superstition, risk-taking and risk perception of accidents among South African taxi drivers. *Accident Analysis & Prevention, 35*(4), 619–623.

Rakauskas, M., Ward, N., Gerberich, S., & Alexander, B. (2007). *Rural and urban safety cultures: Human-centered interventions toward zero deaths in rural Minnesota.* Minnesota Department of Transportation, Mn/DOT 2007-41.

Richard, C. M., Campbell, J. L., Lichty, M. G., Brown, J. L., Chrysler, S., & Lee, J. D. (2013). *Motivations for speeding: Findings report* (Vol. II). NHTSA, US Department of Transportation, DOT HS 811 818.

Royal, D. (2003). *National survey of speeding and unsafe driving attitudes and behaviors: 2002: Findings report* (Vol. II). NHTSA, US Department of Transportation, DOT HS 809 688.

Schroeder, P., Kostyniuk, L., & Mack, M. (2013). *2011 national survey of speeding attitudes and behaviors.* NHTSA, US Department of Transportation, DOT HS 811 865.

Shi, J., Tao, L., Li, X., Xiao, Y., & Atchley, P. (2014). A survey of taxi drivers' aberrant driving behavior in Beijing. *Journal of Transportation Safety and Security, 6*(1), 34–43.

Taxicabs of New York City. (2017). Wikipedia. Available at: https://en.wikipedia.org/wiki/Taxicabs_of_New_York_City. Accessed August 7, 2021.

The Central People's Government of the People's Republic of China. (2012). Provisions on the application and use of motor vehicle driving license. Available at: http://www.gov.cn/flfg/2012-10/09/content_2239595.htm. Accessed August 7, 2021.

Traffic Management Bureau. (2012). *Annual statistical report on road traffic accidents.* Ministry of Public Security.

Tseng, C. M. (2013). Operating styles, working time and daily driving distance in relation to a taxi driver's speeding offenses in Taiwan. *Accident Analysis and Prevention, 52*, 1–8.

Wali, B., Ahmed, A., Iqbal, S., & Hussain, A. (2017). Effectiveness of enforcement levels of speed limit and drink driving laws and associated factors–Exploratory empirical analysis using a bivariate ordered probit model. *Journal of traffic and transportation engineering (English edition), 4(3)*, 272-279.

Williams, A. F., Kyrychenko, S. Y., & Retting, R. A. (2006). Characteristics of speeders. *Journal of Safety Research, 37*(3), 227–232.

Yeh, M. S., Tseng, C. M., Liu, H. H., & Tseng, L. S. (2015). The factors of female taxi drivers' speeding offenses in Taiwan. *Transportation Research Part F: Traffic Psychology and Behavior, 32*, 35–45.

Zhang, G., Yau, K. K. W., & Gong, X. (2014). Traffic violations in Guangdong province of China: Speeding and drunk driving. *Accident Analysis and Prevention, 64*(1), 30–40.

Zhang, K., Sun, D., Shen, S., & Zhu, Y. (2017). Analyzing spatiotemporal congestion pattern on urban roads based on taxi GPS data. *Journal of Transport and Land Use,10(1)*, 675-694.

Chapter 9
Effects of Congestion on Drivers' Speed Choice: Assessing the Mediating Role of State Aggressiveness Based on Taxi Floating Car Data

Abstract Inappropriate cruising speed, such as speeding, is one of the major contributors to the road safety, which increases both the quantitative number and severity of traffic accidents. Previous studies have indicated that traffic congestion is one of the primary causes of drivers' frustration and aggression, which may lead to inappropriate speed choice. In this study, the large taxi floating car data (FCD) was used to empirically evaluate how traffic congestion-related negative moods, defined as state aggressiveness, affected drivers' speed choice. The indirect effect of traffic delay on the cruising speed adjustment through the state aggressiveness was assessed through the mediation analysis. Furthermore, the moderated mediation analysis was performed to explore the effect of driver type, value of time, and working duration on the mediation role of state aggressiveness. The results proved that the state aggressiveness was the mediator of the relationship between travel delays and driving speed adjustment, and the mediation role was different across various driver types. As compared to the aggressive drivers, the normal drivers and the steady drivers tended to behave more aggressively after experiencing non-recurrent congestion during the early stage of the trips. When the value of time was high, steady drivers were more likely to adjust their speed choice although the effect was not statistically significant for other driver types. The validation results indicated that the speed model incorporating state aggressiveness could better predict the travel time than the traditional speed model that only considering the specific expected speed distribution. The prediction results for the manifest indicators of state aggressiveness, such as the maximum speed and the speed deviation, also demonstrated a reasonable reflection of the field data.

Keywords Speed choice · Traffic congestion · State aggressiveness · Mediation analysis · Moderated mediation analysis · Floating car data

1 Introduction

For a long time, speed has become one of the major contributors to road safety, and inappropriate cruising speed, such as speeding, inevitably increases the quantitative number and severity of traffic accidents. In the United States, speeding was

S. Zhong and D. Sun, *Logic-Driven Traffic Big Data Analytics*,
https://doi.org/10.1007/978-981-16-8016-8_9

listed as a contributing factor in nearly one-third of fatal crashes from 2002 to 2011 (National Center for Statistics and Analysis, 2013). Nearly three-quarters of drivers reported that they had driven over the speed limit within the past month (Royal, 2003). In China, more than 90,000,000 speeding offenses were recorded in 2012 and over 7000 victims were dead from speeding-related crashes during the year (Traffic Management Bureau, 2012). Consequently, it is important to understand and predict the speed choice among drivers. Numerous studies have explored the situational determinants of drivers' speed choice, such as road attributions (Giles, 2004; Huang et al., 2018; Yagar & Van Aerde, 1983), traffic conditions (Huang et al., 2018), vehicular parameters (Giles, 2004), and time of day (Oppenlander, 1966; Richard et al., 2013). While these surveys mainly aimed to figure out the exogenous reasons for the speed choice of drivers, it is necessary to quantify how these factors affect the speed choice, so as to predict the speed selection from the psychological process of drivers.

Traffic congestion is one common exogenous source of frustration and aggression (Cœugnet et al., 2013; Shinar, 1998; Shinar & Compton, 2004; Szollos, 2009), which has been frequently overlooked by speed choice studies. According to Cœugnet et al. (2013), congestion-related travel delay can lead to tight time constraints and high time uncertainty, which are two main causal factors for time pressure. Numerous studies have indicated that time pressure is one of the most important factors affecting drivers' speed choice. In a national survey of speeding attitudes and behaviours, the most frequent reason for speeding was "being in a hurry" (Schroeder et al., 2011). The similar conclusion was found in McKenna (2005) that those who claimed frequently exceeding speed limit reported that the time pressure played a significant role in their speed choice. When drivers are under high time pressure, they are more likely to change lanes and increase speeds that can shorten their journey time as much as possible (Peer & Rosenbloom, 2013; Tarko, 2009). A simulator study found that compared to normal drivers, very hurried drivers tended to select higher speeds, accelerate faster after red lights, and accept smaller gaps on left turns (Fitzpatrick et al., 2017). They were more likely to overtake a slow vehicle and run a yellow light (Fitzpatrick et al., 2017). Stern (1998, 1999) first introduced time pressure to explain the dynamics of driver state and choice under different conditions of congestion. Stern (1998) found that when under congestion-related time pressure, many drivers attempt to improve speeds on later sections of the trip. However, Stern (1998, 1999) did not empirically investigate the speed choice after congestion but rather stated behaviour response.

After Stern (1998, 1999), few studies have focused on the effect of traffic congestion on drivers' en-route choice empirically, especially for speed choice. The major reason is the scarcity of reliable driver behaviour-related data. The data sample required to identify and model the speed preference/choice triggered by congestion needs to be sufficiently massive, owing to the randomness of driver state (Hess & Train, 2011). Traditional methods, like traffic survey (Dahlen et al., 2012; Hennessy & Wiesenthal, 1997) and driving simulation (Danaf et al., 2015; Stern, 1999) have limitations to incorporate enough driver sample and represent the real world. Within the last decade, the wide usage of GPS devices allows traffic engineers to collect tremendous amounts GPS derived speed data with high time and space resolution. Some studies have demonstrated the advantages of introducing GPS data into traffic safety research (Huang et al., 2018), named as floating car data (FCD) method. Compared to the survey studies and driving simulator methods, which were frequently used in the speed-safety research, the FCD.method assists to capture the en-route choice of drivers in more details and can represent drivers' speed choice under real scenarios.

1.1 State-Trait Theory

The state-trait theory, first introduced by Cattell and Scheier (1961), points out that the emotional experience of human being can be divided into two parts: a transient mood state and a long-term trait. The theory has been shown especially useful in anxiety research (Spielberger, 1988; Spielberger et al., 1983) and has been widely applied to aggressive driving research since Deffenbacher et al. (1994). The state of driver can be described as emotional feelings of drivers, such as anger or excitement, which can reflect the intra-personal heterogeneity of drivers under different conditions. The trait of driver can be described as a long-term tendency of drivers, such as driving style and the characteristics of drivers, which can reflect the inter-personal heterogeneity of drivers. To date, it has been repeatedly shown in the literatures that both the trait of a driver and the state of a driver have strong relationships with aggressive driving behaviours (Dahlen et al., 2012; Danaf et al., 2015; Hennessy & Wiesenthal, 1997; Huang et al., 2018; Peer, 2010; Peer & Rosenbloom, 2013; Shinar, 1998; Shinar & Compton, 2004; Tseng, 2013). Hennessy and Wiesenthal (1997) applied the state-trait theory to explain the relationship between traffic congestion, driver stress and coping behaviours of drivers. To empirically test and quantify this relationship, this study applies the state-trait theory by including driver state and driver type to account for the effect of congestion on drivers. The congestion-related negative mood is defined as state aggressiveness, which was investigated in detail in Sect. 2.3.1.

When drivers experience heavy congestion condition, they are more likely to exhibit elevated levels of stress, including frustration, irritation, and other negative moods (Feng et al., 2016; Hennessy & Wiesenthal, 1997, 1999; Underwood et al., 1999). In a phone survey, Hennessy and Wiesenthal (1997) found that drivers' level of state stress and aggression in heavy congestion were higher than mild congestion. Using driving anger scale (DAS), Feng et al. (2016) found that traffic congestion was the most anger provoking of the four driving anger subscales for professional drivers in China. Similarly, Underwood et al. (1999) examined the diaries of 100 drivers and found that drivers were more likely to report angry when they were experiencing congestion. They also found that drivers who generally reported high congestion had no tendency to report more driving anger. This can be explained by Stern (1999), as congestion could generally be divided into recurrent congestion and non-recurrent one. Drivers would have less time pressure within recurrent congestion than non-recurrent one because they could arrange the preventive response in advance, such as leaving home earlier. In addition, Shinar and Compton (2004) mentioned that the value of time also played an important role in affecting the feeling of drivers. For example, when the value of time is high, e.g., rush hours, the traffic delay could be very frustrating. Drivers are more likely to drive aggressively during the rush hours than the periods with low value of time, such as non-rush weekdays or weekends. The effect of traffic congestion on the state of drivers and their speed choice may also have significant differences among drivers with various traits, such as cultural norms (Shinar, 1998), sensation-seeking (Peer & Rosenbloom, 2013), time-saving bias, trait stress (Hennessy & Wiesenthal, 1997), and trait anger (Danaf et al., 2015). These personality factors can be well reflected by drivers' daily habitual behaviour, such as preferred speed, which was found to be one of the most important factors in predicting driving anger (Feng et al., 2016).

1.2 Aim and Objective

The present study was to empirically analyse and quantify how did the traffic delay affect drivers' speed choice. The driving style and the state of driver were identified based on the large taxi FCD dataset in Shanghai. While the results of taxi drivers may not apply to the non-professional drivers, the approach proposed in this study can be used to investigate these drivers if the corresponding FCD data were available. Moreover, investigating the speed choice of taxi drivers is also important because taxi drivers tend to have more opportunities to be involved in a speed violation (Huang et al., 2018; Tseng, 2013).

More specifically, the main objective of the current study is to test and model the role of driver state aggressiveness and driver trait in the relationship between congestion-related delay on the first half of the trips and drivers' speed adjustment on the second half of the trips. The three sub-objectives are:

1. To classify drivers into different driver types according to their habitual speeds in different traffic conditions.
2. To test the mediation role of driver state aggressiveness when analysing the influence of traffic delays on driver speed choice.
3. To test and incorporate driver type, the value of time, and working hours through a moderated mediation process, so as to explain and predict drivers' speed choice with travel delays.

The remainder of the chapter is structured as follows: Sect. 2 proposes the methods of classifying drivers along with the mediation analysis process and moderated mediation analysis process. Next, the model implementation and validation process are presented in Sect. 3. Section 4 discusses the results and demonstrates the limitations and the future work. Finally, conclusions are provided in Sect. 5.

2 Statistical Methods

This section presents an empirical data-based approach to detect and model the mediation effect of state aggressiveness on the relationship between traffic congestion and drivers' speed choice on urban roads. The first step relates to the driver type classification, including habitual speed identification and driver type clustering. Next, a mediation analysis is performed to test the mediation role of state aggressiveness when analysing the influence of congestion on drivers' speed choice. Finally, a moderated mediation analysis is carried out to identify whether the mediation role of state aggressiveness is different for various types of drivers. The remainder of this section describes more details related to these three steps.

2.1 Data Description

The FCD data was collected in 2016 from Shanghai Qiang-Sheng Taxi Company, one of the largest taxi companies in Shanghai. The entire dataset contains more than 10,000 taxi drivers' sequential trajectories and operation statuses, including taxi ID, data, measurement time, longitude, latitude, speed (km/h), and operation status identifier (1 for vacant/0 for hired). The daily dataset has over one hundred million items (e.g., 114, 633, 142 items), and records every taxi's ID, speed, location and other operation information in every 10 s interval. In Shanghai, the speed limits of most arterial and secondary roads are 60 km/h and 40 km/h, respectively. As a result, the records with the speeds higher than 120 km/h were removed from the dataset. Additional detailed explanation of the dataset and data cleaning process can be found in Huang et al. (2018).

After the data cleaning process, 10,488 taxi drivers were selected as the research subjects. The records from March 21 to 25, 2016 (Monday to Friday) were chosen to calculate the habitual behaviours of drivers and the trajectories on March 29, 2016 (Tuesday) were used for model calibration. The trips were identified by the operation status identifier and the distance of each trip was calculated by the relative location of each record set. Only trips with distance over 10 km and average speed of the first half trip lower than drivers' habitual speeds were selected. Finally, 16,682 effective trips (nearly 80% of all effective trips) were used to calibrate the model and the other 4151 effective trips (nearly 20% of all effective trips) were used for validation.

2.2 *Driver Type Classification*

2.2.1 Habitual Speed Identification

In general, drivers behave consistency in speed choice (Haglund & Åberg, 2000) and tend to have habitual speeds under different traffic conditions. In this research, the habitual speeds of each taxi driver were firstly calculated according to their historical average speeds from March 21, 2016 to March 25, 2016. Figure 1 presents the distributions of habitual speeds of taxi drivers in six different time periods, representing the various traffic conditions in Shanghai: AM peak (from 07:00 to 09:00), noon off-peak (from 11:00 to 13:00), afternoon off-peak (from 15:00 to 17:00), PM peak (from 17:00 to19:00), evening off-peak (from 19:00 to 21:00) and midnight off-peak (from 21:00 to 23:00).

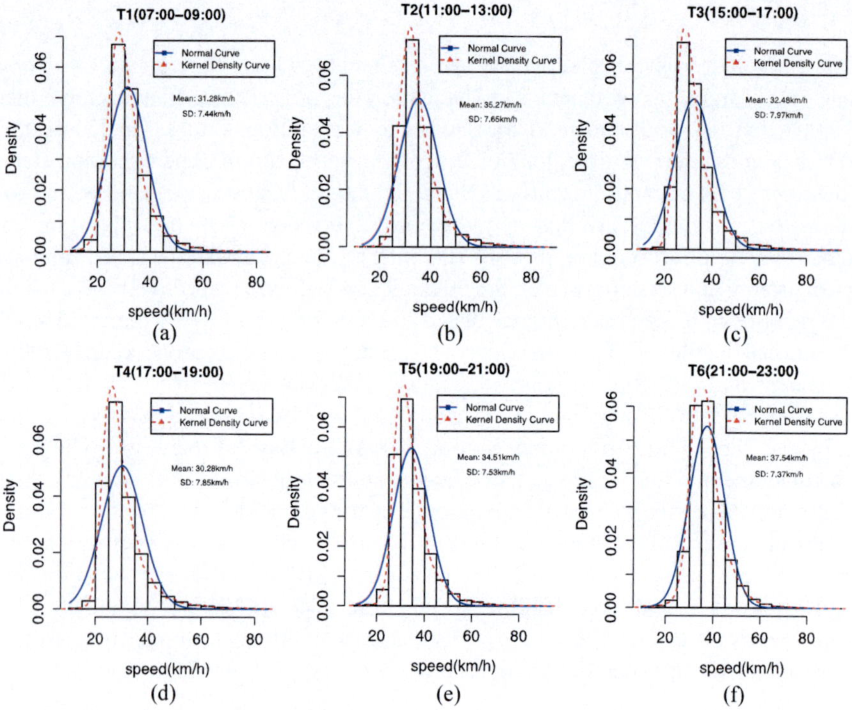

Fig. 1 Speed distributions in six time periods, **a** 07:00–09:00, **b** 11:00–13:00, **c** 15:00–17:00, **d** 17:00–19:00, **e** 19:00–21:00, **f** 21:00–23:00

2.2.2 Gaussian Mixture Model

Previous studies have indicated that the driving anger provoked by congestion was significantly correlated with drivers' preferred speeds (Feng et al., 2016). Some demographics, such as age, driving experience, and driving mileage in the last year, were not significantly correlated with the congestion-related driving anger for professional drivers in China (Feng et al., 2016). In this study, the driving style of each driver was detected and classified according to their habitual speeds from different traffic conditions. To do this, we used Gaussian Mixture Model (GMM), which has been shown very effective in modelling and classifying driving speed data (Koh & Kang, 2015; Park et al., 2010). The habitual speeds of drivers within the six time-periods would be regarded as different dimensions to perform cluster analysis. The expectation–maximization (EM) algorithm, initially named by Dempster et al. (1977), is used to estimate the GMM.

The appropriate number of components and the different type of multivariate mixture in GMM are determined by the information-based criteria. It has been shown that the Bayesian information criterion (BIC) is more reliable for EM algorithm to find the maximum mixture likelihood (Schwarz, 1978) and it can be calculated using:

$$2 \log p(x|M) + \text{constant} \approx 2l_M(x, \widehat{\theta}) - m_M \log(n) \equiv BIC \tag{1}$$

where, $p(x|M)$ is the likelihood of the data for the model M, $l_M(x, \widehat{\theta})$ is the maximized mixture log-likelihood for the model and m_M is the number of independent parameters to be estimated.

While the number of clusters is not considered as an independent parameter to compute the BIC, the integrated complete-data likelihood (ICL) criteria is also proposed for model selection. The ICL can be calculated as:

$$ICL = BIC + 2\sum_{i=1}^{n}\sum_{k=1}^{G} c_{ik} \log(Z_{ik}) \tag{2}$$

where, Z_{ik} is the conditional probability that x_i arises from the kth mixture component, and $c_{ik} = 1$ if x_i is clustered into k and 0 otherwise. ICL penalizes the BIC through an entropy term which measures clusters overlap.

In Gaussian mixture model, clusters are centered at the mean vector μ_i and ellipsoidal with different geometric features, such as volume, shape, and orientation, determined by the covariance matrix $\sum_i$. In the model-based clustering approach, the type of multivariate mixture components with best clustering performance (e.g., using BIC and ICL as criteria) will be chosen.

2.3 *Mediation Analysis*

2.3.1 State Aggressiveness Identification

Several studies have explored the negative moods, such as state stress (Hennessy & Wiesenthal, 1997) and state anger (Underwood et al., 1999), when experiencing heavy congestion. In this research, the congestion-related negative mood is defined as state aggressiveness (SA), a latent variable, which is manifested in risky or aggressive driving behaviour, such as speeding and high acceleration. Based upon existing literature (Danaf et al., 2015), certain measures of SA can be extracted from these actions and used as manifest indicators, which can be expressed as follows:

$$I_l = \alpha_l + \lambda_l SA + w_l \quad (3)$$

where, I_l represents a vector of three different manifest indicators of state aggressiveness: I_1 is the maximum speed indicator, I_2 is the maximum acceleration indicator, and I_3 is the standard deviation of speed indicator; α_l and λ_l are parameters that need to be estimated. One of the λ_l terms need to be normalized to 1 for identification purpose; w_l represents an error term with standard deviation σ_{w_l} to be estimated.

2.3.2 Mediation Analysis Process

To test the role of congestion-related state aggressiveness in drivers' speed choice behaviour, a mediation analysis approach was conducted in this study. Mediation analysis is to establish the indirect effect of independent variables (i.e., congestion-related delay) on an outcome variable (i.e., speed adjustment) through a mediating variable (i.e., state aggressiveness). Since the state aggressiveness is a latent variable, the mediation process needs to be estimated by structural equation modelling (SEM).

In this research, the indirect effect of speed delay on the adjustment of travel speed for the second half trip through the mediation role of drivers' state aggressiveness was tested using the mediation analysis. The treatment variable was the congestion-related travel delay, which was calculated as the speed delay of the first half trip divided by drivers' habitual speed during the same time of day in the past. This treatment variable can reflect the level of the non-recurrent congestion. The dependent variable was the speed adjustment factor, which was calculated as the speed adjustment during the second half trip divided by the average speed during the first half trip.

2.4 *Moderated Mediation Analysis*

To validate and incorporate driver type into our analysis, a moderation analysis was performed after the mediation analysis. Moderation process refers to a sequence of causal relations by which moderator factors (e.g., individual difference or contextual variable) influence the strength and/or direction of the relationship between the treatment variable and the dependent variable (Muller et al., 2005). Moderation process can be combined with mediation process into a moderated mediation model if the moderator factor can significantly influence the mediation process that intervenes between the treatment and the outcome. In other words, the moderated mediation analysis is to estimate the conditional indirect effect of congestion on speed choice through state aggressive at values of moderator, e.g., driver type. The moderated mediation analysis can be conducted using multiple-group SEM. Models with equality constraints across groups would be compared with models with no regression parameters constraints across groups (Hayes, 2009).

The indicators of value of time and working hour were also included to better explain and predict taxi drivers' state aggressiveness. The value of time has been indicated significantly affecting the aggressive behaviour under congestion-related delays for non-professional drivers (Shinar & Compton, 2004). However, the value of time of taxi drivers is different from non-professional drivers. In this research, the average incomes in different time of day were used to estimate the value of time for taxi drivers.

3 Model Implementation and Validation

3.1 Model Implementation

3.1.1 Driver Type Classification

We first fitted the habitual speed data of drivers with increasing number of clusters from K = 1 to 9 using 14 different multivariate mixture models respectively. Using Eqs. (1) and (2), the information-based criteria, BIC and ICL, were calculated for each model. Figure 2 provides the results of analysis for numbers from 2 to 9. A large ascend is found from BIC (2) to BIC (3), which means by splitting 2 clusters to 3 clusters, the value of BIC increases significantly, while by splitting to 4 clusters, the value of BIC does not increase as much. When the cluster number is larger than 3, the value of BIC does not increase too much. According to the ICL criteria, both K = 2 and 3 are preferred. Therefore, using the BIC criteria together with ICL criteria, the appropriate number of clusters used is recommended as 3. The appropriate type of multivariate mixture is recommended as VEE (ellipsoidal, equal shape, and orientation) because both values of BIC and ICL for VEE of cluster number 3 are the largest (−403,288.9 for BIC and −407,055.5 for ICL).

Drivers can be categorized using GMM. Based on the features of habitual speeds within six different periods, as shown in Fig. 3, the following three types of drivers were identified:

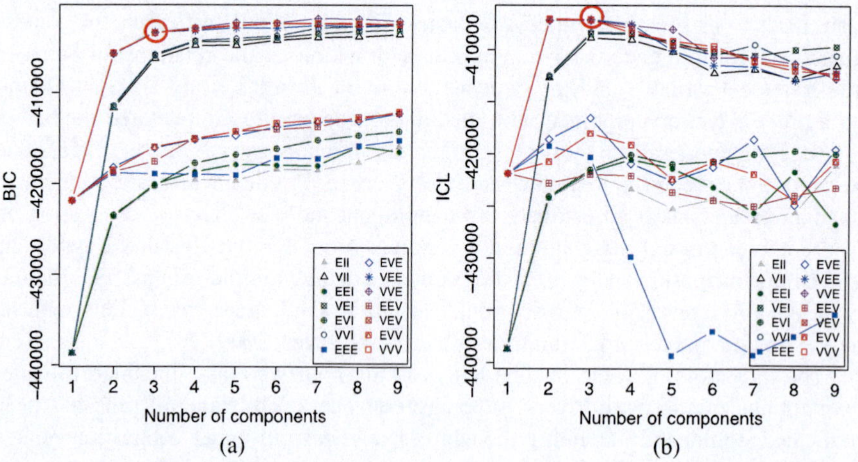

Fig. 2 Results for clustering with different number of clusters and different type of multivariate mixtures, **a** *BIC* criterion, **b** *ICL* criterion

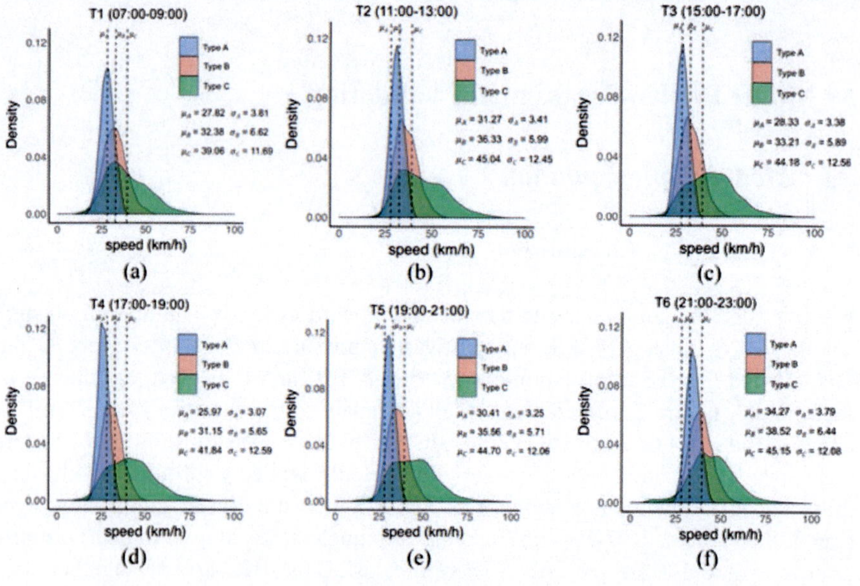

Fig. 3 Speed distributions for different types of drivers during six time periods: **a** 07:00–09:00, **b** 11:00–13:00, **c** 15:00–17:00, **d** 17:00–19:00, **e** 19:00–21:00, **f** 21:00–23:00

- **Type A**: Drivers in this group would not change expected speeds under most situations. They are inclined to keep the habitual speeds (not too fast) in general. We define drivers in this group as steady drivers. The total number of drivers in this group is 4401.

- **Type B**: Drivers in this group would like to get better speed advantages for some situations under low risk. Compared to Type A, drivers in this group are more willing to change lanes to obtain speed advantages during the six different time periods, and consequently are somewhat more aggressive. We define drivers in this group as normal drivers. The total number of drivers in this group is 4824.
- **Type C**: Compared to Type B, drivers in this group have higher habitual speeds and speed variations. They are more ambitious in getting speed advantages and would take increasing risks under good traffic conditions. We define drivers in this group as aggressive drivers. The total number of drivers in this group is 1263.

3.1.2 Mediation Analysis

The mediation process was estimated in R package lavaan using maximum likelihood estimation (Rosseel, 2012). The model converged normally after 107 iterations. The estimation results are presented in Table 1.

The results demonstrated a reasonable model fit according to the SEM goodness-of-fit measures: Comparative Fit Index (CFI) = 0.998; Tucker-Lewis Index (TLI) = 0.994; Root Mean Square Error of Approximation (RMSEA) = 0.035; Root Mean Square Residual (SRMR) = 0.011. Rules of thumb guidelines are that CFI > 0.95, TLI > 0.95, RMSEA < 0.05, and SRMR < 0.05, which represent a good fitting model.

In the regression model without the mediator, the estimate of the causal path from congestion-related delay to speed adjustment was significant (c = 1.548, $p < 0.001$). In the SEM model with the mediator, both estimated paths (path a and path b) for the indirect effect were statistically significant. Besides, compared to the path c of the regression model, the estimate of the direct effect (path c′) in the SEM model was closer to zero. Although path c′ was still significant in the SEM model, a significant deduction (from $c = -1.548$ to $c' = -0.603$) can still demonstrate that the given mediator was indeed potent.

3.1.3 Moderated Mediation Analysis

To model drivers' speed choice considering congestion-related delays across different driver types, the multi-group structural equation modelling was employed. The baseline model (M1), without any regression parameter constraints, yielded $\chi^2(30) = 294.86$, $p < 0.001$, CFI = 0.992, TLI = 0.986, RMSEA = 0.040, SRMR = 0.014. The second model (M2) developed with two regression parameters (SA ~ Congestion-related delays and SA ~ Speed adjustment) constrained to be equal across groups, yielded $\chi^2(36) = 598.84$, $p < 0.001$, CFI = 0.984, TLI = 0.975, RMSEA = 0.053, SRMR = 0.030. Both M1 and M2 have a good model fit. The chi-square difference test demonstrated that there was a significant difference between M1 and M2, with $\Delta\chi^2(6) = 303.97$, $p < 0.001$, indicating that the indirect effect of state aggressiveness differs across driver groups.

Table 1 Unstandardized estimation results for driver state mediation process ($n = 16{,}682$ *trips*)

Parameter/variable	Parameter estimate	Standard error	*t*-test	*p*-value
1. **Regression model: delay—speed adjustment** R^2: 0.1977 $F(1, 16{,}680) = 4111$, $p < 0.001$				
β_0 (Intercept)	0.057	0.006	9.168	<0.001
c (Delay—speed adjustment)	−1.548	0.024	−64.121	<0.001
2. **Structural equation model (SEM): speed reduction-state—speed adjustment**				
Driver state measurement				
α_1 (Constant—Max. speed)	8.218	0.125	65.992	<0.001
α_2 (Constant—Sd. of speed)	2.779	0.045	62.189	<0.001
α_3 (Constant—Max. acceleration)	−0.031	0.018	−1.753	<0.001
λ_1 (Factor loading—Max. speed)	1			
λ_2 (Factor loading—Sd. of speed)	0.373	0.003	138.242	<0.001
λ_3 (Factor loading—Max. acceleration)	0.011	0.001	8.108	<0.001
Mediation process				
β_1 (Intercept)	0			
β_2 (Intercept)	0			
a (Delay—*SA*)	−34.922	0.741	−47.144	<0.001
b (*SA*—speed adjustment)	0.027	0.000	110.025	<0.001
c' (Delay—speed adjustment)	−0.603	0.019	−31.726	<0.001
Variance				
σ_{w1} (Max. speed)	60.339	1.176	53.849	<0.001
σ_{w2} (Standard deviation of speed)	6.183	0.194	41.710	<0.001
σ_{w3} (Max. acceleration)	5.190	0.057	91.286	<0.001
Delay	0.022	0.000	91.329	<0.001
SA (State aggressiveness)	168.344	2.536	66.385	<0.001
Speed adjustment	0.092	0.001	74.126	<0.001
Goodness-of-fit measures				
Comparative fit index (CFI)		0.998		
Tucker-Lewis index (TLI)		0.994		
Root mean square error of approximation (RMSEA)		0.035, 90% CI (0.029, 0.042)		
Root mean square residual (SRMR)		0.011		

Figure 4 presents the multi-group SEM results (M1) for three driver types. From the results, we found that the effect of different indicators varied across different driver types. The effect of congestion-related delay to state aggressiveness was significant for all three driver types ($p < 0.001$). The influence of traffic-related delay for the steady drivers ($\beta = 40.823$, SE = 1.224) and the normal drivers ($\beta = -41.185$, SE = 1.041) was greater than the aggressive drivers ($\beta = -28.534$, SE = 2.146).

The effect of value of time is significant for the steady drivers ($\beta = 0.024$, SE = 0.007, z-value = 3.368, $p = 0.001$) while the effect was not significant enough for the normal drivers ($\beta = -0.013$, SE = 0.007, z-value = 1.878, $p = 0.060$) and even negative for the aggressive drivers ($\beta = -0.043$, SE = 0.017, z-value = −2.536, $p = 0.011$). The results also indicated that working hour would significantly affect aggressive behaviours of the steady drivers ($\beta = -0.312$, SE = 0.037, z-value = 8.369, $p = 0.000$) and the normal drivers ($\beta = 0.109$, SE = 0.035, z-value = 3.149, $p = 0.002$), while it was not significant for the aggressive drivers ($\beta = -0.075$, SE = 0.087, z-value = −0.860, $p = 0.390$).

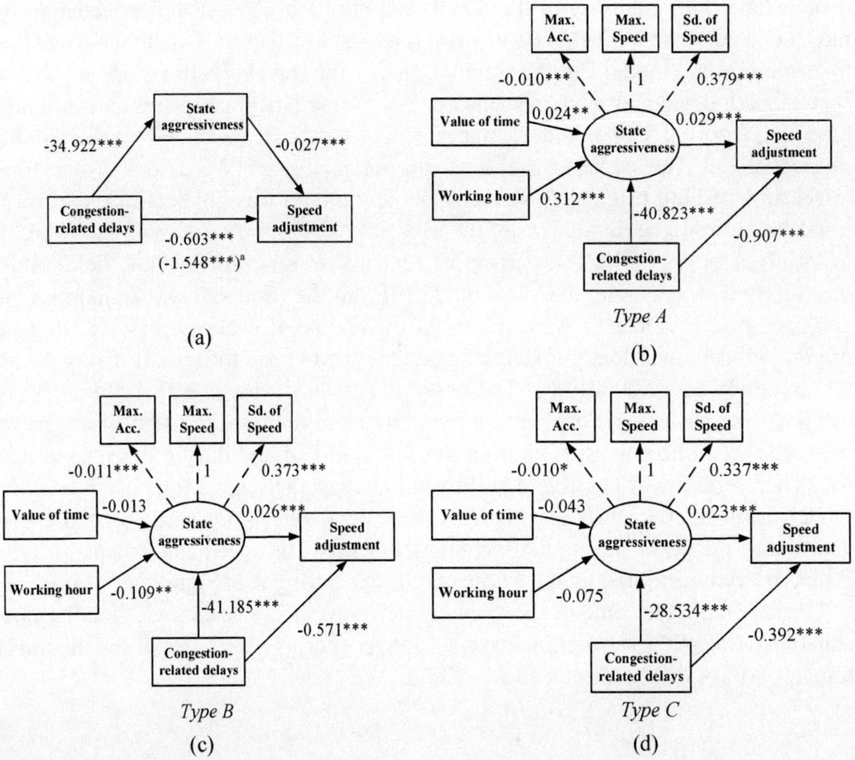

Fig. 4 Mediation effect of "state aggressiveness" for all drivers (**a**) and the results of multi-group SEM (**b–d**) ($^*p < 0.01$; $^{**}p < 0.005$; $^{***}p < 0.001$; numbers on arrows are unstandardized regression coefficients; [a] coefficient of the regression model)

3.2 Model Validation

The purpose of validation is to test and determine the extent to which the new model replicates the real system under different traffic conditions (Sun & Kondyli, 2010). This section presents the comparisons between the model result and the field data. The models were implemented in 4151 trips from different time periods (07:00–09:00, 11:00–13:00, 15:00–17:00, 17:00–19:00, 19:00–21:00, and 21:00–23:00). All trips used to validate the results had speed delays during the first half of trip. Four measures of performance, average travel time, standard deviation of speed, maximum speed, and maximum acceleration were selected to evaluate the model performance.

3.2.1 Comparison of the Overall Travel Time

In most traffic simulation packages, the intra-individual variability of drivers is kept constant or has tiny oscillation based on the desired speed of drivers. To compare and demonstrate the robustness of the new speed model proposed, the expected speed model embedded in VISSIM (PTV, 2014) was employed as the benchmark (referred to as the original model). In the original model, the travel speeds of drivers within specific period were normally distributed. The average habitual speeds and variance, identified from each driver's daily traces, were used as the mean and variance in the original model. The average travel time was obtained from GPS field data. The total travel time of both original model and the new model was predicted by the travel information from the first halves of the trips.

Figure 5 presents the comparison of total travel time among the "field data", the "original model", and the "new model". From the results, it was found that the "original model" tended to overestimate the travel time. However, it appeared that the original model consistently underestimated the travel time for type B drivers (with the exception of 19:00–21:00). The possible reason of this was that although the trip length of each single trip was more than 10 km, there still existed a correlation between the traffic conditions during the first half trip and during the second half trip. Therefore, it was possible that the predicted average travel time using original model could be frequently less than the field data, since congestion-related speed delays may also exist during the second half of trip. The speed adjustments of type B drivers were restricted by the traffic conditions during the second half of trip.

Overall, the travel time of the "new model" was closer to the field data in most situations, except the travel time for type B drivers during 19:00–21:00 and the travel time for type A drivers during 11:00–13:00.

3.2.2 Comparison of the Aggressive State

The three indicators of aggressiveness state were obtained from the GPS data and were compared with the "new model." The comparative results of the "new model"

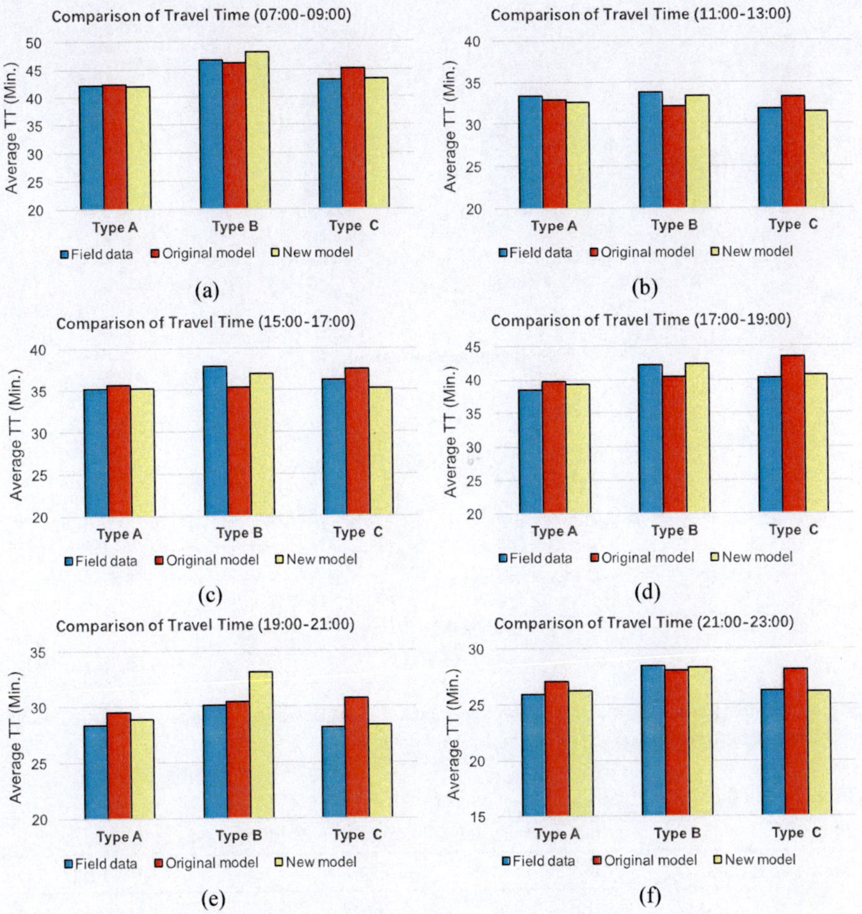

Fig. 5 Comparison of the average travel time, **a** 07:00–09:00, **b** 11:00–13:00, **c** 15:00–17:00, **d** 17:00–19:00, **e** 19:00–21:00, **f** 21:00–23:00

and the "field data" are provided in Fig. 6, in terms of the maximum speed, the standard deviation of speed, and the maximum acceleration. The prediction results were similar to the field data. The "new model" tended to slightly overestimate the maximum speed and underestimate the standard deviation of speed.

The goodness-of-fit measures were also employed to evaluate the performance of the new model. In this study, mean error (ME), mean percent error (MPE), root mean square error (RMSE), and root mean square percent error (RMSPE) were used to quantify the overall error of the model results (Sun & Kondyli, 2010; Toledo et al., 2007), as presented in Table 2. It was shown that the MPEs for both indicators of maximum speed and standard deviation of speed were less than 5% and the RMSPEs for both were less than 50%, which indicated a reasonable presentation of the field data. However, the RMSPEs for maximum acceleration were larger than 50% for all

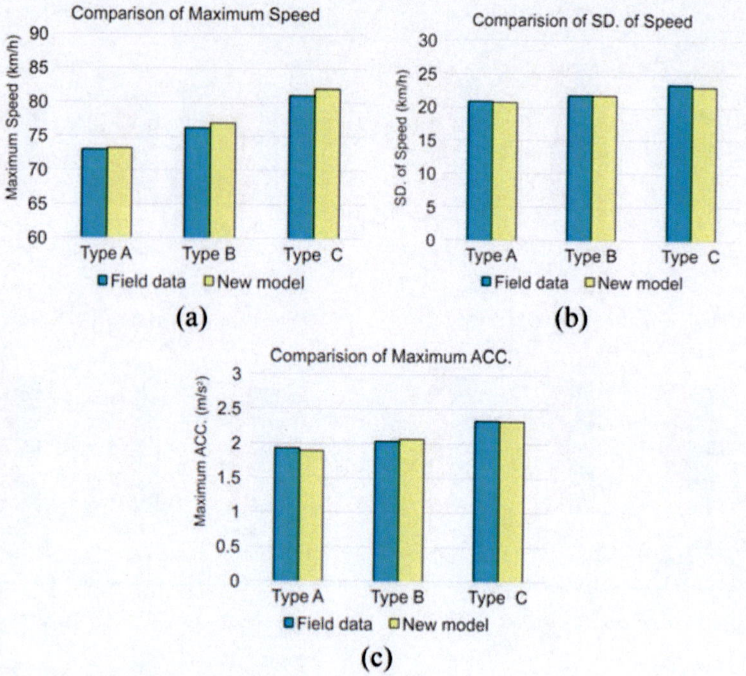

Fig. 6 Comparisons of the **a** maximum speed, **b** standard deviation of speed, and **c** maximum acceleration between the new model and the field data

Table 2 Goodness-of-fit statistics for indicators of driver state

Indicator	Statistical measures	Driver type A	Driver type B	Driver type C
Max. speed (km/h)	ME	0.21	0.71	1.02
	MPE	2.74%	3.78%	4.75%
	RMSE	14.18	15.46	16.15
	RMSPE	21.60%	23.64%	24.97%
SD. of speed (km/h)	ME	−0.14	−0.05	−0.34
	MPE	3.73%	4.79%	3.51%
	RMSE	4.70	5.44	6.12
	RMSPE	26.22%	30.42%	30.51%
Max. acceleration (m/s^2)	ME	−0.04	0.03	−0.01
	MPE	36.31%	44.05%	50.75%
	RMSE	1.99	2.33	2.89
	RMSPE	137.33%	149.67%	192.04%

three types of drivers, which indicated an inaccurate prediction performance. The calculation of acceleration was based on the GPS data recorded in every 10 s, which may result in the poor presentation of aggressiveness-related acceleration.

4 Discussion

The results of this study, with respect to the effect of congestion on drivers' speed choice, were expected. Drivers' speed response to congestion would be mediated by state aggressiveness and the mediation role was different across different driver types. These empirical findings supported the survey findings by Underwood et al. (1999) and the simulation findings by Stern (1999). One point should be mentioned in this study was that the steady drivers and the normal drivers tended to behave more aggressively and increase their driving speeds after experiencing the non-recurrent congestion. However, influence of congestion-related delay on the aggressive drivers was not as great as the steady drivers and the normal drivers. Besides, when the value of time was high, the steady drivers tended to behave more aggressively while the value of time would not significantly affect the normal drivers and the aggressive drivers. A possible explanation for these results was that the aggressive drivers had the predisposition to act aggressive behaviours regardless of the environmental triggers. They would get better speed advantages if they had a chance. For the normal drivers and steady drivers, their behaviours were more likely to be triggered by negative emotions, such as anger and frustration (Berdoulat et al., 2013). These negative emotions could be easily provoked by non-recurrent congestion (Stern, 1999). The steady drivers were not likely to behave aggressively in most situations and they had tendency to keep steady average speeds during the trips. However, if they experienced non-recurrent congestion in the early stage of the trips and the value of time was high for them, they would also act aggressively and adjust their speed choice, especially when they already had long working hours. This was in line with Zhong et al. (2012) that steady drivers could better judge the traffic conditions and were more likely to adjust their previous behaviours and adapt to the new conditions.

Although the results are promising, the proposed approach indeed has some theoretical and practical limitations. First, the state aggressiveness is a dynamic variable, which can change continuously with time. However, in this study, the trips were divided into two sections and only the variations between these two sections were tested. Further modelling needs to shorten the interval and use dynamic hybrid approach to model drivers' speed choice. Secondly, this study did not incorporate the information of road condition and other environmental factors, such as speed limit, road type, weather, traffic condition, etc. Besides, although the trip length of each single trip was more than 10 km, there still existed a correlation between the traffic condition during the first half trip and during the second half trip, which may lead to a bias to predict the speed adjustment. Further studies could empirically analyse the speed choice or speeding on specific scenarios after experiencing the non-recurrent congestion. Thirdly, due to the limitation of data, this study only considered taxi drivers in Shanghai and the results may only apply to taxi drivers in Shanghai. Future research could focus on applying this approach to analyse and model the speed choice of other un-occupational drivers. Finally, the model defined an adjusted definition of state aggressiveness and used three manifest indicators to measure drivers' state aggressiveness, which was considered as a latent variable in this study. These three indicators also have been used in Danaf et al. (2015) to indicate the dynamic state anger of drivers. However, as mentioned in Danaf et al. (2015), these indicators appeared to be directly correlated with the observed speed adjustments of the drivers. The problem could be improved if using more advanced technologies with higher location accuracy. More indicators of drivers' aggressiveness state could be measured, such as lane changing frequency, gap acceptance, and lateral speeds.

5 Conclusions

The general objective of this study was to test and model the role of driver state aggressiveness and driver type in the relationship between congestion-related delay on the first half of trips and drivers' speed adjustment on the second half of trips. The results of mediation process showed that the congestion-related state aggressiveness was the mediator of the influence of travel delay on driver speed adjustment. The mediation role of the state aggressiveness was different across driver types, which was indicated by the moderated mediation process. The validation results indicated that by incorporating state aggressiveness, the new speed model has the ability to represent the travel time more realistically than the method that only considering the specific expected speed distribution. The proposed model can also reasonably predict indicators of aggressive behaviours, such as maximum speed and standard deviation of speed.

This research is one of the first studies that has empirically demonstrated the effect of congestion on drivers' speed choice through the mediation effect of state aggressiveness. The intra-personal heterogeneity of drivers may modify the relationship between the independent variables and behavioural outcome in driving behaviour research. Future research is suggested to add the intra-personal heterogeneity of drivers, affected by the dynamic driving environment, into the behavioural framework, instead of simply treating it as a constant variable or random variable.

References

Berdoulat, E., Vavassori, D., & Sastre, M. T. M. (2013). Driving anger, emotional and instrumental aggressiveness, and impulsiveness in the prediction of aggressive and transgressive driving. *Accident Analysis and Prevention, 50*, 758–767.

Cattell, R. B., & Scheier, I. H. (1961). The meaning and measurement of neuroticism and anxiety. *Psychosomatics, 135*(4), 495–496.

Cœugnet, S., Naveteur, J., Antoine, P., & Anceaux, F. (2013). Time pressure and driving: Work, emotions and risks. *Transportation Research Part F: Traffic Psychology and Behavior, 20*, 39–51.

Dahlen, E. R., Edwards, B. D., Tubré, T., Zyphur, M. J., & Warren, C. R. (2012). Taking a look behind the wheel: An investigation into the personality predictors of aggressive driving. *Accident Analysis and Prevention, 45*(1), 1–9.

Danaf, M., Abou-Zeid, M., & Kaysi, I. (2015). Modeling anger and aggressive driving behavior in a dynamic choice-latent variable model. *Accident Analysis and Prevention, 75*, 105–118.

Deffenbacher, J. L., Oetting, E. R., & Lynch, R. S. (1994). Development of a driving anger scale. *Psychological Reports, 74*(1), 83–91.

Dempster, A. P., Laird, N. M., & Rubin, D. B. (1977) Maximum likelihood from incomplete data via the EM algorithm. *Journal of the Royal Statistical Society. Series B (Methodological), 39*(1), 1–38.

Feng, Z., Lei, Y., Liu, H., Kumfer, W. J., Zhang, W., Wang, K., & Lu, S. H. (2016). Driving anger in china: A case study on professional drivers. *Transportation Research Part F: Traffic Psychology and Behavior, 42*, 255–266.

Fitzpatrick, C. D., Samuel, S., & Knodler, M. A., Jr. (2017). The use of a driving simulator to determine how time pressures impact driver aggressiveness. *Accident Analysis and Prevention, 108*, 131–138.

Giles, M. J. (2004). Driver speed compliance in Western Australia: A multivariate analysis. *Transport Policy, 11*(3), 227–235.

Haglund, M., & Åberg, L. (2000). Speed choice in relation to speed limit and influences from other drivers. *Transportation Research Part F: Psychology and Behavior, 3*(1), 39–51.

Hayes, A. F. (2009). Beyond Baron and Kenny: Statistical mediation analysis in the new millennium. *Communication Monographs, 76*(4), 408–420.

Hennessy, D. A., & Wiesenthal, D. L. (1997). The relationship between traffic congestion, driver stress and direct versus indirect coping behaviors. *Ergonomics, 40*(3), 348–361.

Hennessy, D. A., & Wiesenthal, D. L. (1999). Traffic congestion, driver stress, and driver aggression. *Aggressive Behavior, 25*(6), 409–423.

Hess, S., & Train, K. E. (2011). Recovery of inter- and intra-personal heterogeneity using mixed logit models. *Transportation Research Part B: Methodological, 45*(7), 973–990.

Huang, Y., Sun, D., & Tang, J. (2018). Taxi driver speeding: Who, when, where and how? A comparative study between Shanghai and New York City. *Traffic Injury Prevention, 19*(3), 311–316.

Koh, D. W., & Kang, H. B. (2015). Smartphone-based modeling and detection of aggressiveness reactions in senior drivers. In *Intelligent Vehicles Symposium (IV)*, Seoul. Korea.

McKenna, F. P. (2005). What shall we do about speeding-education? In *International Conference of Traffic and Transport Psychology*, Nottingham, England.

Muller, D., Judd, C. M., & Yzerbyt, V. Y. (2005). When moderation is mediated and mediation is moderated. *Journal of Personality and Social Psychology, 89*(6), 852.

National Center for Statistics and Analysis. (2013). *Speeding: Traffic safety facts 2011 data.* NHTSA, Report No. DOT HS-811-751.

Oppenlander, J. C. (1966). *Variables influencing spot speed characteristics.* National Academy of Sciences. Highway Research Board Special Report No. 89.

Park, B. J., Zhang, Y., & Lord, D. (2010). Bayesian mixture modeling approach to account for heterogeneity in speed data. *Transportation Research Part B: Methodological, 44*(5), 662–673.

Peer, E. (2010). Speeding and the time-saving bias: How drivers' estimations of time saved in higher speed affects their choice of speed. *Accident Analysis and Prevention, 42*(6), 1978–1982.

Peer, E., & Rosenbloom, T. (2013). When two motivations race: The effects of time-saving bias and sensation-seeking on driving speed choices. *Accident Analysis and Prevention, 50*, 1135–1139.

PTV. (2014). VISSIM 8.0 user manual. www.ptvgroup.com, Accessed November 30, 2017.

Richard, C. M., Campbell, J. L., Lichty, M. G., Brown, J. L., Chrysler, S., & Lee, J. D. (2013). *Motivations for speeding: Findings report* (Vol. II). NHTSA, U.S. Department of Transportation. DOT HS-811-818.

Rosseel, Y. (2012). lavaan: An R package for structural equation modeling. *Journal of Statistical Software, 48*(2), 1–36.

Royal, D. (2003). *National survey of speeding and unsafe driving attitudes and behaviors: 2002: Findings report* (Vol. II). NHTSA. Report No. DOT HS-809-688.

Schroeder, P., Kostyniuk, L., & Mack, M. (2011) *National survey of speeding attitudes and behaviors*. NHTSA, 2013. Report No. DOT HS-811-865.

Schwarz, G. (1978). Estimating the dimension of a model. *The Annals of Statistics, 6*(2), 461–464.

Shinar, D. (1998). Aggressive driving: The contribution of the drivers and the situation. *Transportation Research Part F: Traffic Psychology and Behavior, 1*(2), 137–160.

Shinar, D., & Compton, R. (2004). Aggressive driving: An observational study of driver, vehicle, and situational variables. *Accident Analysis and Prevention, 36*(3), 429–437.

Spielberger, C. D. (1988). *Manual for the state-trait anger expression inventory (STAXI)*. Psychological Assessment Resources.

Spielberger, C. D., Jacobs, G., Russell, S., & Crane, R. S. (1983). Assessment of anger: The state-trait anger scale. *Advances in Personality Assessment, 2*, 159–187.

Stern, E. (1998). Choice behavior in congested situations: Modeling and research needs. In T. Garling, T. Laitila, & W. Kerstin (Eds.), *Theoretical foundations of travel choice modeling* (pp. 173–200). Pergamon.

Stern, E. (1999). Reactions to congestion under time pressure. *Transportation Research Part C: Emerging Technologies, 7*(2–3), 75–90.

Sun, D., & Kondyli, A. (2010). Modeling vehicle interactions during lane-changing behavior on arterial streets. *Computer-Aided Civil and Infrastructure Engineering, 25*(8), 557–571.

Szollos, A. (2009). Toward a psychology of chronic time pressure: Conceptual and methodological review. *Time and Society, 18*(2–3), 332–350.

Tarko, A. P. (2009). Modeling drivers' speed selection as a trade-off behavior. *Accident Analysis and Prevention, 41*(3), 608–616.

Toledo, T., Koutsopoulos, H. N., & Ben-Akiva, M. (2007). Integrated driving behavior modeling. *Transportation Research Part C: Emerging Technologies, 15*(2), 96–112.

Traffic Management Bureau, Ministry of Public Security. (2012). *Annual statistical report on road traffic accidents*. Beijing.

Tseng, C. M. (2013). Operating styles, working time and daily driving distance in relation to a taxi driver's speeding offenses in Taiwan. *Accident Analysis and Prevention, 52*, 1–8.

Underwood, G., Chapman, P., Wright, S., & Crundall, D. (1999). Anger while driving. *Transportation Research Part F: Traffic Psychology and Behavior, 2*(1), 55–68.

Yagar, S., & Van Aerde, M. (1983). Geometric and environmental effects on speeds of 2-lane highways. *Transportation Research Part A: Policy and Practice, 17*(4), 315–325.

Zhong, S., Zhou, L., Ma, S., & Jia, N. (2012). Effects of different factors on drivers' guidance compliance behaviors under road condition information shown on VMS. *Transportation Research Part A: Policy and Practice, 46*(9), 1490–1505.

Chapter 10
Analysis of the Spatio-temporal Distribution of Traffic Accidents Based on Urban Built Environment Attributes and Microblog Data

Abstract Previous studies have proven that urban built environments influence people's travel behaviors, such as travel mode, travel routes, and travel frequency, which subsequently influences the spatio-temporal distribution of traffic flow and accidents. In addition, spatial heterogeneity means that the effects of urban built environments on accidents vary between regions. This chapter, therefore, uses a geographically weighted regression model and microblog data to analyze the spatio-temporal distribution of accidents from the perspective of urban built environments. In terms of spatial aspects, the results show that proximity to a subway can effectively reduce accidents, whereas proximity to primary schools, kindergartens, and hospitals is positively associated with accidents. In terms of temporal aspects, the number of accidents on weekdays is much higher than that on weekends. As the spatial effects of factors such as subways, schools, and hospitals on weekdays are quite different from those on weekends, transportation planning and management departments should devise traffic control plans that are based on the spatio-temporal effects of different variables. This study provides approaches for the government to use to adjust the attributes of urban built environments to reduce the number of traffic accidents.

Keywords Microblog · Traffic accidents · Urban built environment · Geographically weighted regression · Spatial heterogeneity

1 Introduction

In recent years, rapid urbanization worldwide has led to increased road safety problems, such as traffic accidents. Accidents threaten peoples' lives and property and the efficiency of a road transportation system. Thus, understanding the factors that led to accidents is the key to reducing the number of accidents. For decades, research has focused on developing this understanding by examining the factors influencing accidents, such as illegal parking, the layout of intersections, and speed limits (Poch & Mannering, 1996). However, no studies have considered the urban built environment as an influencing factor, or the spatial heterogeneity of the effects of urban built environment on accidents.

S. Zhong and D. Sun, *Logic-Driven Traffic Big Data Analytics*,
https://doi.org/10.1007/978-981-16-8016-8_10

Urban built environment attributes (e.g., land use density, diversity, design, destination accessibility, and distance to public transit) are variables that characterize the spatial features and attributes of different regions of a city, and thus affect the spatio-temporal distribution of accidents by influencing residents' travel behaviour in the following two ways. First, the spatial layout in urban areas of different types of land use, such as workplaces and residences, determines travel destinations (Pan et al., 2020). Second, distance to transit and destination accessibility affect travel frequency, modes, and routes. Researchers have shown that urban built environments influence the effects of implementing transportation policies (Acker & Witlox, 2011; Frank et al., 2008; Zhong & Bushell, 2017; Zhong et al., 2015, 2020). However, few researchers have analyzed the effects of urban built environments on the spatio-temporal distribution of traffic accidents. Moreover, the effect of a given built environment attribute on accidents varies between areas, and the quantitation of this spatial heterogeneity of attributes remains to be determined (Zhong et al., 2021).

Accordingly, this study uses a geographically weighted regression (GWR) model and microblog message data, land use data, and road network data in Beijing, China, to explore the effects of urban built environments on the spatio-temporal distribution of accidents. Microblog data with continuous spatio-temporal distribution characteristics are used as a proxy for accident data, as traffic accident data collected by previous investigators has been limited by sparse sensor coverage (e.g., Gu et al., 2016).

The remainder of this chapter is organized as follows. The second section reviews and summarizes previous research. The third section presents the methods used in this chapter. A case study is discussed in the fourth section to demonstrate the effects of urban built environments on the spatio-temporal distribution of accidents. The fifth section concludes the chapter.

2 Literature Review

2.1 Research on Methods for Accident Detection

Existing methods for accident identification can be divided into two categories. The first category comprises methods based on detector data, such as traffic flow parameter data collected by fixed detectors and trajectory data obtained from vehicles. Researchers have analyzed traffic flow parameter data using a neural network (Samant & Adeli, 2001), a support vector machine (Xiao et al., 2014; Yuan & Cheu, 2003), wavelet theory (Teng & Qi, 2003), and nonlinear time-series analysis (Fragkou et al., 2018). Similarly, researchers have examined floating car data using hierarchical analysis (Kamran & Haas, 2007) and real-time spatio-temporal analysis (D'Andrea & Marcelloni, 2017) to identify accident locations. However, as sensing coverage and the budgets for detector installation are usually limited, such approaches struggle to comprehensively and accurately identify the spatio-temporality of traffic flow.

The other approach to extracting accident information is based on the use of low-cost and efficient Internet-transmitted text, which comprises massive amounts of long-term data and requires neither the installation and maintenance of detector equipment nor the use of data processing machines. For example, Zhang et al. (2015) used spatial-point pattern analysis to evaluate the spatial dependence between accident-related tweets and accidents, which proved that social media feeds functioned as sensors. Sakaki et al. (2012) pioneered a method for detecting the level and location of congestion by solving the problems of text-based classification and transforming geographically related terms into geographical coordinates. Vallejos et al. (2021) extracted accident information from tweets by combining machine learning and natural language processing techniques.

2.2 Research on the Law of Accidents

Methods for analyzing accidents are divided into two groups. The first group of methods comprises statistical regression methods based on a single factor or multiple factors. For example, Poch and Mannering (1996) used negative binomial regression analysis to reveal the significant effects of traffic-related and geometric elements on the frequency of accidents at intersections, based on 7-year history data of 63 intersections in Bellevue, Washington (United States). The other group of methods comprises systematic causation theory-based approaches. For example, Ozbay and Noyan (2006) created dynamic accidents duration estimation trees by using a Bayesian network to clearly describe accidents clearance patterns.

Single-factor theory and multiple-factor theory focus on the independent effects of the factors influencing accidents and thus do not sufficiently characterize the mechanisms of accidents and how their influencing factors interact. Specifically, studies have not explored the factors influencing traffic accidents from the perspective of urban built environments. Systematic causation theory does examine the interactions between various influencing factors, but only at a macro level; thus, it cannot explore the effect of the spatial heterogeneity of urban built environments on traffic accidents. To solve these problems, this study uses a GWR model to quantitatively analyze the spatial heterogeneity of the effect of urban built environments on the spatio-temporal distribution of accidents.

3 Methodology

3.1 Selection of Study Region

An area within the 5th Ring Road of Beijing is selected as the research object, which encloses a total area of 668.8 km^2 and 8.75 million people. The study area is divided

Fig. 1 Study route

into 97 traffic analysis zones (TAZs) according to its township-level administrative divisions, as presented in Fig. 1.

3.2 Data Collection

Microblog data, land use data, and road network data are collected for this study. The microblog data contains traffic accident data from 2014 to 2016, which includes data on 1217 traffic accidents in 2014, 773 traffic accidents in 2015, and 705 traffic accidents in 2016, comprising a total of 2695 accidents, with accurate occurrence times and coordinates. The land use data are the geographic data of various land use types in Beijing, including the spatial location and attribute information of facilities such as subways, schools, hospitals, and bus stations. The road network data is the geographic road network data and Baidu Maps data for Beijing.

3.2.1 Collection and Processing of Microblog Data

We use the Chinese microblog platform Sina Weibo as the data source, which is a Twitter-like emerging social platform that integrates rapid publishing, sharing, and dissemination of messages. Users can post text, pictures, links, and other information anytime and anywhere via smartphones.

Web crawlers are used to capture data on accidents from Sina Weibo, with accidents-related terms such as "car crash" and "traffic accident" used as search terms. Then, regular expressions are used to remove symbols, users' names, and other irrelevant information from the textual data.

Word segmentation is used to identify the location of each piece of textual data. Unlike English, Chinese sentences lack a middle delimiter. Thus, we use the Chinese Academy of Sciences' word segmentation system, natural language processing information retrieval, to segment the denoised text. This affords the TAZs of each accident based on the coordinate selection program of the Baidu advanced programming interface.

3.2.2 Built Environment Variables of Traffic Analysis Zone

This case study aims to determine the influence of urban built environments on the spatio-temporal distribution of traffic accidents, so the evaluation variables of urban built environment elements must be clarified. Common built environment variables include density, diversity, design, destination accessibility, and the distance to public transportation. On this basis, and considering the availability of data, we select population, the number of hospitals, the number of primary schools and kindergartens, the number of restaurants, the number of bus stops, the number of subway stations, the number of traffic lights, the length of expressway, and the length of arterial road as built environment variables.

3.2.3 Statistical Analysis of the Built Environment Variables in Different Traffic Analysis Zones

The above built environment variables are used as the main independent variables in an ordinary least squares (OLS) regression model. Table 1 presents the statistical results of each built environment variable in the 97 TAZs. The spatial distribution of incidents identified from the microblog is presented in Fig. 2.

3.3 GWR Model

GWR models are an extension of the general global linear regression model that fit spatial data effectively due to their low spatial dependence (Torun et al., 2020). In

Table 1 Statistic results of built environment variables

Variables	Min	Max	Mean	Standard deviation
Incidents (item)	0	116	27.463918	30.209257
Area (km^2)	0.092	54	6.894990	8.952745
Population density (ten thousand/km^2)	3.05	24	9.020928	4.841273
Hospital density (count/km^2)	0	43	6.104431	7.059674
Subway density (count/km^2)	0	18	5.644227	4.489701
Signal light density (count/km^2)	0	6	1.012651	1.061969
Restaurant density (count/km^2)	0	46	3.803204	6.710241
Arterial road density (km/km^2)	0	34	3.745857	3.704526
Bus stops density (count/km^2)	0	112	24.216495	21.436661
Density of primary schools and kindergartens (count/km^2)	0	65	12.659794	11.617518
Expressway density (km/km^2)	0	59	8.478787	11.183789

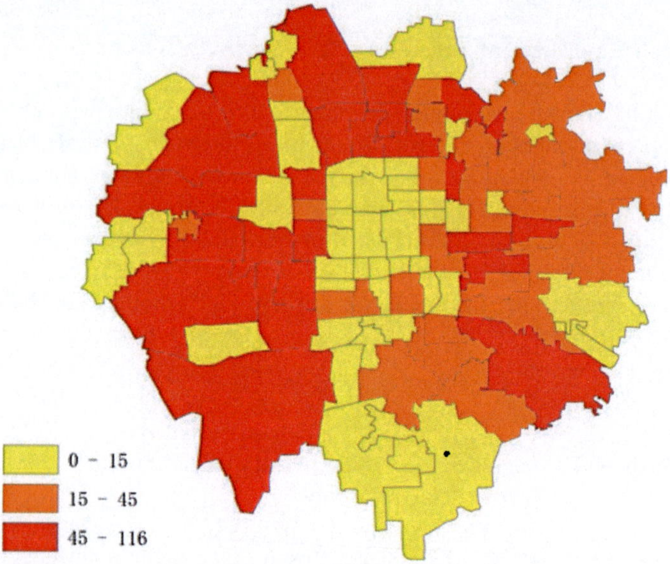

Fig. 2 Spatial distribution of incidents in Beijing

GWR models, spatial heterogeneity is described by adjusting the estimated parameter of each variable to geographic space. Thus, the GWR model we use employs local regression and variable parameters; that is, it embeds the spatial position of variables into the regression parameters, and then uses the local weighted least-squares method to perform point-by-point estimation of the parameters. The model structure is as follows:

$$y_i = \beta_0(\mu_i, \upsilon_i) + \sum_{p=1}^{K} \beta_{ip}(\mu_i, \upsilon_i)x_{ip} + \varepsilon_i, \quad i = 1, 2, \ldots n, \tag{1}$$

where y_i is the number of incidents in TAZ i; n is the number of TAZs; K is the number of affecting factors; β_{ip} is the regression coefficient; (μ_i, υ_i) is coordinates of TAZ i; x_{ip} is the pth independent variable associated with TAZ i; ε_i is the random error term; $\beta_0(\mu_i, \upsilon_i)$ is the constant of TAZ i.

In practice, although the parameters estimated for TAZ i from its own set of data are unbiased, they have large standard deviations and low accuracy. Thus, to reduce the standard deviation of parameter estimation and improve the estimation accuracy, samples around TAZ i should be considered; however, this inevitably leads to deviations from the actual values for TAZ i. Therefore, appropriate data must be sampled around TAZ i to balance the deviation and standard deviation of parameter estimation.

To determine which TAZs around TAZ i to select for estimating parameters, the spatial weight matrix W_{ij} is introduced. This matrix assigns a spatial weight to all TAZs around TAZ i, which establishes a functional distance relationship between the regression of TAZ and its neighboring TAZs; this is the basis of the GWR model. Essentially, the spatial weight matrix selects a function whose value decreases as d_{ij} increases. This study selected the commonly used Gauss function as the spatial weight matrix.

$$W_{ij} = e^{-(d_{ij}/b)^2}, \quad b \geq 0, \tag{2}$$

where b is the bandwidth which is an attenuation parameter of the weight matrix as a function of spatial distance; d_{ij} is the distance between TAZ i and TAZ j. The larger the bandwidth, the slower rate of the decrease of the weight with distance. When the bandwidth is 0, the weight reaches the maximum value at 1 (Fig. 3).

Fig. 3 Gauss function

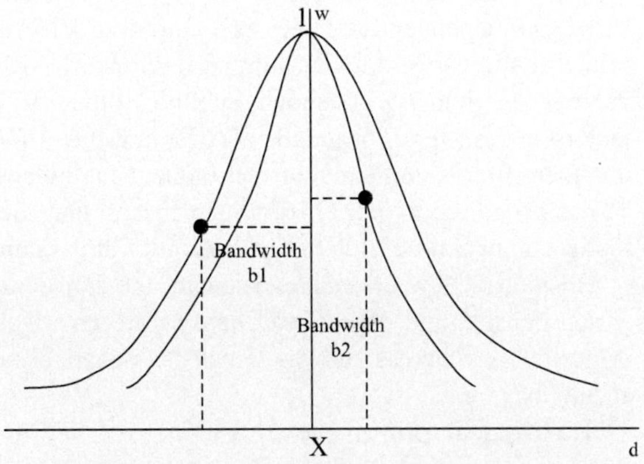

The estimations generated by the GWR model are insensitive to the types of weight functions but are sensitive to the bandwidth of specific weight functions. If the bandwidth is too small, the spatial weight decreases too rapidly with increasing distance, and thus the standard deviation increases. If the bandwidth is too large, the spatial weight decreases too slowly with increasing distance, which increases the deviation of parameter estimation. This study uses the Akaike information criterion (AIC) to select the bandwidth (Fotheringham et al., 2002).

4 Results and Analysis

The average travel speed on the road segments is estimated using the global regression model and the GWR model. Local estimated coefficients may not be consistent with global estimated results, because some estimated global parameters may take a sign opposite to that of the local parameters at some locations. The results are presented in the following two sections, together with figures depicting the spatial variations.

4.1 OLS Model Results and Analysis

The GWR model allows the structural relationships between variables in various locations and the spatial heterogeneity to be easily established. Before applying the GWR model, an OLS regression model is used to determine the significance of each influencing factor and understand the interconnections between the urban built environment attributes and the spatio-temporal distribution of traffic accidents. The results shown in Table 2 are obtained using the OLS toolbox in ArcGIS 10.2, which is a family of client software, server software, and online geographic information system (GIS) services developed and maintained by Esri (Scott & Janikas, 2010).

VIF is the variance inflation factor, which is used to test the multicollinearity between independent variables; the larger the VIF value, the higher the probability of multicollinearity. ArcGIS stipulates that there is no redundant variable if the VIF value is less than 7.5. As shown in Table 2, the VIF values of all of the influencing factors are less than 7.5, and the maximum value is not greater than 2.5, which means that there is no significant correlation between independent variables in this model. The adjusted R^2 is 0.831176, which means that the independent variables in the model can explain 83.1% of the variability in the number of accidents.

Hospital density, arterial road density, bus stop density, primary school and kindergarten density, and expressway density are positively correlated with the number of accidents, whereas subway density is negatively correlated with the number of accidents.

The residual error of the OLS model is listed in Table 3. Koenker's degree of freedom is less than 0.05, which is statistically significant. The chi-square value is used to show the significance of the whole model. The joint F value (53.515460)

Table 2 Estimated results of the OLS regression model (the global regression model)

Variable	Coefficient	Standard deviation	t-value	Robust t-value	Robust *p*-value	VIF
Constant	−9.333599	3.738221	−2.496802	−3.058266	0.002961***	–
Population density	0.157121	0.342480	0.441826	0.008748	0.993039	1.712953
Hospital density	2.984933	0.191919	15.553090	9.343356	0.000000***	1.143831
Subway density	−1.509533	0.428440	−3.523322	−3.933036	0.000173***	2.305536
Signal light density	−0.529054	1.354227	−0.390668	−0.650246	0.517246	1.288733
Restaurant density	−0.160010	0.202710	−0.789352	−0.856693	0.393962	1.152879
Arterial road density	0.000668	0.000406	1.743702	2.488163	0.014741**	1.411192
Bus stops density	0.534516	0.086724	6.163414	5.251892	0.000001***	2.153526
Density of primary schools and kindergartens	0.728758	0.145126	5.021566	4.25572	0.000055***	1.771214
Expressway density	0.000417	0.000149	2.802672	2.898542	0.004745***	1.724855
R^2: 0.847003		Adjusted R^2: 0.831176		AIC: 779.459000		

Note *** Significant at 99%; ** Significant at 95%

Table 3 Statistic results of the residual errors of the OLS model

	Joint F	Joint chi-square	Koenker (BP)	Jarque–Bera
Value	53.515460	318.358751	31.328623	4.672199
Freedom degree	0.000000	0.000000	0.000260	0.096704

indicates that the linear relationship between the number of accidents and the independent variables is significant. The Jarque–Bera value is greater than 0.05, which indicates that the residual errors follow the normal distribution.

4.2 GWR Model Results and Analysis

4.2.1 Model Estimation

The spatial weight and bandwidth are determined by the GWR toolbox in ArcGIS 10.2, using an adaptive bandwidth method and the bandwidth with the minimum AIC value. As indicated by the OLS regression result, hospital density, subway density, arterial road density, bus stop density, primary school and kindergarten density, and expressway density are selected as independent variables. The estimation results of the GWR model are presented in Table 4, and the regression coefficients of each independent variable and its average *t*-value are shown in Table 5.

As can be seen in Table 4, the adjusted R^2 is 0.837048, indicating that the GWR model explains 83.7% of the variability in the number of accidents. Compared with the OLS model, the GWR model has a higher adjusted R^2 and a lower AIC value,

Table 4 Estimation results of the GWR model

	Minimum value	Maximum value	Mean	Standard deviation
Constant	−10.927644	−7.546074	−9.268660	0.874280
Residual error	−31.387381	37.165172	0.250579	11.157586
Standard deviation	3.783607	12.086405	11.102990	1.130536
Standard residual error	−3.407577	3.317880	0.024036	1.048192
AIC	778.180405			
R^2	0.863517			
Adjusted R^2	0.837048			

Table 5 Regression coefficients and average t-values of independent variables

	Minimum value	Maximum value	Mean	Standard deviation	Average t-value
Hospital density	2.900881	3.069552	2.988250	0.038616	15.297060
Subway density	−1.659483	−1.328427	−1.563054	0.080180	−3.533156
Arterial road density	0.000509	0.000780	0.00064	0.000059	1.809434
Bus stops density	0.586224	0.701678	0.625256	0.023073	4.081439
Density of primary schools and kindergartens	0.453513	0.615776	0.536448	0.035882	5.981602
Expressway density	0.000275	0.000740	0.000460	0.000117	2.893082

which indicates that the fitting degree of the GWR model is better than that of the OLS model.

As can be seen in Table 5, the *t*-values of the independent variables in the GWR model are greater than 1.7, meaning that these six independent variables have significant effects on the number of accidents, with that of hospital density being the most significant. The regression coefficients indicate that hospital density, arterial road density, bus stop density, primary school and kindergarten density, and expressway density have positive effects on the number of accidents in all TAZs, whereas subway density has negative effects on the number of accidents in all TAZs. The results of the GWR model are thus similar to those of the OLS model.

4.2.2 Analysis of the GWR Model Results

The regression coefficients and the *t*-values of 97 TAZs are obtained from the GWR model, and are used to explore the effect of each built environment attribute on the spatio-temporal distribution of accidents and to determine the characteristics of these effects. Moreover, the results of the GWR model can help urban transportation planners and managers to understand the factors that affect the occurrence of traffic accidents in different areas and provide guidance and decision support for improving traffic safety.

- Spatial characteristics of accidents

Hospital density has the highest *t*-value (15.29), indicating that it has the most significant effect on the number of accidents. The regression coefficient of hospital density in all TAZs is positive (see Fig. 4), indicating that hospital density is positively related to the number of accidents. The TAZs in southwest Beijing, i.e., Fengtai district and Shijingshan district, are most affected by hospital density. This is because there are many residential areas and few business areas in these two districts, and because hospitals are non-business-related gathering places. Thus, the more northeast a TAZ, the higher is its proportion of gatherings of business people, and the less the effect hospital density has on the number of accidents in the TAZ.

The spatial distribution of the regression coefficients of arterial road density is presented in Fig. 5. As can be seen, the number of accidents increases with arterial road density; this is because the longer a road, the higher the probability that accidents will occur on the road. Furthermore, the arterial road length has the most significant positive effect on the number of accidents in the east and south of the 4th and 5th Ring Roads, because the subway beyond the 4th Ring Road is not conveniently accessible for people in these areas, and the south mostly contains residential areas where people usually travel by personal vehicles.

The expressway density *t*-value (2.8) is higher than the arterial road density *t*-value (1.8), indicating that the expressway length is more positively related to the number of accidents. In addition, the difference value of expressway density is higher than that of arterial road density, which means that the effect of the length of the expressway

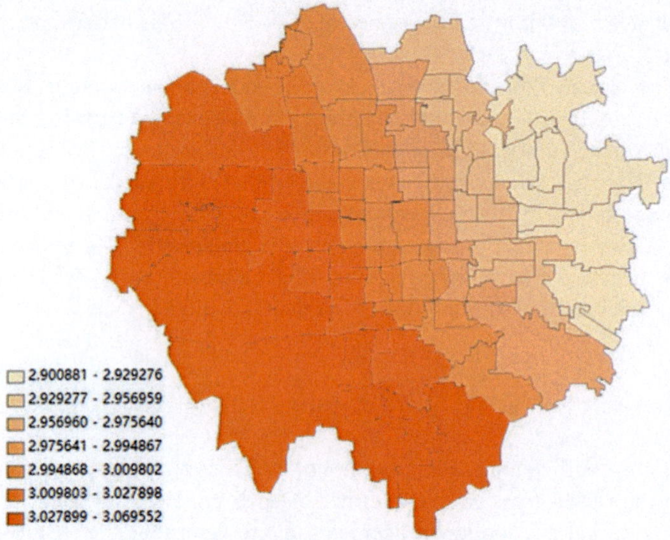

Fig. 4 Spatial distribution of the regression coefficients of hospital density

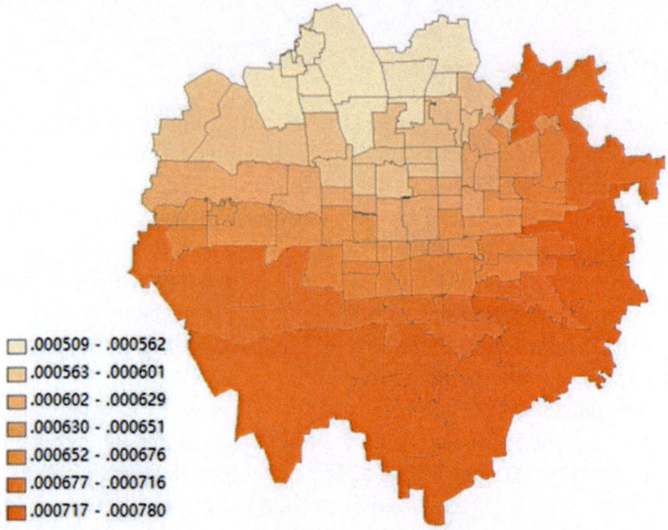

Fig. 5 Spatial distribution of the regression coefficients of arterial road density

has much greater spatial heterogeneity than that of arterial road density. As shown in Fig. 6, the influence of expressway density in the western region, especially in the middle western region (i.e., the area between Dinghuiqiao in the 4th Ring Road area, east of Nanshawo, and the 5th Ring Road), is significantly higher than that in the

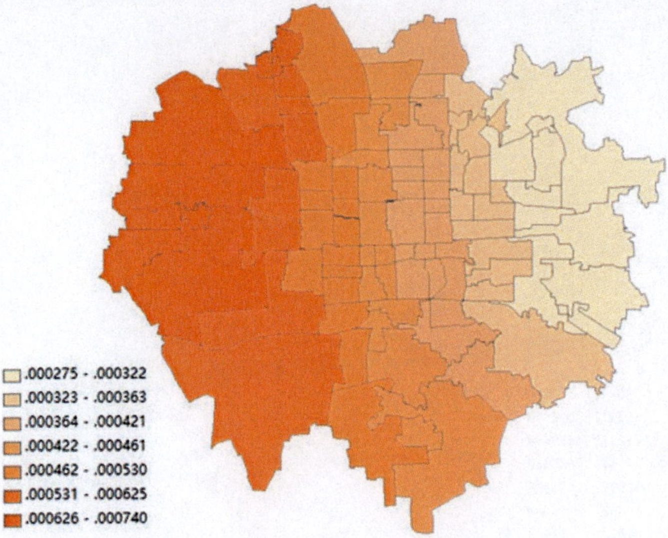

Fig. 6 Spatial distribution of the regression coefficients of expressway density

eastern region. This is because the expressway in the western region is more crowded in the morning and evening than the expressway in the eastern region, and the high speeds and large traffic volumes on expressways lead to a high accident rate.

Figure 7 indicates that primary school and kindergarten density has a positive effect on the number of accidents. The effect in the northern and southern regions is higher than that in the middle region, and the number of accidents in the northeastern regions is least influenced by this variable. This is because there are few business centers in the northern and southern regions, and because primary schools and kindergartens are one of the main places where people gather. Moreover, parents usually use personal vehicles to pick up and drop off their children, which leads to more accidents.

The spatial distribution of the regression coefficients of subway density is presented in Fig. 8. This is the only independent variable that has a negative correlation with the number of accidents, and the average t-value is -3.533156, which indicates that the negative correlation is significant. Overall, as regions with more subway stations offer more convenient public rail transportation services, people will travel by subway more often and by personal vehicles less often in these areas, resulting in a lower number of traffic accidents in these regions. In addition, subways are almost always underground and so they rarely conflict with ground transportation.

Figure 9 shows that bus stop density is positively correlated with the number of accidents, and the t-value shows that the effect of bus stops on the number of accidents is very high. This is because bus stops are located in areas with high traffic demand and high traffic volumes, where accidents are more likely to occur (as pedestrians gather around bus stops, and many pedestrian crossing activities occur when passengers

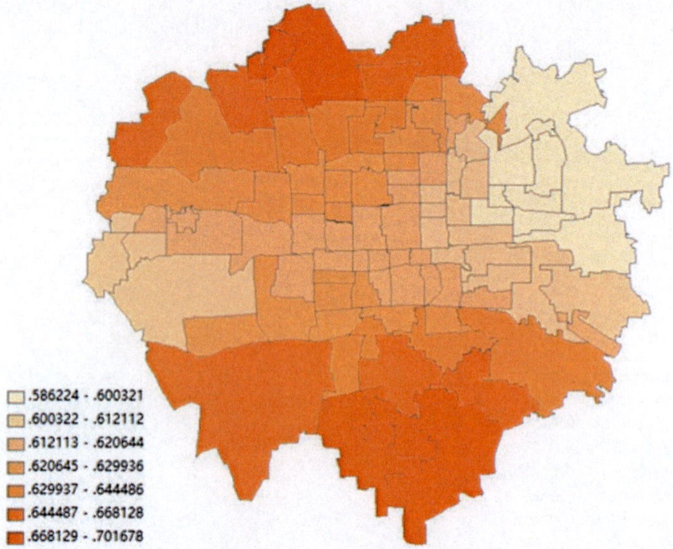

Fig. 7 Spatial distribution of the regression coefficients of the density of primary schools and kindergartens

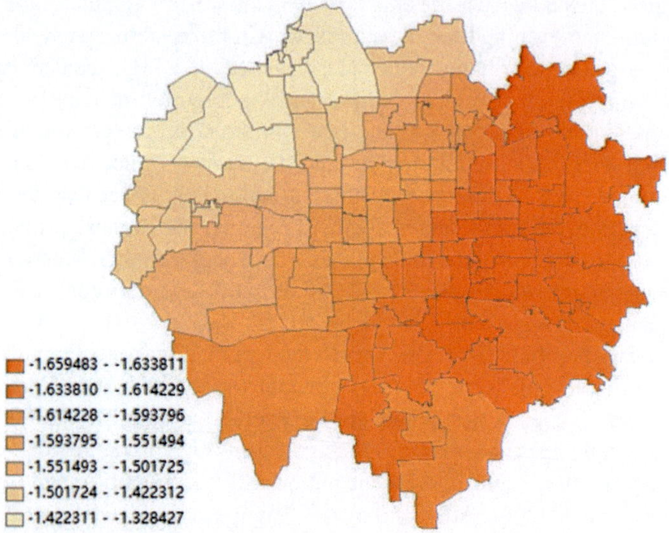

Fig. 8 Spatial distribution of the regression coefficients of subway density

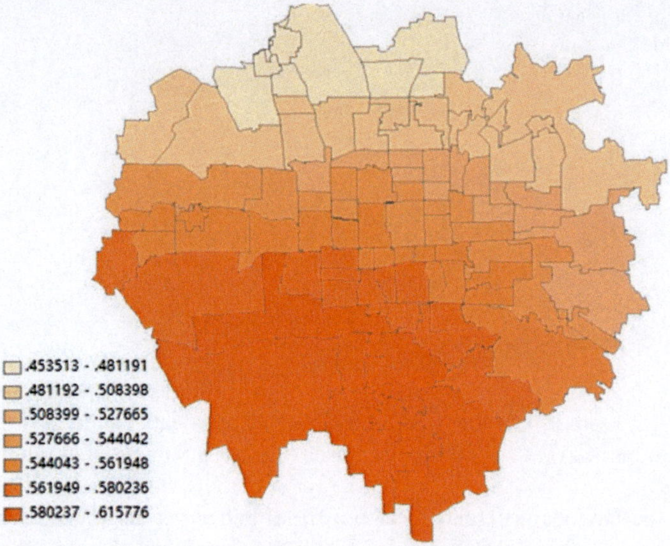

Fig. 9 Spatial distribution of the regression coefficients of bus stops density

enter and exit bus stops). Buses entering and exiting a stop also cause inconvenience to the normal passage of other vehicles, which increases the conflict between buses, private cars, and pedestrians, and further increases the likelihood of accidents.

- Temporal characteristics of accidents

The hourly distribution map shown in Fig. 10 is consistent with reality. The number of accidents in the morning and evening peak hours is significantly higher than that in other periods, which is attributable to a large number of travelers and high traffic volumes during peak hours. The number of accidents increases sharply after 6:30 a.m. and reaches a maximum (295) at 7:30 a.m. Thereafter, the number of accidents decreases to 49 at 9:30 a.m. and stabilizes at this level. The evening peak period

Fig. 10 Hourly distribution map of incidents

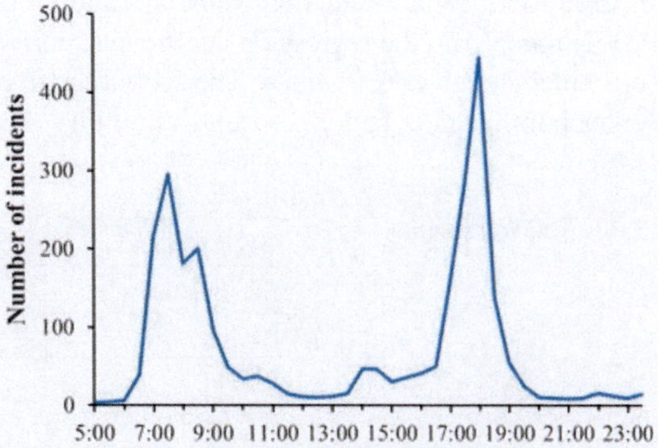

Fig. 11 Daily distribution map of incidents

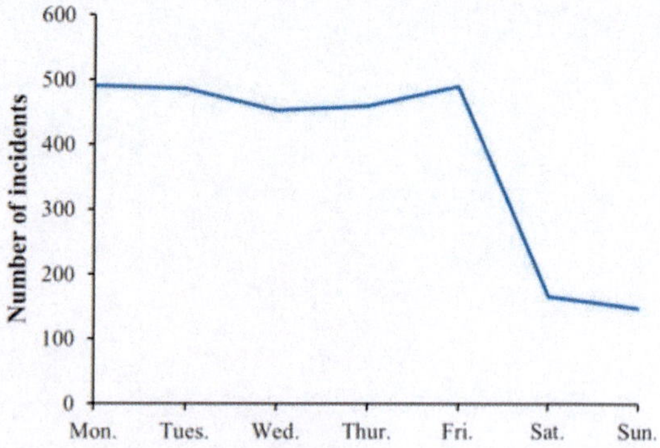

begins at 4:30 p.m.; from then on, the number of accidents increases significantly to a maximum (445) at 6:00 p.m., decreases to 10 at 8:00 p.m., and remains stable thereafter.

Given the obvious difference between travel behaviors on weekdays and those on weekends, the daily distribution of accidents is analyzed. As shown in Fig. 11, Monday saw 491 accidents; the number of accidents slowly declines to a weekday minimum on Wednesday, which is 453 accidents, and then increases to 490 on Friday. In contrast, there are fewer accidents on weekends (~150), and the daily number of accidents on weekdays is significantly higher than that on weekends. This is because the traffic demand is large on weekdays, with most travelers on weekdays commuting in the morning and evening, during which time drivers are rushing, which leads to a high number of accidents during these periods.

- Spatial distribution of accidents on weekdays and on weekends

From the foregoing results, we obtain the six influencing factors that affect the spatial distribution of accidents: hospital density, subway density, bus stop density, primary school and kindergarten density, expressway density, and arterial road density. In addition, we find that the number of accidents on weekdays and that on weekends are vastly different. Therefore, this subsection presents a spatial comparison of the accidents in these two periods. The GWR model containing six influencing factors is used to model accidents on weekdays and those on weekends. The results of the GWR model and the regression coefficients corresponding to these two periods are presented in Tables 6, 7, and 8. The spatial distribution of the regression coefficients of each influencing factor is visualized in Figs. 12, 13, 14, 15, 16 and 17.

Table 6 GWR results

GWR results	Workdays	Weekends
Bandwidth	8715.618676	360.626239
AIC	756.269559	444.483846
R^2	0.877503	0.711651
Adjusted R^2	0.845455	0.642093

Table 7 Regression coefficients of incidents that happened on workdays

Independent variable	Min	Max	Mean	Standard deviation
Hospital density	2.600695	2.871137	2.682239	0.06485
Subway density	−1.578099	−1.011878	−1.373215	0.139614
Arterial road density	0.00029	0.000721	0.000439	0.000086
Bus stops density	0.317975	0.58099	0.451727	0.069836
Density of primary schools and kindergartens	0.084958	0.629562	0.393933	0.116714
Expressway density	0.00022	0.000913	0.000518	0.000242

Table 8 Regression coefficients of incidents that happened on weekends

Independent variable	Min	Max	Mean	Standard deviation
Hospital density	0.290090	0.3100590	29.368360	0.302767
Subway density	−0.285150	−0.157266	−21.870500	−0.225470
Arterial density	0.000103	0.000318	0.018179	0.000187
Bus stops density	0.061872	0.092393	7.507887	0.077401
Density of primary schools and kindergartens	0.000103	0.000318	0.018179	0.000187
Expressway density	0.000017	0.000105	0.005318	0.000055

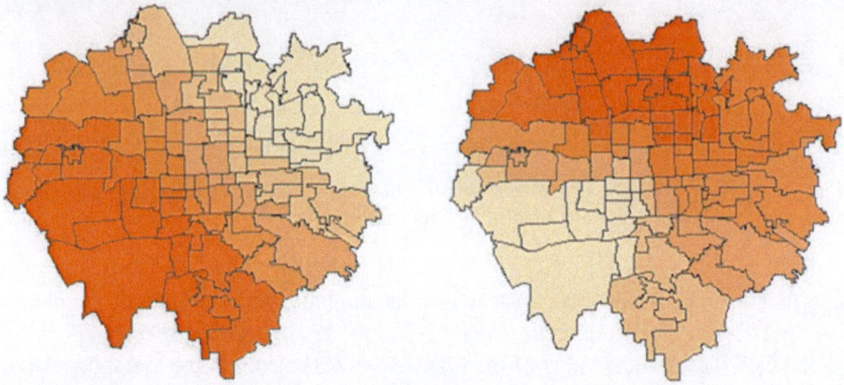

Fig. 12 Spatial distribution of the regression coefficient of hospital density on workdays (left) and that on weekends (right)

The number of accidents that occur on weekends and the number that occur on weekdays are 325 and 2370, respectively. Due to this vast difference, the regression coefficients of the same variable in a different period are also different. Thus, this subsection analyzes the difference between the regression coefficients on weekends and those on weekdays, rather than analyzing the difference between regression

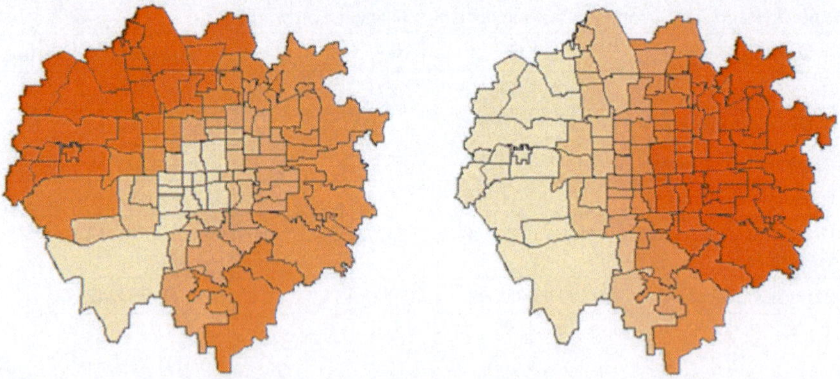

Fig. 13 Spatial distribution of the regression coefficient of subway density on workdays (left) and that on weekends (right)

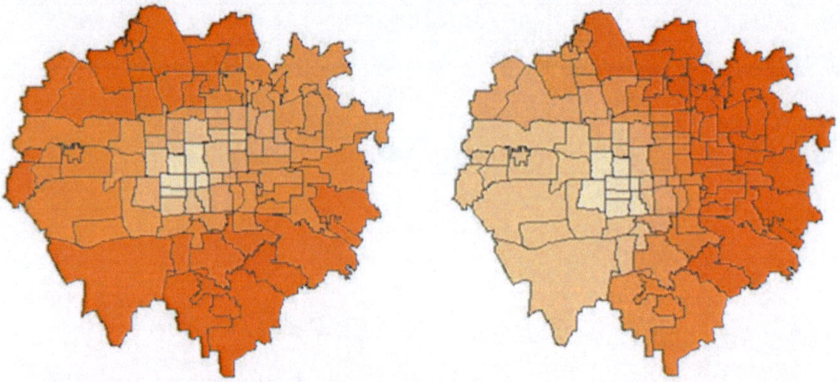

Fig. 14 Spatial distribution of the regression coefficient of the density of primary schools and kindergartens on workdays (left) and that on weekends (right)

coefficients in different areas. Instead of indicating the specific values of the regression coefficient of each variable, Figs. 12, 13, 14, 15, 16 and 17 use different colors to indicate the absolute value of the regression coefficients of the variables, where darker colors indicate larger absolute values. Figures 12, 13, 14 and 15 show that the spatial effects of hospital density, subway density, primary school and kindergarten density, and arterial road density on the number of accidents on weekdays and that on weekends are substantially different.

Hospital density has a greater positive effect in the southwestern region during weekdays, whereas it has a greater positive effect in the northern region on weekends (Fig. 12). This may be because people are more likely to go to hospitals on weekdays than on weekends, as fewer doctors work on weekends. Therefore, accidents in the

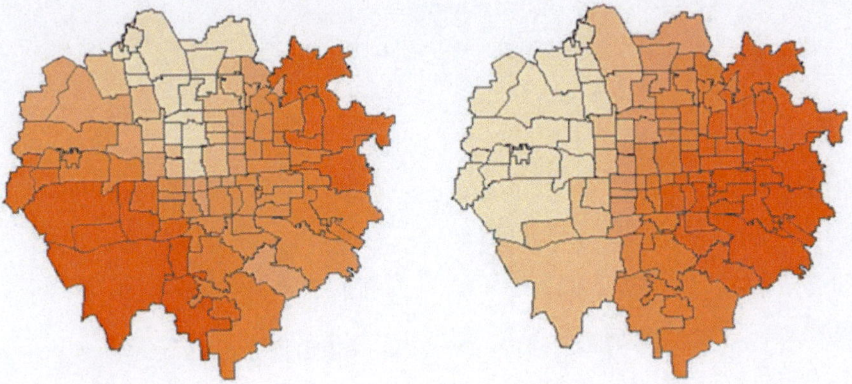

Fig. 15 Spatial distribution of the regression coefficient of arterial road density on workdays (left) and that on weekends (right)

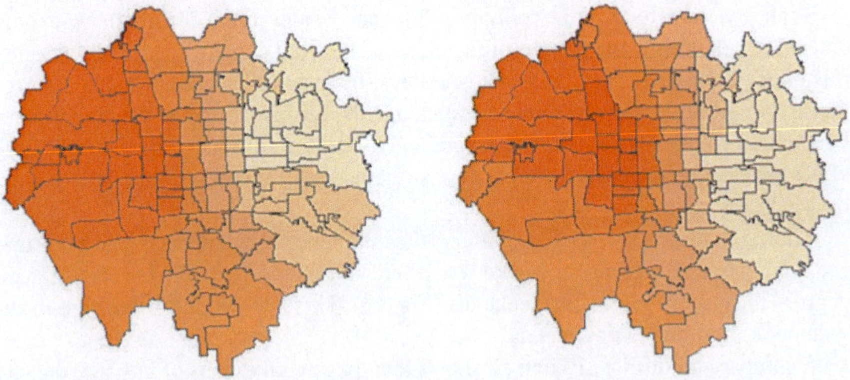

Fig. 16 Spatial distribution of the regression coefficient of expressway density on workdays (left) and that on weekends (right)

southwestern region where people live are easily influenced by hospital density on weekdays.

The negative correlation between subway density and the number of accidents in the entire peripheral region is significant on weekdays, whereas it only significantly affects the number of accidents in the eastern region on weekends (Fig. 13). In this case, the darker the color, the lower the number of traffic accidents. On weekdays, residents usually commute by subway; thus, as subways in the peripheral region are more developed, fewer accidents occur in these regions. On weekends, there are fewer commutes; moreover, as there are many subway stations in the eastern business districts, there are fewer driving trips and a small number of accidents in these regions on weekends.

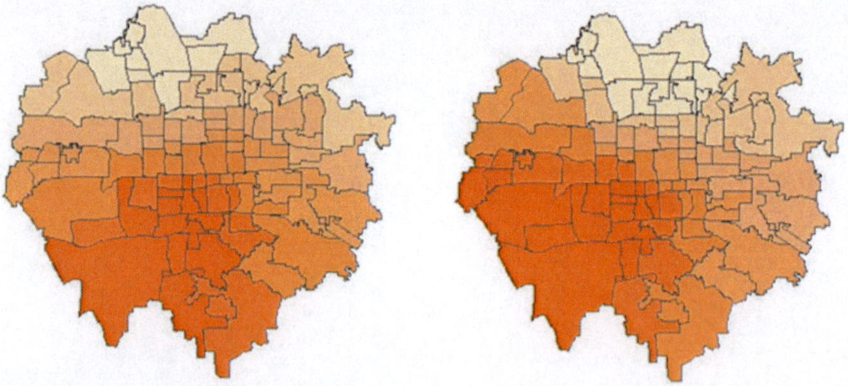

Fig. 17 Spatial distribution of the regression coefficient of the bus stops density on workdays (left) and that on weekends (right)

Similarly, the effects of the primary school and kindergarten density on weekdays and weekends follow the pattern of the effect of subway density, but the direction of this effect is opposite (Fig. 14). On weekdays, the primary school and kindergarten density are significantly positively correlated with accidents in the peripheral regions, whereas on weekends the primary school and kindergarten density only significantly affect the eastern region. This is because students go to school on weekdays and not on weekends.

Arterial road density positively affects the number of accidents in the northeastern and southwestern regions on weekdays, whereas it only affects the number of accidents in the eastern region on weekends (Fig. 15). This may be due to the high traffic volume of commuters on weekdays.

A comparison of Figs. 16 and 17 shows that the spatial effects of bus stop density and expressway density on accidents on weekdays and on weekends are similar; this shows that the spatial effects of these parameters have no temporal characteristics. Specifically, there is always a significantly positive correlation between expressway density and the number of accidents in the northwest region, and between bus stop density and the number of accidents in the southwestern region.

5 Conclusion

Most studies on the factors influencing accidents have been based on statistical regression methods or systematic causation theory. This has prevented a comprehensive understanding of the factors contributing to accidents and how these factors influence each other, as single-factor theory and multifactor theory focus only on the independent effects of factors that influence accidents. Moreover, although systematic causation theory examines the interactions between various influencing factors, it only does so at a macro level. Furthermore, these methods cannot analyze the intrinsic relationship between urban built environments and the spatio-temporal distribution of traffic accidents and cannot explore the spatially heterogeneous effect of urban built environments on traffic accidents. Thus, this study uses an innovative GWR model to analyze the urban built environment factors that influence the spatio-temporal distribution of traffic accidents.

First, microblog messages, land use data, and road network data in Beijing are used to build a general OLS model that contains nine built environment attributes as independent variables. The results from this model indicate that accidents have spatial heterogeneity in the study area, as the effects of each variable on the number of accidents differ substantially between regions. Accordingly, a GWR model is used to quantify the effects of urban built environment factors on the spatial distribution of accidents. The regression coefficients of hospital density, arterial road density, bus stop density, primary school and kindergarten density, and expressway density are different, and they only exert significant effects on the numbers of accidents in some TAZs.

In terms of spatial characteristics, hospital density and primary school and kindergarten density have the greatest positive effect on the number of accidents. Therefore, traffic management and control around densely populated areas such as hospitals and schools must be strengthened to optimize road environments and reduce accidents. Analogously, in the expressways in the western region of the study area, other measures should be taken to reduce accidents, such as by diverting some of the traffic volumes of the arterial road to other trunk or branch roads, and fining drivers for speeding and randomly changing lanes. In addition, the government should pay attention to the layout of bus stops. For example, bus bays should be used to reduce the negative influence of bus stops on the passage of other vehicles. Subway coverage should also be strengthened to decrease the use of private vehicles, and reduce pedestrian exposure and vehicle conflicts on roads, which can prevent accidents.

In terms of temporal characteristics, the number of accidents on weekdays is much higher than that on weekends, with most occurring on Mondays and Fridays, and during the morning and evening peak hours. Therefore, road management in the morning and evening rush hours should be strengthened, especially on Mondays and Fridays. The spatial structure characteristics of accidents on weekdays and on weekends show that the effects of hospital density, subway density, primary school and kindergarten density, and arterial road network density on the spatial distribution of accidents vary over time, whereas the effects of bus stop density and expressway density are largely stable over time. Therefore, the government should strengthen traffic management on weekdays and pay more attention to arterial roads and commuting areas, such as hospitals and primary schools, than to expressways.

The results of this study will assist transportation planners and managers to understand the relationship between urban built environments and the spatio-temporal distributions of accidents. The results will serve as a useful reference when developing specific urban planning and management strategies to reduce accidents and improve traffic safety. In addition, the general relationship between urban built environments and the spatio-temporal distribution of accidents that is revealed in this work can be easily and fruitfully applied to other areas of urban planning.

References

D'Andrea, E., & Marcelloni, F. (2017). Detection of traffic congestion and incidents from GPS trace analysis. *Expert Systems with Applications, 73*, 43–56.

Fotheringham, A. S., Brunsdon, C., & Charlton, M. (2002). *Geographically weighted regression: The analysis of spatially varying relationships*. Wiley.

Fragkou, A. D., Karakasidis, T. E., & Nathanail, E. (2018). Detection of traffic incidents using nonlinear time series analysis. *Chaos: An Interdisciplinary Journal of Nonlinear Science, 28*(6), 063108.

Frank, L., Bradley, M., Kavage, S., Chapman, J., & Lawton, T. K. (2008). Urban form, travel time, and cost relationships with tour complexity and mode choice. *Transportation, 35*(1), 37–54.

Gu, Y., Qian, Z. S., & Chen, F. (2016). From Twitter to detector: Real-time traffic incident detection using social media data. *Transportation Research Part C: Emerging Technologies, 67*, 321–342.

Kamran, S., & Haas, O. (2007, June). A multilevel traffic incidents detection approach: Identifying traffic patterns and vehicle behaviours using real-time gps data. In *2007 IEEE Intelligent Vehicles Symposium* (pp. 912–917). IEEE.

Ozbay, K., & Noyan, N. (2006). Estimation of incident clearance times using Bayesian Networks approach. *Accident Analysis and Prevention, 38*(3), 542–555.

Pan, Y., Chen, S., Niu, S., Ma, Y., & Tang, K. (2020). Investigating the impacts of built environment on traffic states incorporating spatial heterogeneity. *Journal of Transport Geography, 83*, 102663.

Poch, M., & Mannering, F. (1996). Negative binomial analysis of intersection-accident frequencies. *Journal of Transportation Engineering, 122*(2), 105–113.

Sakaki, T., Matsuo, Y., Yanagihara, T., Chandrasiri, N. P., & Nawa, K. (2012, May). Real-time event extraction for driving information from social sensors. In *2012 IEEE International Conference on Cyber Technology in Automation, Control, and Intelligent Systems (CYBER)* (pp. 221–226). IEEE.

Samant, A., & Adeli, H. (2001). Enhancing neural network traffic incident-detection algorithms using wavelets. *Computer-Aided Civil and Infrastructure Engineering, 16*(4), 239–245.

Scott, L. M., & Janikas, M. V. (2010). Spatial statistics in ArcGIS. In *Handbook of applied spatial analysis* (pp. 27–41). Springer.

Teng, H., & Qi, Y. (2003). Application of wavelet technique to freeway incident detection. *Transportation Research Part C: Emerging Technologies, 11*(3–4), 289–308.

Torun, A. Ö., Göçer, K., Yeşiltepe, D., & Argın, G. (2020). Understanding the role of urban form in explaining transportation and recreational walking among children in a logistic GWR model: A spatial analysis in Istanbul, Turkey. *Journal of Transport Geography, 82*, 102617.

Vallejos, S., Alonso, D. G., Caimmi, B., Berdun, L., Armentano, M. G., & Soria, Á. (2021). Mining social networks to detect traffic incidents. *Information systems frontiers, 23*(1).

Van Acker, V., & Witlox, F. (2011). Commuting trips within tours: How is commuting related to land use? *Transportation, 38*(3), 465–486.

Xiao, J., Gao, X., Kong, Q. J., & Liu, Y. (2014). More robust and better: A multiple kernel support vector machine ensemble approach for traffic incident detection. *Journal of Advanced Transportation, 48*(7), 858–875.

Yuan, F., & Cheu, R. L. (2003). Incident detection using support vector machines. *Transportation Research Part C: Emerging Technologies, 11*(3–4), 309–328.

Zhang, S., Tang, J., Wang, H., & Wang, Y. (2015). Enhancing traffic incident detection by using spatial point pattern analysis on social media. *Transportation Research Record, 2528*(1), 69–77.

Zhong, S., & Bushell, M. (2017). Built environment and potential job accessibility effects of road pricing: A spatial econometric perspective. *Journal of Transport Geography, 60*, 98–109.
Zhong, S., Li, X., Jiang, Y., Cheng, R., & Wang, Z. (2020). Identifying the combined effect of shared autonomous vehicles and congestion pricing on regional job accessibility. *Journal of Transport and Land Use, 13*(1), 273–297.
Zhong, S., Wang, S., Jiang, Y., Yu, B., & Zhang, W. (2015). Distinguishing the land use effects of road pricing based on the urban form attributes. *Transportation Research Part A: Policy and Practice, 74*, 44–58.
Zhong, S., Wang, Z., Wang, Q., Liu, A., & Cui, J. (2021). Exploring the spatially heterogeneous effects of urban built environment on road travel time variability. *Journal of Transportation Engineering, Part A: Systems, 147*(1), 04020142.

Chapter 11
Taxi Hailing Choice Behavior and Economic Benefit Analysis of Emission Reduction Based on Multi-mode Travel Big Data

Abstract During the passing decade, taxi floating car data (FCD) has become an important tool to investigate urban trip choice behaviors and activities. The corresponding taxi exhaust reduction issue is also with rather significance for traffic emission mitigation in urban areas. By taking Shanghai as an empirical case, this chapter analyzed the spatiotemporal characteristics of multimode travelers by combining the taxi FCD (from Qiangsheng Inc.), the metro smartcard data and the GPS trajectories of Mobike, one of the most popular shared bicycles in China, 2018. Binomial logit models (BNL) were proposed to estimate mode choices for both peak and off-peak periods by incorporating socio-economic, demographic, urban morphology, land use properties, and various trip-related variables. The choices between metro and taxi, Mobike and taxi were analyzed, respectively, with the corresponding influential factors identified. The results indicated that the percentage of residential and commercial land uses, the number of educational facilities have significant impacts on travel mode choice during peak hours, while the percentage of commercial land, the number of hospitals and bus lines are more prominent during off-peak periods. To quantify the emission reduction benefits, localized calculation of automobile exhaust was established according to the Vehicle Specific Power (VSP) based measurements obtained from the Portable Emission Measurement System (PEMS) experiments. Then, five corresponding emission mitigation schemes were proposed based on the model findings, and the cost–benefit of each countermeasure was further analyzed. Comparing with releasing the peak-hour crowdedness of metro stations, increasing Mobike supply, updating taxis into electric vehicles, and equipping taxis with catalytic converters, the scheme of removing non-motor vehicle restrictions was found with the shortest payback period and was consequently recommended as accordance with the proposal of urban eco and non-motorized transportation. Findings of this study is useful for transportation management in improving the mode share of metro and bicycles, thus to alleviate the congestion and auto emissions in urban areas.

Keywords Urban travel big data · Travel mode selection · Logit regression model · Exhaust emission model · Cost–benefit analysis · Removing the non-motor vehicle restrictions (on road)

S. Zhong and D. Sun, *Logic-Driven Traffic Big Data Analytics*,
https://doi.org/10.1007/978-981-16-8016-8_11

1 Introduction

With the increasing development of urbanization and motorization, transportation has become a key component of social and economic lives (Huang et al., 2020). In 2018, the annual average daily traffic volume and the overall travelling distance for Chinese National Highways attain 14,179 vehicles per road, and 3099.39 million kilometers per day, respectively, increasing by 3.5% and 2.5% comparing with those of 2017 (MOTPRC, 2019). In Shanghai, the average daily taxi passenger volume accounts for 15% of the total public transportation ridership. As an important part of passenger transportation with the characteristics of both public and private sectors, taxi floating car data (FCD) has become an important tool to investigate urban trip choice behaviors and activities (Ge et al., 2020). Meanwhile, transportation has taken an important potion in urban fuel consumption and atmospheric environmental pollutions (Figliozzi et al., 2020). Globally, carbon dioxide from motor vehicles exhaust has attained up to 25% of the entire CO_2 emission, and the traffic fuel consumption makes around 50% of the total consumption (Yutaka et al., 2006). Motor vehicle emissions have become one of the largest sources of CO, NO_X, HC, and CO_2 (Benoit, 2009; Richardson, 2005), in which taxis also contribute to a rather large amount of the exhaust emissions because the extensive utilization and frequent brakes, launches and idle movements tend to produce more pollutants (Bischoff & Maciejewski, 2014; Cai et al., 2014). The Chinese government has taken various measures to deal with urban transportation-related pollutions (Luo et al., 2017), including improving vehicle emission standard level, optimizing industrial energy structure, promoting public transportation, implementing low emission policies, and so on (Cipriani et al., 2019). Nowadays, Global Positioning Systems (GPS) installed on the shared bicycles and taxies provides an automated way to obtain massive order and trajectory data, which may be matched with road networks in Geographic Information Systems (GIS) software, so as to analyze and exhibit passenger-related traffic statistics. Meanwhile, the metro Integrated Circuit (IC) Card, road loop detectors, and on street surveillance cameras have also provided multi-source traffic information. To this end, various aggregated statistics may be obtained based on the taxi GPS and transit IC card travelling data, so that the taxi emission and reduction could be approximated.

For urban traveler choice behavior studies, Wang et al. (2013a) revealed the social and economic characteristics, travel habits, "the last one-mile" demand for the transit and the illegal taxi usage through field investigation and questionnaire surveys. With the ten influencing factors, including travel distance, age, job, and education background, the choice between shuttle bus transit and illegal taxis was approximated using a binomial logit model. Then, prediction was carried out to improve the quality and efficiency of the shuttle bus trips. Zhan et al. (2013) retrieved the passenger-based Origin Destination (OD) information and travel time from the taxicab GPS data in New York City, integrating with the minimum expected travel time and the variances of each traveler, to estimate hourly average of urban link travel times. Alghuraybi et al. (2006) used the nest and machine learning algorithms to reveal the relationship between traveler choice and tolling, and found that the travelers' willingness to pay was affected by fare, location, time, as well as other factors. Chavis and Gayah (2017) established three multinomial logit models with different levels and structures to study the choice between public and private transit systems, including both fixed and variable routes. Zheng and Gan (2013) conducted a willingness investigation in Shanghai and built up an MNL (Multi-nomial Logit) model with travel time, individual characteristics and trip fare, and proposed corresponding transportation and transit optimization advices to the municipal government. In summary, the traditional studies on travel choice behavior generally utilized disaggregate models, with data mainly obtained through questionnaires and surveys. Few have quantified the environmental benefit from the public transit (e.g. metro) and bicycle sharing (e.g. Mobike), although massive taxi GPS and public transit IC cards data have been used at the aggregated level (Sun et al., 2014; Szeto et al., 2013).

This chapter starts by proposing a disaggregate model by investigating relationship between taxis, and metros (or shared bicycles) using the trip data from the three modes, together with land use and built environment factors. Then, a localized emission model for Shanghai was built using the experimental Portable Emission Measurement System (PEMS) based field data, in which Vehicle Specific Power (VSP) was introduced to establish the relationship between cruising condition and emission rate. Furthermore, five traffic emission reduction schemes were proposed based on the significant influence factors revealed in the mode choice studies. The economic cost–benefit analysis for each scheme was conducted from the environment perspective, thus to demonstrate the potential applicability of the metro and sharing bikes related schemes. The remainder of the chapter is structured as follows: Sect. 2 explains the binomial logit (BNL) model for mode choice purpose, and proposes two BNL models, i.e., metro-taxi model, and Mobike-taxi model, in which the significance of different influence factors is exhibited. Section 3 deals with the localized emission model based on VSP, and Sect. 4 analyzes the benefits of each emission reduction scheme, considering both the construction and maintenance costs involved. Finally, conclusions and recommendations for future works are provided in Sect. 5.

2 Data and Taxi Choice Models

2.1 Data and Attributes

The time horizon of the data used in this study is April 2016, which includes daily taxi floating car data (FCD) from Qiangsheng Inc., daily IC card swiping data of metro stations, GPS trajectories of Mobikes (one of the most popular shared bikes in China), population density and land use, road bicycle lane infrastructure and the vehicle emission statistics from the PEMS field experiment. Qiangsheng Inc., the largest taxi company in Shanghai, holds about 30% of the total taxi ownership. The study mainly focused on the center city of Shanghai (within the outer ring), where the majority of taxis is concentrated. The measuring interval of the Qiangsheng taxi FCD trajectories is 10-s, and each record includes six attributes: vehicle ID, longitude and latitude, occupancy state, time, instantaneous speed, and the moving direction. The metro IC card data collected the swiping location and time, including card ID, date, entry/exit time, entry/exit station ID and trip fare. The Mobike data includes bicycle ID, user ID, check-out time, check-in time, starting and ending locations (longitude and latitude), as well as the real-time position coordinates of the bicycle cruising trajectory.

As both metros and sharing bicycles may be chosen to replace taxis trips, two separate binary logit models were introduced to analyze the mode choice between metro and taxi trips and sharing bicycle and taxi trips, respectively. Three categories of attributes were selected as the potential influencing factors, including population density, urban morphology and land use properties (i.e. land use type, construction environment, non-motor vehicle infrastructure, sharing bicycle delivery amount), and trip schedule related variables (i.e. travel time, trip distance and fare) (Hochmair, 2016; Sun et al., 2018; Zhong & Bushell, 2017). Figure 1a–c present the hourly number of trips (from 6:00 to 22:00) for the three modes, respectively.

2.2 Binary-Nomial (BNL) Model and Related Attributes

As a typical disaggregate logit model, BNL deals with choices between two alternatives (Taxi and Mobike), assuming each individual traveler is rational and chooses the mode with the largest utility (Cao, 2019). The probability of a traveler q choosing taxi is calculated as follows.

$$P_{taxi,q} = exp\left(V_{taxi,q}\right)/\left(\exp\left(V_{taxi,q}\right) + \exp\left(V_{mobike,q}\right)\right) \tag{1}$$

where,

$P_{taxi,q}$ represents the probability of a traveler q choosing taxi.
$V_{taxi,q}$ represents the deterministic utility value of mode Taxi to a traveler q.
$V_{mobike,q}$ represents the deterministic utility value of mode Mobike to a traveler q.

To assess the mode choice between metro and taxi, or sharing bicycle and taxi, variables used in BNL models to reveal the influence factors of mode choice are as follows.

Dependent variables: In the metro-taxi model, choice between metro and taxi travel was used as the dependent variable. The outcome becomes 0, if a traveler chooses taxi. Otherwise, if the traveler takes metro, the outcome is 1. Similarly, in the Mobike-taxi model, the choice between Mobike and taxi was chosen as the dependent variable. If a traveler chooses taxi, the value is 0. Otherwise, Mobike will be chosen.

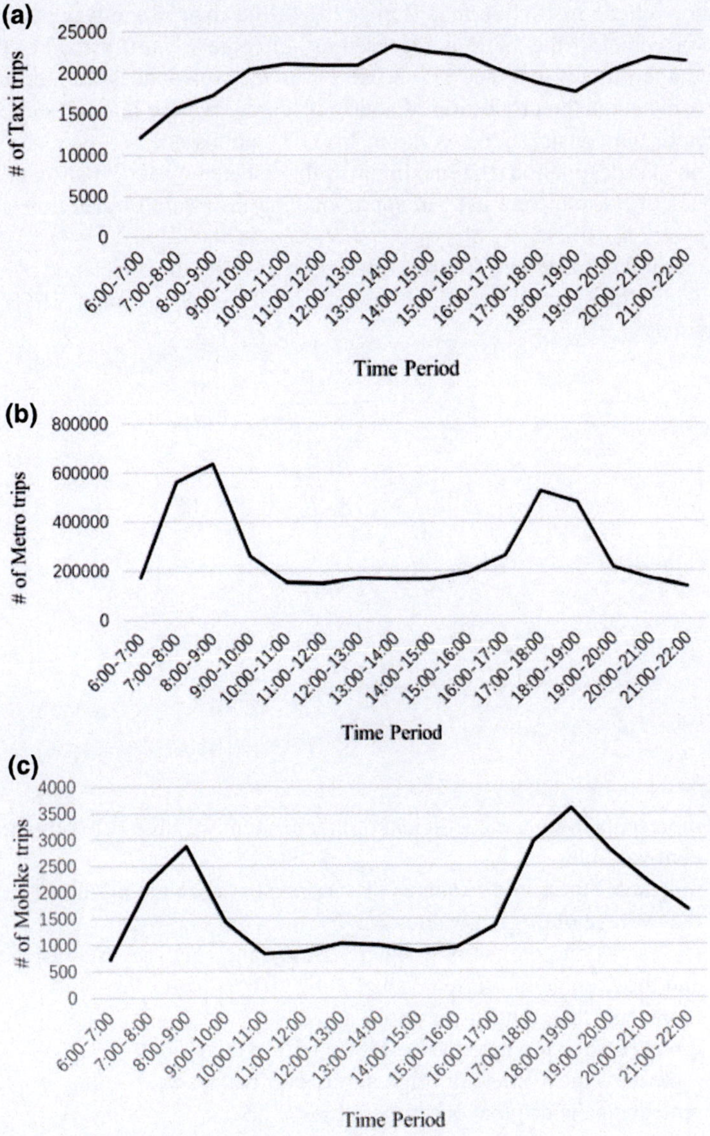

Fig. 1 Hourly ridership of taxi, metro and Mobike. **a** Hourly ridership of taxi. **b** Hourly ridership of metro. **c** Hourly ridership of Mobike

Independent variables: Taking the maximal acceptable walking distance (500 m) as the radius of buffer zone, the average value of population density and land use variables within the buffer zone were obtained (Sun and Ding 2019; Sun et al., 2016; Xu et al., 2019). In particular, the value of non-motor vehicle restriction, defining as the road facilities prohibit bicyclists to enter in, was interpolated from the Shanghai non-motor vehicle restriction map (Lang, 2018). If a road is one-way restricted for non-motor vehicles, the value is recorded as −1; otherwise, if a road is two-way restricted, the value is recorded as −2. Moreover, the crowdedness of metro stations is approximated as the proportion of hourly IC cards swiping into the station to the floor area of the construction. As the metro OD stations doesn't represent the true trip origin and destination, the maximum walk distance of 500 m and the average walk speed of 5 km/h were used in approximating the overall travel time of metro trips.

Variance inflation factor (VIF) was introduced to verify the correlation of independent variables, thus to ensure no multicollinearity. The calculation of VIF is defined as follows.

$$\mathrm{VIF}_i = \frac{1}{1 - R_i^2} \tag{2}$$

$$\mathrm{R}_i = \frac{SSR_i}{SST_i} \tag{3}$$

$$\mathrm{SSR}_i = \sum_{j=1}^{n} (\widehat{y_j} - \overline{y_j})^2 \tag{4}$$

$$\mathrm{SST}_i = \sum_{j=1}^{n} (y_j - \overline{y_j})^2 \tag{5}$$

where,

R_i represents R-square coefficient of independent variable x_i to other independent variables;

SSR_i represents the sum of squares of variance between the predicted value and the average of original values;

SST_i represents the total sum of squares of variance between the original value and the average of original values;

n represents the number of samples;

$\widehat{y_j}$ represents the predicted value by the regression model;

$\overline{y_j}$ represents the average of original observed values y_j;

y_j represents the original observed values.

The selected independent variable values (for both peak and off-peak hours), including minimum, maximum and mean, and the corresponding VIF values of each variable calculated using R software package, are presented in Table 1. If VIF exceeds 4, the choice of independent variables needs further considerations. If VIF exceeds 10, the independent variables have large multicollinearity and the highly correlated variables should be removed. As presented in Table 1, the VIFs of all independent variables range from 1.05 to 1.81 (<4), indicating no multicollinearity within the independent variables.

Table 1 Description of independent variables and VIF

Metro-taxi					Mobike-taxi					Description
Variable	Min	Max	Mean	VIF	Variable	Min	Max	Mean	VIF	
pop_1	22	85,214	19,627	1.27	pop_3	25	87,468	25,142	1.38	Origin population density, people/km^2
pop_2	22	85,229	21,276	1.24	pop_4	25	85,848	24,554	1.39	Destination population density, people/km^2
resi_1	0	100	22.97	1.43	resi_3	0	100	29.82	1.50	Origin residential land use ratio, %
resi_2	0	100	32.08	1.56	resi_4	0	100	30.76	1.53	Destination residential land use ratio, %
farm_1	0	100	26.67	1.41	farm_3	0	100	12.52	1.75	Origin agricultural land use ratio, %
farm_2	0	100	13.90	1.54	farm_4	0	100	12.70	**1.81**	Destination agricultural land use ratio, %
comm_1	0	39.31	1.39	1.14	comm_3	0	39.92	2.49	1.18	Origin commercial land use ratio, %
comm_2	0	34.52	2.77	1.15	comm_4	0	42.29	2.40	1.13	Destination commercial land use ratio, %
admin_1	0	9.49	0.26	1.06	admin_3	0	11.58	0.42	1.11	Origin administrative land use ratio, %
admin_2	0	9.49	0.61	1.06	admin_4	0	9.97	0.39	1.10	Destination administrative land use ratio, %
trans_1	0	100	12.36	1.26	trans_3	0	100	11.54	1.31	Origin transportation land use ratio, %
trans_2	0	100	13.77	1.37	trans_4	0	100	11.40	1.33	Destination transportation land use ratio, %
bus_1	1	106	20.06	1.31	bus_3	1	105	25.66	1.24	Origin bus line amount
bus_2	1	106	25.69	1.31	bus_4	1	106	27.03	1.28	Destination bus line amount
hosp_1	1	38	1.93	1.21	hosp_3	1	38	3.46	1.11	Origin medical institution amount
hosp_2	1	38	3.00	1.24	hosp_4	1	38	3.69	1.10	Destination medical institution amount
sch_1	1	13	2.71	1.40	sch_3	1	13	3.56	1.28	Origin school amount
sch_2	1	13	3.25	1.53	sch_4	1	13	3.45	1.36	Destination school amount
off_1	1	36	2.95	1.18	off_3	1	40	5.29	1.39	Origin office building amount
off_2	1	42	4.93	1.21	off_4	1	41	5.52	1.39	Destination office building amount

(continued)

Table 1 (continued)

Metro-taxi					Mobike-taxi					Description
Variable	Min	Max	Mean	VIF	Variable	Min	Max	Mean	VIF	
TTime	−322	323.12	38.36	1.12	–					Travel time difference within the same OD, minute
fee	−218	221.60	−4.26	1.17	–					Trip fare difference within the same OD, RMB
cong	0.01	4.12	1.16	1.10	–					Origin metro station crowdedness (for peak hours only), passenger/m^2
–					sup	1	198	80	1.08	Origin Mobike supply, veh
–					restrict_3	−21	0	−1.14	1.44	Origin non-motor vehicle restriction, lane
–					restrict_4	−20	0	−1.31	1.52	Destination non-motor vehicle restriction, lane
–					TDist	0.15	54.34	3.13	**1.05**	Travel distance, km

2.3 Metro-Taxi BNL Choice Model

Urban taxi trips may be divided into three types according to the functional relationship with metro, namely substitutable type, complementary type, and extended type with the definitions as below.

Substitutable type. The taxi trips can be completely replaced by metro. In this study, a taxi trip with metro stations located within 500 m for both the origin and destination was recognized as a substitutable type.

Complementary type. The taxi trips can't be accomplished by the metro, as they have no metro stations within 500 m for both origin and destination.

Extended type. The taxi trips may be substituted by taking metro and then followed with a shorter taxi travel, which are defined as there exists metro stations near either the origin or the destination (within 500 m).

As the study mainly focused on the mode choice between metro and taxi, only the substitutable trips were taken as the subject for further investigation. Because of the apparent difference in mode choice during peak and off-peak hours, the probabilities were modeled separately, by taking 6:00–9:00, 16:00–19:00 as the peak hours, and 9:00–16:00, 19:00–22:00 as the off-peak hours. The morning and afternoon peak hours were not separated and the purpose of trips, e.g. commuting trips, was identified by the land use types at the origin or destination. Results of the BNL models for both peak and off-peak hours are shown in Table 2.

As presented in Table 2, during the off-peak hours, destination population density (**pop_2**), travel time (**TTime**) and trip fare (**fee**) are the most significant during the binary mode choice, indicating higher destination population density and more travel time difference tend to improve the probability of taking taxis during off-peak hours. The higher trip fare difference in between increases the probability for travelers to take metros. Moreover, origin population density (**pop_1**), origin agricultural land use ratio (**farm_1**) and origin office building amount (**off_1**) all have large significance in the mode choice, which may be related to incomes and travel purposes. Travelers whose origin has more agricultural land use tend to take metro because most of them may be farmers, with comparable low income. On the contrary, travelers whose origin has more official buildings tend to take taxi because many business trips tend to be more concerned about comfort and convenience than cost.

However, during the peak hours, the origin residential land use ratio (**resi_1**), travel time (**TTime**) and trip fare (**fee**) have the largest significance during the mode choice. Travel time and trip fare are still important influencing factors. With the enlarging of travel time difference, the possibility of switching to taxis increases. Meanwhile, as the trip fare difference grows, the ratio changing to metro increases. During peak hours, travel time is more important compared to trip fare, and crowdedness (**cong**) is an extremely important factor of metro trips, which may deter metro usage during peak hours.

Table 2 Results of BNL models for choice between metro and taxi

Variable	Off-peak hours		Peak hours	
	Coefficient	Significance	Coefficient	Significance
(Intercept)	6.146	***	1.493e + 01	***
pop_1	−1.023e−05	**	−1.067e−05	•
pop_2	−1.389e−05	***	−6.737e−06	
resi_1	2.326e−01		6.808e−01	***
resi_2	3.667e−01		2.620e−02	**
farm_1	7.552e−01	**	5.354e−01	
farm_2	5.052e−01	•	4.269e−02	
comm_1	−1.001	•	4.800e−01	
comm_2	−7.720e−01	*	3.053	•
admin_1	−2.028		−6.617	
admin_2	−4.034	•	3.508	
trans_1	1.195e−01		4.065e−01	•
trans_2	6.888e−02		6.979e−02	*
bus_1	7.319e−03	•	1.650e−03	*
bus_2	7.893e−03	*	1.105e−03	•
hosp_1	−2.170e−02	*	−1.269e−02	*
hosp_2	−1.303e−02	•	−1.603e−02	
sch_1	2.314e−02		4.883e−02	*
sch_2	3.322e−02		6.327e−02	**
off_1	−2.391e−02	**	3.893e−03	•
off_2	−8.312e−03	•	2.654e−02	•
TTime	−1.161e−01	***	−3.140	***
Fee	7.979e−02	***	1.214e−02	***
Cong	–	–	−7.207e−01	**

Note *** indicates $p < 0.001$; ** indicates $p < 0.01$; * indicates $p < 0.05$; • indicates $p < 0.1$

2.4 *Mobike-Taxi Choice Model*

Because Mobike and taxi do not have strict operational schedules as metro, the division of peak and off-peak hours is different, in which peak hours were defined as ranging from 6:00 to 9:00 and 16:00 to 19:00, and the rests are off-peak hours. Substitutable trips for Mobike were used, defined as taxi trips whose origin and destination are both within the maximal acceptable walking distance (500 m) of the Mobike (bicycles) supply sites. Results of the BNL model for choice between Mobike and taxi are shown in Table 3.

As presented in Table 3, during the off-peak hours, destination administrative land use ratio (**admin_4**), destination number of office building (**off_4**), origin number

Table 3 Results of BNL model for choice between mobike and taxi

Variable	Off-peak hours		Peak hours	
	Coefficient	Significance	Coefficient	Significance
(Intercept)	1.092	***	2.804	***
pop_3	7.764e−06	*	3.832e−06	
pop_4	1.088e−05	**	1.124e−05	***
resi_3	−3.495e−01		−5.183e−01	**
resi_4	2.710e−01		−5.149e−01	**
farm_3	8.060e−03		3.172e−01	•
farm_4	3.366e−01		1.471e−01	
comm_3	−2.786	*	−1.795	
comm_4	−8.970e−01		−3.432	***
admin_3	−1.238e+01	**	−1.653e+01	***
admin_4	−1.971e+01	***	−2.345e+01	***
trans_3	8.558e−01	**	7.271e−01	**
trans_4	6.021e−02		3.680e−01	•
bus_3	6.189e−03	•	1.706e−02	***
bus_4	−2.107e−03		5.940e−03	*
hos_3	−2.510e−02	**	−6.974e−02	***
hos_4	−2.685e−02	**	−4.504e−02	***
sch_3	−9.713e−03	***	4.230e−02	*
sch_4	−1.615e−02		3.699e−02	*
off_3	−4.461e−01	***	−1.075e−01	***
off_4	−6.495e−02	***	−5.366e−02	***
sup	2.033e−03	**	7.908e−04	*
restrict_3	2.897e−02	*	1.805e−01	***
restrict_4	6.406e−03	*	7.614e−02	***
TDist	−7.724e−01	***	−5.437e−01	***

Note *** indicates $p < 0.001$; ** indicates $p < 0.01$; * indicates $p < 0.05$; • indicates $p < 0.1$

of school (**sch_3**) and origin number of office building (**off_3**) have the largest significance for the mode choice. The coefficients are all negative, indicating all have positive effects in choosing taxis. These may be explained that during off-peak hours, trips to or from schools and office buildings are mostly with the purposes of non-commuting official business, during which travelers are more concerned about comfort and convenience, and consequently tend to choose taxis. However, during the peak hours, more factors are significant, including administrative land use ratio (**admin_3**), number of bus line (**bus_3**), number of medical institution (**hosp_3**), number of office building (**off_3**), non-motor vehicle restriction (**restrict_3**) of the origin, population density (**pop_4**), commercial land use ratio (**comm_4**), administrative land use ratio (**admin_4**), number of medical institution (**hosp_4**), number of office building (**off_4**), non-motor vehicle restriction (**restrict_4**) of the destination and travel distance (**TDist**). Coefficients of destination population density (**pop_4**), origin bus line amount (**bus_3**), origin and destination non-motor vehicle restriction (**restrict_3, restrict_4**) are positive. This indicates that the rise of these variables will promote using Mobike. Other variables, such as origin and destination land use ratio (**resi_3, resi_4**), travel distance (**TDist**), are with negative coefficients, indicating that the rise of these variables will promote taking taxis.

3 Development of Localized Emission Model

Vehicle specific power (VSP), a commonly used parameter reflecting the travel condition of vehicles, was introduced to build up a localized emission model, defined as the required power to pull the mass of 1 ton (unit: kw/t). The influence factors of VSP include gradient, acceleration and average speed. Assuming the gradient is 0, the calculation of VSP is shown as follows (Martinetz, 1989).

$$\mathrm{VSP} = v \times (1.1 \times a + 0.132) + 0.000302 \times v^2 \tag{6}$$

where, v represents the average speed of the vehicle (unit: m/s), and a represents the acceleration of the vehicle (unit: m/s^2).

For the VSP-based emission model, each VSP section has to correspond to a certain emission rate. To this end, PEMS field experiment was introduced to assess the field emission rate of vehicles, so as to localize the emission model and determine the pollutant emission rate of different VSP sections. During the field experiment, the acceleration and average speed of the instrumented vehicle were obtained through GPS equipment, to calculate the VSP values. Then, the emissions corresponding to each acceleration rate and average speed were collected by the pollutant collection instrument OBEAS-3000 during the field experiment. Finally, the relationship between VSP and emission rate was established.

The local street experimental vehicle is a light-duty gasoline-fueled car, Skoda Superb 2010, with an engine capacity 2 L, subjected the emission standard China IV. The test vehicle for highways is a light-duty gasoline-fueled car, Buick Regal 2013, with an engine capacity 2 L, also subjected the emission standard China IV. The experiments were carried out on November 25 and 26, 2014 from 9:00 to 18:00, including peak hours (17:00–18:00) and off-peak hours (15:00–16:00), on Jianchuan Road and Dongchuan Road (urban roads), and Gushan section of S19 Tongxi Highway (highway). Field data collected from both urban roads and highways were used to estimate the emission rates for different VSP sections. A detailed description of the experiment can be found from Sun et al. (2017).

As found in PEMS experiments, the VSP section of [−25, 25] takes up 93.46% and the VSP section of [−30, 30] takes up 97.17%. To make the distribution of VSP section more accurate, VSP sections within [−25, 25] were further divided with the step of 1 kW/t, while the sections out of the range of [−25, 25] were divided with steps of 3 kW/t or larger. Results of the localized VSP model is shown in Table 4.

Emission model based on VSP can be approximated by the following equation.

$$\mathrm{M} = \sum_{i=1}^{n}\sum_{j=1}^{m} VSP_{ij} * t_{ij} \tag{7}$$

where,

M	represents the emission of one single pollutant from the taxis;
n	represents the number of travels;
m	represents the number of uniform sections and acceleration sections.
VSP_{ij}	represents the emission rate of travel i related to corresponding vehicle specific power, VSP, calculated based on the average speed and acceleration on road section j;
t_{ij}	represents the travel time of travel i on road section j;

With the VSP emission model, the hourly emissions of Qiangsheng taxis for a typical workday were calculated, as presented in Fig. 2. Then, the overall daily emissions of CO_2, CO, HC and NO_X were aggregated as 2371.92 t, 3296.53 kg, 114.59 kg and 647.26 kg, respectively, and the daily total fuel consumption was approximated as 395,420.13 L. These would be taken as the benchmark for the emission reduction analyses in Sect. 4.

4 Emission Reduction Schemes and Assessment

As mentioned, the crowdedness of the metro stations (**cong**) during peak hours significantly affected the choice between metro and taxi. Meanwhile, Mobike supply (**sup**) and non-motor vehicle restriction (**restrict_3, restrict_4**) also played important roles in choosing between Mobike and taxi. Therefore, three schemes were proposed based on the BNL models as follows

- Releasing the peak-hour crowdedness of metro stations.
- Increasing the Mobike supply (to 200%).
- Removing the non-motor vehicle restrictions (on road).

Moreover, according to previous studies (Shaheen et al., 2020; Wang et al., 2013b), updating to electric vehicles, and equipping with catalytic converters are also popular countermeasures to reduce exhaust emissions. In China, cities, such as Shenzhen, Taiyuan and Nanjing, have promoted electric taxis since 2016, and the Beijing government has forced all taxis to equip with catalytic converters. Consequently, two additional schemes were also considered as below.

- Updating taxis into electric vehicles.

Table 4 Emission rate of light-duty vehicles of VSP interval based on PEMS test

NO	VSP range	CO_2 (g/s)	CO (mg/s)	HC (mg/s)	NOx (mg/s)
1	<−30	2.5760	5.5983	0.1870	1.3131
2	[−30, −26]	4.4714	7.9783	0.1762	1.0257
3	−25	1.8209	8.0494	0.0530	0.9517
4	−24	4.2887	7.5925	0.1138	1.1114
5	−23	2.9193	6.5674	0.1899	1.1025
6	−22	2.5712	3.7649	0.2729	1.4416
7	−21	2.5960	3.6113	0.1626	1.3313
8	−20	4.4569	7.5783	0.1999	0.8209
9	−19	2.4734	8.2688	0.2085	1.1409
10	−18	6.9313	8.6350	0.2095	1.3302
11	−17	5.5723	8.6226	0.2229	1.3528
12	−16	4.1717	7.2805	0.2600	1.2943
13	−15	2.0338	9.4853	0.2851	1.2241
14	−14	4.9154	6.9929	0.2568	1.2282
15	−13	4.3437	7.3696	0.2711	1.4590
16	−12	3.9303	9.1183	0.3694	1.7694
17	−11	2.9862	4.8153	0.2277	1.2867
18	−10	2.7490	8.2690	0.1431	0.9439
19	−9	5.9139	10.5527	0.3011	1.2043
20	−8	4.9285	8.8614	0.3202	1.2986
21	−7	9.0655	11.6479	0.5199	0.6438
22	−6	4.6304	5.0686	0.7231	1.5588
23	−5	4.0270	5.6186	0.4352	1.2939
24	−4	7.2600	14.5153	0.4503	1.5747
25	−3	3.6531	13.0153	0.6525	1.7610
26	−2	8.0341	11.8943	0.5766	1.6254
27	−1	6.7724	8.9078	0.3679	1.4046
28	0	4.6100	6.4022	0.1487	1.2503
29	1	1.3870	4.0901	0.1065	1.0630
30	2	5.8668	3.1637	0.2500	1.5185
31	3	7.0617	6.2841	0.2983	2.0373
32	4	4.3502	8.1183	0.3644	2.2336
33	5	6.3980	10.0948	0.3845	1.8260
34	6	8.4799	14.5480	0.4169	2.2575
35	7	8.5223	12.4159	0.6018	3.1275
36	8	7.8936	9.6506	0.6499	3.1109

(continued)

Table 4 (continued)

NO	VSP range	CO_2 (g/s)	CO (mg/s)	HC (mg/s)	NOx (mg/s)
37	9	8.4361	13.0065	0.4999	3.3647
38	10	7.6381	10.9906	0.4561	3.0492
39	11	10.4020	13.0062	0.4998	2.4078
40	12	11.6154	14.7883	0.4508	2.7380
41	13	12.3404	14.8126	0.6562	2.9178
42	14	8.3483	11.2054	0.5566	2.7509
43	15	8.5596	12.3012	0.3968	3.1299
44	16	12.8856	13.7315	0.6125	4.7936
45	17	10.7433	18.2405	0.8941	3.6384
46	18	11.8845	17.4825	1.0617	3.3372
47	19	9.2394	16.0921	0.8409	3.3490
48	20	10.0347	13.3682	0.6214	3.4167
49	21	8.2103	9.5976	0.6118	4.9003
50	22	7.9887	7.9354	0.7423	4.5568
51	23	6.3683	8.3816	0.5956	4.2449
52	24	9.1096	9.9409	0.5855	3.6502
53	25	8.1562	12.6573	0.7372	3.1142
54	26	10.3993	14.7639	0.5412	4.7668
55	[27,31]	16.0764	15.9627	0.6381	3.3259
56	[32,36]	10.1752	16.5965	0.7226	5.0590
57	[37,40]	8.1053	12.3570	0.6619	4.6963
58	>40	10.2360	11.2702	0.8939	6.6456

Note In **VSP range**, −25 indicates the value belongs to the range of [−25.5, −24.5], so as the rests (Sun et al., 2017)

- Equipping taxis with catalytic converters.

The cost–benefit analyses of the five emission reduction methods were conducted, by considering both the construction and maintenance costs included (Laurino et al., 2019). The benefit consists of energy-saving (fuel) and emission reduction. The latter is further divided into the reduction of environmental cost by harmful pollutants (NO_X, CO, HC) and the reduction of the carbon taxes by greenhouse gases (CO_2e). Other parameters, e.g. feasibility, applicability, and system connections were not considered. Benefits of the schemes were calculated using the following equation.

$$Q = \sum_{i=1}^{n} P_i * m_i / e_i \tag{8}$$

where,

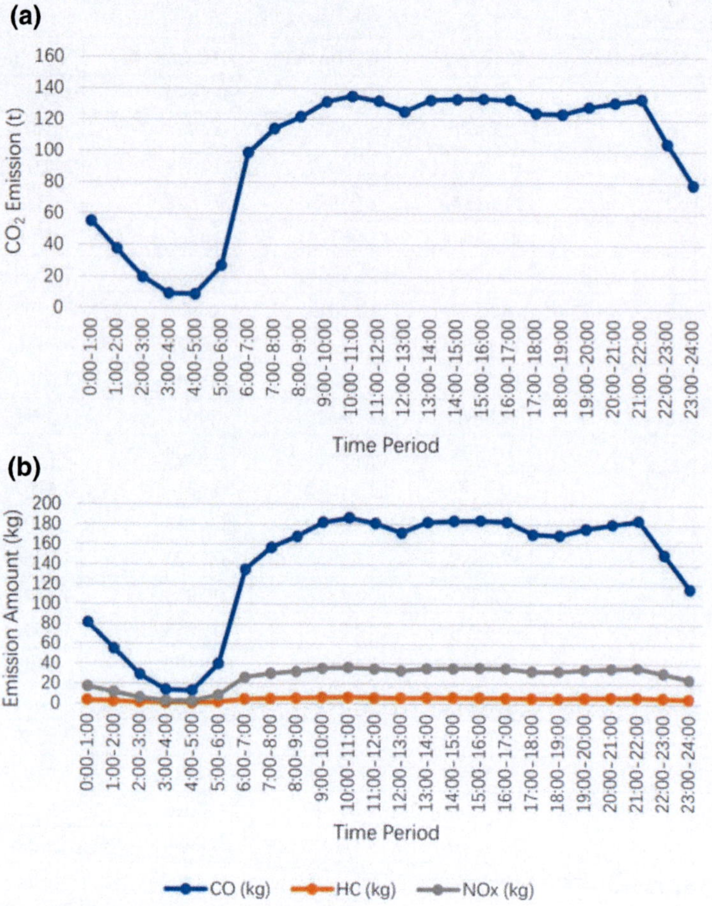

Fig. 2 Hourly variation of the overall Qiangsheng taxi emissions. **a** Hourly variation of CO_2 emission. **b** Hourly variation of CO, HC, NO_x emissions

Q represents the daily benefit of one scheme;
n (=5) represents the five components, i.e., CO_2, NO_X, CO, HC and fuel;
P_i represents the price or cost of each component (unit: RMB/t for CO_2, RMB/kg for NO_X, CO and HC, RMB/L for fuel);
m_i represents the mass weight of reduction (unit: t for CO_2, kg for NO_X, CO and HC, L for fuel);
e_i represents the equivalent coefficient of each component (1 for CO_2 and fuel, 0.95,16.7 and 0.27 for NO_X, CO and HC, respectively).

According to the average carbon price (P_1) of December 24, 2018 by the China Carbon Trade Net website (www.tanjiaoyi.com/tanshichang/), 33 RMB per ton was used as the price of CO_2. The environmental cost of harmful gases is decided by the "Schedule of Environmental Protection Tax" of China, in which the tax of atmospheric pollutant is at least 1.2 RMB per pollution equivalent. Pollution treatment cost is generally 5–20 times of the tax. To simplify the calculation, 10 RMB per pollution equivalent was used as the environmental cost of atmospheric pollutants (Long et al., 2018) for NO_X, CO and HC (P_2, P_3, and P_4). According to "Standard Measures for the Administration of the Collection of Pollutant Discharge Fees" of China, the equivalent coefficients for NO_X, CO and HC are 0.95, 16.7 and 0.27 kg/equivalent (Chen et al., 2005), respectively. Moreover, the fuel price (P_5) was set as 6.71 RMB/L, the average price of 92# gasoline in Shanghai in June 2019.

The total costs consist of **fixed costs** (C_{fixed}) and **annual maintenance costs** (C_{mtn}), on which the recycling period of the costs can be calculated by Eq. (9):

$$t = C_{fixed}/(Q - C_{mtn}) \tag{9}$$

where,

t	represents the recycling period (unit: year);
C_{fixed}	represents the fixed investment cost (unit: RMB);
C_{mtn}	represents the annual maintenance cost (unit: RMB/year);
Q	represents the benefit (unit: RMB/year).

Scheme 1: Releasing the Peak-Hour Crowdedness of Metro Stations

The crowdedness of metro stations may be mitigated from several perspectives, including increasing dispatching frequency or enhancing the capacity of subway trains, controlling passenger entrance, and renovating metro stations, etc. The expenses of such strategies range from rather low staff stipend (labor cost) to the extremely high infrastructure construction cost. As a result, only the benefits were considered by assuming the crowdedness is reduced to 50%, which were calculated as follows.

$$Q(sch\_1) = \sum_{i=1}^{5} P_i * m_{sch\_1,i}/e_i \tag{10}$$

$$m_{sch\_1,i} = M_i - M_{sch\_1,i} \tag{11}$$

$$M_{sch\_1,i} = \sum_{i=1}^{n'} \sum_{j=1}^{m} VSP_{ij} * t_{ij} \tag{12}$$

$$P_{taxi,q}(sch\_1) = exp(V_{taxi,q})/(\exp(V_{taxi,q}) + exp(V_{metro,q})) \tag{13}$$

$$V_{metro,q} = \beta_0 + \beta_1 * pop\_1 + \cdots + \beta_{n-1} * fee + \beta_n * (\mathbf{0.5} * \boldsymbol{cong}) \tag{14}$$

where,

$m_{sch\_1,i}$ is the emission/fuel reduction of component i in Scheme 1;
M_i is the original mass weight of component i with no scheme conducted;
$M_{sch\_1,i}$ is the mass weight of component i with Scheme 1 conducted;
n' is the amount of taxi travels with Scheme 1 conducted;
$\beta_0, \beta_1, \ldots, \beta_{n-1}, \beta_n$ are the coefficients of influencing factors.

The updated mode choice for the mitigation scheme (Scheme 1) was estimated by Eqs. (13) and (14), and then the number of travelers switching to metro was approximated. The results revealed that an overall 26.98% of trips used to take taxis during peak hours would switch to metro. Then, the updated emissions from the taxi trips were calculated by Eq. (12). As a result, the environmental treatment savings were calculated as the economic benefit and assessed by Eqs. (10) and (11) as follows.

$$\begin{aligned} \mathrm{Q(CO_2)} &= P_1 * \frac{m_{sch\_1,1}}{e_1} = P_1 * \frac{M_1 - M_{sch\_1,1}}{e_1} = 33 * \frac{2371.92 - 2178.92}{1} \\ &= 6369\,\mathrm{RMB} \end{aligned}$$

$$\begin{aligned} \mathrm{Q(CO)} &= P_2 * \frac{m_{sch\_1,2}}{e_2} = P_2 * \frac{M_2 - M_{sch\_1,2}}{e_2} = 10 * \frac{3296.53 - 3031.36}{16.7} \\ &= 158.78\,\mathrm{RMB} \end{aligned}$$

$$\begin{aligned} \mathrm{Q(HC)} &= P_3 * \frac{m_{sch\_1,3}}{e_3} = P_3 * \frac{M_3 - M_{sch\_1,3}}{e_3} = 10 * \frac{114.59 - 105.86}{0.27} \\ &= 323.33\,\mathrm{RMB} \end{aligned}$$

$$\begin{aligned} \mathrm{Q(NO_X)} &= P_4 * \frac{m_{sch\_1,4}}{e_4} = P_4 * \frac{M_4 - M_{sch\_1,4}}{e_4} = 10 * \frac{647.26 - 596.12}{0.95} \\ &= 538.32\,\mathrm{RMB} \end{aligned}$$

$$\begin{aligned} \mathrm{Q(fuel)} &= P_5 * \frac{m_{sch\_1,5}}{e_5} = P_5 * \frac{M_5 - M_{sch\_1,5}}{e_5} = 6.71 * \frac{395{,}420.13 - 363{,}245.98}{1} \\ &= 538.32\,\mathrm{RMB} \end{aligned}$$

$$\begin{aligned} \mathrm{Q(sch\_1)} &= \mathrm{Q(CO_2)} + \mathrm{Q(CO)} + \mathrm{Q(HC)} + \mathrm{Q(NO_X)} + \mathrm{Q(fuel)} \\ &= 223{,}277.93\,\mathrm{RMB} \end{aligned}$$

To this end, the annual benefit of releasing the peak-hour crowdedness of metro stations is approximated as 223,277.93 * 365 days = 81.395 million RMB.

Scheme 2: Increasing the Mobike Supply (to 200%)

The total number of Mobikes in Shanghai attains 1.5 million by 2018. Assuming the major cost of increasing the supply is the manufacture expenses, which is about 600 RMB per bike, with the total fixed cost of 900 million RMB. If a maintenance ratio of 10% should be carried out (0.3 million), with a cost of 400 RMB per vehicle per year, the total maintenance cost becomes 120 million RMB/year. Consequently, if Mobike supply doubles, the total daily benefit is approximated as follows.

$$Q(\text{sch\_2}) = \sum_{i=1}^{5} P_i * m_{sch\_2,i}/e_i \tag{15}$$

$$m_{sch\_2,i} = M_i - M_{sch\_2,i} \tag{16}$$

$$M_{sch\_2,i} = \sum_{i=1}^{n''}\sum_{j=1}^{m} VSP_{ij} * t_{ij} \tag{17}$$

$$\mathrm{P}_{\text{taxi,q}}(\text{sch\_2}) = exp(\mathrm{V}_{\text{taxi,q}})/(exp(\mathrm{V}_{\text{taxi,q}}) + exp(\mathrm{V}_{\text{mobike,q}})) \tag{18}$$

$$\mathrm{V}_{mobike,q} = \beta_0 + \beta_1 * pop\_3 + \cdots + \beta_j * (\mathbf{2} * \boldsymbol{sup}) + \cdots + \beta_{n-1} * restrict_4 + \beta_n * TDist \tag{19}$$

where,

$m_{sch\_2,i}$	is the emission/fuel reduction of component i in Scheme 2;
M_i	is the original mass weight of component i with kno scheme conducted;
$M_{sch\_2,i}$	is the mass weight of component i with Scheme 2 conducted;
n''	is the amount of taxi travels with Scheme 2 conducted;
$\beta_0, \beta_1, \ldots, \beta_{n-1}, \beta_n$	are the coefficients of influencing factors.

The updated mode choice for this scheme was estimated by Eqs. (18) and (19), and then the number of travelers switching to Mobike was approximated. Finally, an overall 27.94% and 34.12% of taxi trips during off-peak and peak hours would switch to Mobike, respectively. The economic benefit of Scheme 2 is assessed as follows.

$$Q(CO_2) = P_1 * \frac{m_{sch\_2,1}}{e_1} = P_1 * \frac{M_1 - M_{sch\_2,1}}{e_1} = 33 * \frac{2371.92 - 1672.61}{1} = 23077.23\,\text{RMB}$$

$$Q(CO) = P_2 * \frac{m_{sch\_2,2}}{e_2} = P_2 * \frac{M_2 - M_{sch\_2,2}}{e_2} = 10 * \frac{3296.53 - 2325.91}{16.7} = 581.21\,\text{RMB}$$

$$\text{Q(HC)} = P_3 * \frac{m_{sch\_2,3}}{e_3} = P_3 * \frac{M_3 - M_{sch\_2,3}}{e_3} = 10 * \frac{114.59 - 81.05}{0.27}$$
$$= 1242.22\,\text{RMB}$$

$$\text{Q(NO}_\text{X}) = P_4 * \frac{m_{sch\_2,4}}{e_4} = P_4 * \frac{M_4 - M_{sch\_2,4}}{e_4} = 10 * \frac{647.26 - 457.11}{0.95}$$
$$= 2001.58\,\text{RMB}$$

$$\text{Q(fuel)} = P_5 * \frac{m_{sch\_2,5}}{e_5} = P_5 * \frac{M_5 - M_{sch\_2,5}}{e_5} = 6.71 * \frac{395{,}420.13 - 278{,}838.93}{1}$$
$$= 782{,}259.8\,\text{RMB}$$

$$\text{Q(sch\_2)} = \text{Q}(CO_2) + \text{Q(CO)} + \text{Q(HC)} + \text{Q}(NO_X) + \text{Q}(fuel)$$
$$= 809{,}162.04\,\text{RMB}$$

To this end, the annual benefit of doubling Mobike supply is approximated as 809,162.04 * 365 days = 295.28 million RMB. As the fixed cost attains 900 million RMB and the annual maintenance cost attains 120 million RMB/year, the payback period of costs is approximated as follows:

$$\text{t} = \frac{\text{C}_{fixed}}{\text{Q} - \text{C}_{mtn}} = \frac{900}{295.28 - 120} = 5.13\ \text{(years)}$$

Thus, the benefit will compensate for the overall cost in about 5 years.

Scheme 3: Removing the Non-motor Vehicle Restrictions (on Road)

In general, the presence of exclusive bicycle lanes may contribute to a higher likelihood of cycling activities (Moudon et al., 2005). Assuming to remove all the non-motor vehicle restrictions in the city central areas, which requires to provide bike lanes and electrostatic spray guardrails within the area, the construction cost includes three parts as follows (Zeng et al., 2014).

- Asphalt layer. The two-layer asphalt will be used, with the thickness of the asphalt layer 10 cm each, and the one-side width is 2 m, with the unit cost of 250 RMB/m^3. Then, the overall cost of asphalt layer is 2 * 0.1 * 2 * 250 * 1000 = 100 thousand RMB/km.
- Road paintings. The thickness is 2 cm per layer, with the cost of 50 RMB/m^3. Then, the cost of road painting is approximated as 0.02 * 2 * 1000 * 50 = 2000 RMB/km.
- Electrostatic spray guardrail, which costs 300 thousand RMB/km for two-way (height is 0.5 m, and the unit cost is 600 RMB/m^2).

As a result, the average cost of building one-way non-motor roads is 402 thousand RMB/km and 804 thousand RMB/km for two-way roads. According to the existing situation, Shanghai requires to build 50.87 km one-way non-motor roads and 12.34 km two-way non-motor roads to remove all non-motor vehicle restrictions (from Baidu maps). The total fixed cost is calculated as:

$$\begin{aligned} C_{fixed} &= 402 * 50.87 + 804 * 12.34 \\ &= 30371.1\text{ (thousand RMB)} = 30.37\text{ (million RMB)} \end{aligned}$$

Assuming the maintenance cost is 10% of the fixed investment (Zeng et al., 2014), which was calculated as 3.037 million RMB/year. Consequently, the total daily benefit for removing all restrictions of non-motor vehicles is approximated as follows.

$$\mathrm{Q(sch\_3)} = \sum_{i=1}^{5} P_i * m_{sch\_3,i}/e_i \tag{20}$$

$$m_{sch\_3,i} = M_i - M_{sch\_3,i} \tag{21}$$

$$M_{sch\_3,i} = \sum_{i=1}^{n'''}\sum_{j=1}^{m} VSP_{ij} * t_{ij} \tag{22}$$

$$\mathrm{P_{taxi,q}(sch\_3)} = exp(\mathrm{V_{taxi,q}})/(exp(\mathrm{V_{taxi,q}}) + exp(\mathrm{V_{mobike,q}})) \tag{23}$$

$$\mathrm{V}_{mobike,q} = \beta_0 + \beta_1 * pop_3 + \cdots + (\beta_{n-2} * restrict_3 + \beta_{n-1} * restrict_4) + \beta_n * TDist \tag{24}$$

where,

$m_{sch\_3,i}$	is the emission/fuel reduction of component i in Scheme 3;
M_i	is the original mass weight of component i with no scheme conducted;
$M_{sch\_3,i}$	is the mass weight of component i with Scheme 3 conducted;
n'''	is the amount of taxi travels with Scheme 3 conducted;
$\beta_0, \beta_1, \ldots, \beta_{n-1}, \beta_n$	are the coefficients of influencing factors.

The updated mode choice for the scheme was estimated by Eqs. (23) and (24), and then the number of travelers switching to Mobike was approximated. Finally, an overall 26.21% and 22.79% of taxi trips during off-peak and peak hours would switch to Mobike, respectively. The economic benefit is assessed as follows.

$$\begin{aligned} \mathrm{Q(CO_2)} &= P_1 * \frac{m_{sch\_3,1}}{e_1} = P_1 * \frac{M_1 - M_{sch\_3,1}}{e_1} = 33 * \frac{2371.92 - 1781.84}{1} \\ &= 19{,}472.64\text{ RMB} \end{aligned}$$

$$Q(CO) = P_2 * \frac{m_{sch\_3,2}}{e_2} = P_2 * \frac{M_2 - M_{sch\_3,2}}{e_2} = 10 * \frac{3296.53 - 2476.60}{16.7}$$
$$= 490.98\,\text{RMB}$$

$$Q(HC) = P_3 * \frac{m_{sch\_3,3}}{e_3} = P_3 * \frac{M_3 - M_{sch\_3,3}}{e_3} = 10 * \frac{114.59 - 86.11}{0.27}$$
$$= 1054.81\,\text{RMB}$$

$$Q(NO_X) = P_4 * \frac{m_{sch\_3,4}}{e_4} = P_4 * \frac{M_4 - M_{sch\_3,4}}{e_4} = 10 * \frac{647.26 - 486.35}{0.95}$$
$$= 1693.79\,\text{RMB}$$

$$Q(fuel) = P_5 * \frac{m_{sch\_3,5}}{e_5} = P_5 * \frac{M_5 - M_{sch\_3,5}}{e_5} = 6.71 * \frac{395{,}420.13 - 297{,}048.53}{1}$$
$$= 660{,}073.40\,\text{RMB}$$

$$Q(sch\_3) = Q(CO_2) + Q(CO) + Q(HC) + Q(NO_X) + Q(fuel)$$
$$= 682{,}785.62\,\text{RMB}$$

To this end, the annual benefit of removing non-motor vehicle restriction area is approximated as 682,785.62 * 365 days = 249.29 million RMB. As the fixed cost attains 30.37 million RMB and the annual maintenance cost is 3.037 million RMB/year, the payback period is approximated as follows:

$$t = \frac{C_{fixed}}{Q - C_{mtn}} = \frac{30.37}{249.29 - 3.037} = 0.12\,(\text{years})$$

Thus, the benefit will compensate for the cost within 1 year.

Scheme 4: Updating Taxis into Electric Vehicles

The cost for updating taxis into electric vehicles generally includes three parts, acquisition cost, maintenance cost, and recycle cost. Assuming the acquisition cost is 160,000 RMB per vehicle, the maintenance cost (recharging and repair expense) is 0.7 RMB/km per vehicle and the recycle cost of vehicle plus battery is 50,000 RMB per vehicle, the overall cost of updating taxis is approximated as follows (Ren et al., 2009).

$$C = (160{,}000 - 50{,}000) * N + 0.7 * s = 110{,}000 * N + 0.7 * s \tag{25}$$

where,

C is the overall cost for updating taxis into electric vehicles (unit: RMB);
N is the total number of taxis.

s is the daily travel distance for all Qiangsheng taxis (unit: km);

According to the GPS trajectories, the total number of Qiangsheng taxis is approximately 15,000, and the overall travel distance is 3,412,000 km/day. Then, the total cost is calculated as:

$$\begin{aligned}\mathrm{C} &= 110{,}000 * 15000 + 0.7 * 3{,}412{,}000\ (\mathrm{km/day}) = 1.65\ (\text{billion RMB}) \\ &\quad + 2.39\ (\text{million RMB/day})\end{aligned}$$

Thus, the overall fixed cost 1.65 billion RMB, and the maintenance cost is 2.39 million RMB/day.

As electric vehicles don't consume gasoline and discharge exhausts, the emission reduction equals fully to the original pollutant and fuel amount with no scheme conducted.

$$\mathrm{Q(sch\_4)} = \sum_{i=1}^{5} P_i * M_i / e_i \tag{26}$$

where,

M_i is the original mass weight of component i with no scheme conducted.

The individual daily benefit of each component is assessed as follows.

$$\mathrm{Q(CO_2)} = P_1 * \frac{M_1}{e_1} = 33 * \frac{2371.92}{1} = 78{,}273.36\,\mathrm{RMB}$$

$$\mathrm{Q(CO)} = P_2 * \frac{M_2}{e_2} = 10 * \frac{3296.53}{16.7} = 1973.97\,\mathrm{RMB}$$

$$\mathrm{Q(HC)} = P_3 * \frac{M_3}{e_3} = 10 * \frac{114.59}{0.27} = 4244.07\,\mathrm{RMB}$$

$$\mathrm{Q}(NO_X) = P_4 * \frac{M_4}{e_4} = 10 * \frac{647.26}{0.95} = 6813.26\,\mathrm{RMB}$$

$$\mathrm{Q}(fuel) = P_5 * \frac{M_5}{e_5} = 6.71 * \frac{395{,}420.13}{1} = 2{,}653{,}269.07\,\mathrm{RMB}$$

$$\begin{aligned}\mathrm{Q(sch\_4)} &= \mathrm{Q}(CO_2) + \mathrm{Q(CO)} + \mathrm{Q(HC)} + \mathrm{Q}(NO_X) + \mathrm{Q}(fuel) \\ &= 2{,}744{,}573.73\,\mathrm{RMB}\end{aligned}$$

To this end, the annual benefit of updating taxis into electric vehicles is approximated as 2,744,573.73 * 365 days = 1001.77 million RMB. As the fixed cost is 1650 million RMB and the annual maintenance cost attains 2.39 * 365 days = 872.35 million RMB/year, the payback period is calculated as follows:

$$t = \frac{C_{fixed}}{Q - C_{mtn}} = \frac{1650}{1001.77 - 872.35} = 12.75\ (\text{years})$$

Thus, the benefit will compensate for the cost in almost 13 years (12.75 years).

Scheme 5: Equipping Taxis with Catalytic Converters

As then fixed cost of equipping taxis with catalytic converters is 610 RMB per vehicle, the total amount of fixed cost for 15 thousand taxis is 610 * 15,000 = 9.15 million RMB. Catalytic converters transform harmful pollutants, such as CO, HC and NO_X into CO_2, H_2O, and N_2, which reduces CO, HC and NO_X emissions while enhances CO_2e emissions. The benefit is approximated as follows.

$$Q(sch\_5) = \sum_{i=1}^{5} P_i * m_{sch\_5,i}/e_i \tag{27}$$

$$m_{sch\_5,1} = -(m_{sch\_5,2} * h_{CO} + m_{sch\_5,3} * h_{HC}) \tag{28}$$

where,

$m_{sch\_5,i}$ is the reduction of component i in Scheme 5;
$m_{sch\_5,1}$ is the reduction of CO_2 in Scheme 5;
$m_{sch\_5,2}$ is the reduction of CO in Scheme 5;
h_{CO} is the transformation index of CO to CO_2;
$m_{sch\_5,3}$ is the reduction of HC in Scheme 5;
h_{HC} is the transformation index of HC to CO_2.

According to the conservation of mass, $h_{CO} = 44{:}28$ and $h_{HC} = 44 : 13$. As Scheme 5 doesn't contribute to fuel consumption, the benefit of fuel-saving was not calculated and the individual daily benefit is assessed as follows.

$$Q(CO) = P_2 * \frac{m_{sch\_5,2}}{e_2} = P_2 * \frac{M_2}{e_2} = 10 * \frac{3296.53}{16.7} = 1973.97\,\text{RMB}$$

$$Q(HC) = P_3 * \frac{m_{sch\_5,3}}{e_3} = P_3 * \frac{M_3}{e_3} = 10 * \frac{114.59}{0.27} = 4244.07\,\text{RMB}$$

$$Q(NO_X) = P_4 * \frac{m_{sch\_5,4}}{e_4} = P_4 * \frac{M_4}{e_4} = 10 * \frac{647.26}{0.95} = 6813.26\,\text{RMB}$$

$$m_{sch\_5,1} = -(m_{sch\_5,2} * h_{CO} + m_{sch\_5,3} * h_{HC}) = -(M_2 * h_{CO} + M_3 * h_{HC})$$
$$= -\left(3296.53 * \frac{44}{28} + 114.59 * \frac{44}{13}\right) = -5568.10\,\text{kg} = -5.5681\,\text{ton}$$

$$Q(CO_2) = P_1 * \frac{m_{sch\_5,1}}{e_1} = 33 * \frac{-5.5681}{1} = -183.74\,\text{RMB}$$

$$Q(sch\_5) = Q(CO_2) + Q(CO) + Q(HC) + Q(NO_X) = 12847.56\,RMB$$

To this end, the annual benefit of equipping taxis with catalytic converters is approximated as 12,847.56 * 365 days = 4.69 million RMB. As the fixed cost is 9.15 million RMB with no maintenance cost in each year, the payback period is calculated by Eq. (9) as follows:

$$t = \frac{C_{fixed}}{Q - C_{mtn}} = \frac{9.15}{4.69 - 0} = 1.95\,(\text{years})$$

Thus, the benefit will compensate for all costs in two years. As the average life span of each catalytic converter is 3.6 years, equipping taxis with catalytic converters achieves much shorter cost recycle length.

In summary, the recycling lengths for Schemes 2–5 were approximated as 5.13, 0.12, 12.75 and 1.95 years, respectively, while only the annual benefit of Scheme 1 was approximated as the various investments, including both infrastructure and management costs, may be involved. Removing non-motor vehicle restrictions (Scheme 3) has the shortest cost recycling length within one year (0.12) and was highly recommended as accordance with the proposal of urban non-motorized transportation. Updating taxis into electric vehicles (Scheme 4) has the longest cost recycling length (12.75), which however has eliminated CO_2 emission in response to mitigate global warming. Meanwhile, equipping all taxis with catalytic converters (Scheme 5) also has a cost recycling length within two years (1.95), which unfortunately generates additional CO_2 emissions. To this end, removing non-motor vehicle restrictions (Scheme 3) is recommended as the most efficient strategy although equipping all taxis with catalytic converters (Scheme 5) also has rather outstanding performance.

5 Conclusions

This study investigated two sets of urban mode choice behavior, metro-taxi and Mobike-taxi, proposed and evaluated corresponding emission reduction strategies. Two binary logit models were constructed to investigate the mode choice and the influencing factors based on the trajectory and order data of taxis and Mobikes, IC card swiping data and PEMS experiments. A VSP-based localized emission model for taxi vehicles was established to quantify the benefit of three emission reduction schemes proposed according to the significant influencing factors identified within the mode choice model (**cong, sup, restrict_3, restrict_4**) and two other popular countermeasures. The main contributions are listed as below.

1. Multi-source data from taxis and substitution modes, i.e. metro and Mobike were used, and the corresponding data preprocessing procedure may establish a pilot for similar studies.

2. Two BNL models were constructed to investigate the mode choice for metro-taxi, and Mobike-taxi, by incorporating socio-economic and demographic variables, urban morphology and land use variables, and travel schedule related variables as the influencing factors. Peak/off-peak hours trips were used to build up models respectively, with the results revealed that trip distance, fare and time differences have the largest significance, compared with other attributes. Results of peak/off-peak hours have considerable differences. In peak hours, residential land use ratio and number of office building near the origin and destination promote the metro usage, strengthening metro as the main commuting mode. The improvements of transportation land use ratio and bus line number have a positive effect in choosing metro and Mobike, while commercial land use ratio and the number of medical hospitals promote taxi usage.
3. To further investigate the taxi emissions, a VSP-based emission model was constructed to estimate the exhaust pollutant from taxis. PEMS field experiments were carried out to obtain the localized emissions. Then, the benefit assessment of five emission reduction schemes were calculated based on the localized VSP emission model. The results demonstrate that removing non-motor vehicle restrictions has the shortest payback period within one year (0.12), compared with other schemes.

Although the results are promising, additional works may be conducted to improve the performance of study. For example, incorporating additional independent variables, applying advanced choice models, and using more than one day data, should be effective expansions for the future works. Moreover, the current field data based VSP emission model is rather straightforward, which is not appropriate for precise emission calculations. Advanced emission models, such as CORPERT and MOBILE, may be incorporated for the pollutant emission estimation (Sun et al. 2021). Lastly, assuming that the trip substitution can reduce 100% of the emissions from taxis is actually unrealistic, as the proposed substitute modes, including Metro (e.g., fuel/electricity consumption), Mobike (e.g., bike rebalancing using automobiles), and electric vehicles (e.g., electricity generated from coal), may also generate emissions. In fact, metro emissions may be approximated by the schedules of given metro lines, and the automobile emissions for Mobike rebalancing may be approximated with the daily vehicle mile travel (VMT). However, as this study mainly concentrates on the on-road emissions, the pollutants for operating metro, Mobike and electric vehicles were reasonably not considered, and may be investigated in future studies.

References

Alghuraybi, B., Marvaniya, K., & Xia, G. (2006). Analyze NYC taxi data using hive and machine learning. *International Journal of Database Theory and Application, 9*, 191–198.

Benoit, L. (2009). Long-term energy consumptions of urban transportation: A prospective simulation of "transport–land uses" policies in Bangalore. *Energy Policy, 37*, 940–953.

Bischoff, J., & Maciejewski, M. (2014). Agent-based simulation of electric taxicab fleets. *Transportation Research Procedia, 4*, 191–198.

Cai, H., Jia, X. P., Chui, S. F., Hu, X. J., & Xu, M. (2014). Siting public electric vehicle charging stations in Beijing using big-data informed travel patterns of the taxi fleet. *Transportation Research Part D: Transport and Environment, 33*, 39–46.

Cao, X. J. (2019). Examining the effect of the Hiawatha LRT on auto use in the twin cities. *Transport Policy, 81*, 284–292.

Chavis, C., & Gayah, V. V. (2017). Development of a mode choice model for general purpose flexible-route transit systems. *Transportation Research Record: Journal of the Transportation Research Board, 2650*, 133–141.

Chen, X. X., Wang, W. B., Chen, H. T., Chen, X., & Li, Z. J. (2005). Application of pollutant equivalent in regional pollution source evaluation. *The Administration and Technique of Environmental Monitoring, 17*, 41–43.

Cipriani, E., Mannini, L., Montemarani, B., Nigro, M., & Petrelli, M. (2019). Congestion pricing policies: Design and assessment for the city of Rome, Italy. *Transport Policy, 80*, 127–135.

Figliozzi, M., Saenz, J., & Faulin, J. (2020). Minimization of urban freight distribution lifecycle CO2e emissions: Results from an optimization model and a real-world case study. *Transport Policy, 86*, 60–68.

Ge, W., Shao, D., Xue, M., Zhu, H., & Cheng, J. (2020). Urban taxi ridership analysis in the emerging metropolis: Case study in Shanghai. *Case Studies on Transport Policy, 8*(1), 173–179.

Hochmair, H. H. (2016). Spatiotemporal pattern analysis of taxi trips in New York City. *Transportation Research Record: Journal of the Transportation Research Board, 2542*, 45–56.

Huang, H.-J., Xia, T., Tian, Q., Liu, T.-L., Wang, C., & Li, D. (2020). Transportation issues in developing China's urban agglomerations. *Transport Policy, 85*, A1–A22.

Lang, Y. S. (2018). Research on cycle traffic and dedicated cycle network. *Journal of Transportation Engineering, 18*, 18–23.

Laurino, A., Beria, P., Debernardi, A., Ferrara, E., 2019. Accessibility to Italian remote regions: comparison among different transport alternatives. *Transport Policy, 83*, 127–138.

Long, F., Yang, Q. J., Ge, C. Z., & Dong, Z. F. (2018). Analysis on the influence of environmental protection tax on enterprises economic burden. *Environmental Protection, 46*, 82–85.

Luo, X., Dong, L., Dou, Y., Zhang, N., Ren, J. Z., Sun, L., & Yao, S. Y. (2017). Analysis on spatial-temporal features of taxis' emissions from big data informed travel patterns: A case of Shanghai, China. *Journal of Cleaner Production, 142*, 926–935.

Martinetz, D. (1989). U.S. Environmental protection agency (US-EPA). *Umweltwissenschaften Und Schadstoff-Forschung, 1*, 6–6.

Ministry of Transport of the People's Republic of China (MOTPRC). (2019). Development statistics bulletin of transportation industry in 2018. http://xxgk.mot.gov.cn/jigou/zhghs/201904/t20190412_3186720.html. Accessed on January 30, 2020 (in Chinese)

Moudon, A. V., Lee, C., Cheadle, A. D., Collier, C. W., Johnson, D., Schmid, T. L., & Weather, R. D. (2005). Cycling and the built environment, A US perspective. *Transportation Research Part D: Transport and Environment, 505*, 245–261.

Ren, Y. L., Li, H. F., Sun, R., & Guan, L. (2009). Life-cycle cost model and analysis of electric vehicles based on the consumers' perspective. *Technology Economics, 28*, 54–58.

Richardson, B. C. (2005). Sustainable transport: Analysis frameworks. *Journal of Transport Geography, 13*, 29–39.

Shaheen, S., Martin, E., & Totte, H. (2020). Zero-emission vehicle exposure within U.S. carsharing fleets and impacts on sentiment toward electric-drive vehicles. *Transport Policy, 85*, A23–A32.

Sun, D. J., Chen, S., Zhang, C., & Shen, S. (2016). A bus route evaluation model based on GIS and super-efficient data envelopment analysis. *Transportation Planning and Technology, 39*(4), 407–423.

Sun, D. J., & Ding, X. (2019). Spatiotemporal evolution of ridesourcing markets under the new restriction policy: A case study in Shanghai. *Transportation Research Part A: Practice and Policy, 130*, 227–239.

Sun, D. J., Zhang, C., Zhang, L., Chen, F., & Peng, Z.-R. (2014). Urban travel behavior analyses and route prediction based on floating car data. *Transportation Letters: The International Journal of Transportation Research, 6*(3), 118–125.

Sun, D. J., Zhang, K., & Shen, S. (2018). Analyzing spatiotemporal traffic line source emissions based on massive Didi online car-hailing service data. *Transportation Research Part D: Transport and Environment, 62*, 699–714.

Sun, D. J., Zhang, Y., Xue, R., & Zhang, Y. (2017). Modeling carbon emissions from urban traffic system using mobile monitoring. *Science of the Total Environment, 599–600*, 944–951.

Sun, D. J., Shi, X., Zhang, Y., & Zhang, L. (2021). Spatiotemporal distribution of traffic emission based on wind tunnel experiment and computational fluid dynamics (CFD) simulation. *Journal of Cleaner Production, 282*, 124495.

Szeto, W. Y., Wong, R. C. P., Wong, S. C., & Yang, H. (2013). A time-dependent logit-based taxi customer-search model. *International Journal of Urban Sciences, 17*(2), 184–198.

Wang, B. Y., Zhao, L. Y., Pang, Y. Q., Zhang, D., & Yang, X. G. (2013a). Analysis of passenger's choice between shuttle bus and illegal taxi. *Procedia—Social and Behavioral Sciences, 96*, 1948–1960.

Wang, M., Liu, X., Guo, D. D., & Cao, H. L. (2013b). Investigation on emission of high mileage taxis in Beijing. *Vehicle and Power Technology, 2*, 51–53 (in Chinese).

Xu, W., Long, Y., & Zhang, W. (2019). Prioritizing future funding and construction of the planned high-speed rail corridors of China—According to regional structure and urban land development potential indices. *Transport Policy, 81*, 381–395.

Yutaka, T., Liu, J. P., Yasuhiko, K., Ning, Y. D., & Oki, F. (2006). A survey on energy consumption in rural households in the fringes of Xian city. *Energy and Buildings, 38*, 1335–1342.

Zeng, Y., Min, Z. R., Yu, Z. Y., & Xu, W. (2014). Study of Nanchang city public bicycle system based on the perspective of low carbon. *Technology and Economy in Areas of Communications, 16*, 56–60.

Zhan, X. Y., Hasan, S., Ukkusuri, V., & Kamga, C. (2013). Urban link travel time estimation using large-scale taxi data with partial information. *Transportation Research Part C: Emerging Technologies, 33*, 37–49.

Zheng, X. L., & Gan, H. C. (2013). Analysis on influential factors of travel mode choice behavior of residents. *Journal of University of Shanghai for Science and Technology, 35*, 563–566 (in Chinese).

Zhong, S., & Bushell, M. (2017). Impact of the built environment on the vehicle emission effects of road pricing policies: A simulation case study. *Transportation Research Part A: Policy and Practice, 103*, 235–249.

Chapter 12
Assessing Built Environment and Land Use Strategies from the Perspective of Urban Traffic Emissions: An Empirical Analysis Based on Massive Didi Online Car-Hailing Data

Abstract Nowadays, the massive car-hailing data has become a popular source for analyzing traffic operation and road congestion status, which unfortunately has seldom been extended to capture detailed on-road traffic emissions. This study aims to investigate the relationship between road traffic emissions and the related built environment factors, as well as land uses. The Computer Program to Calculate Emissions from Road Transport (COPERT) model from European Environment Agency (EEA) was introduced to estimate the 24-h NO_x emission pattern of road segments with the parameters extracted from Didi massive trajectory data. Then, the temporal Fuzzy C-Means (FCM) Clustering was used to classify road segments based on the 24-h emission rates, while *Geographical Detector* and *MORAN's I* were introduced to verify the impact of built environment on line source emissions and the similarity of emissions generated from the nearby road segments. As a result, the spatial autoregressive moving average (SARMA) regression model was incorporated to assess the impact of selected built environment factors on the road segment emission rate based on the probabilistic results from FCM. It was found that short road length, being close to city center, high density of bus stations, more ramps nearby and high proportion of residential or commercial land would substantially increase the emission rate. Finally, the 24-h atmospheric NO_2 concentrations were obtained from the environmental monitor stations, to calculate the time variational trend by comparing with the line source traffic emissions, which to some extent explains the contribution of on-road traffic to the overall atmospheric pollution. Result of this study could guide urban planning, so as to avoid transportation related built environment attributes which may contribute to serious atmospheric environment pollutions.

Keywords On-line car-hailing service · NO_x emission · Fuzzy C-means clustering · Spatial analysis · Built environment

1 Introduction

With the increasing development of motorization and urbanization, transportation has gradually become one of the main sources of air pollution in many cities throughout the world. For example, according to statistics, on-road transportation accounts for

S. Zhong and D. Sun, *Logic-Driven Traffic Big Data Analytics*,
https://doi.org/10.1007/978-981-16-8016-8_12

about 47% of CO, 42% of NO_x, and 18.4% of PM emissions, respectively in Europe (Kousoulidou & Sarmaras, 2008); and 27.5% of total greenhouse gas in U.S., in which 41.9% results from passenger cars (Mausami & Weitz, 2017). It also contributes to about 19% of nitrogen oxide emissions in Canada, rising up to 85% in Montreal (IJC, 2012). In China, it is estimated that transportation related activities account for approximately 46% of NO_x, 78% of CO, and 83% of HC in metropolises (Wang et al., 2008). Consequently, it's important to identify the hotspot and the mechanism of on-road emissions to provide guidance for built environment and land use during the planning stage.

Traditional emission measurements are based on laboratory-similar bench test (Silva et al., 2015), remote sensing (Ning & Chan, 2007), and tunnel test (Zhang et al., 2015). However, the emitting characteristics of motor engines extracted from these studies were only applicable to experimental environment. They are expensive and impossible to be used for dynamic monitoring or multi-vehicle test. In recent years, the portable emission measurement system (PEMS) has been widely used in mobile monitoring, which is easy to assemble. Defries et al. (2014) found that PEMS was more compatible to chassis testing than other mobile methods in detecting common pollutants. Gallus et al. (2016) also disclosed that the correlation tests between particle number PEMS and chassis dynamometer tests are in good agreement. However, a set of PEMS could just be used for one-vehicle at one time, taking months or even years to complete the regional data collection.

In such case, simulation models and detailed estimation formulas are developed to simplify the emission estimation. For example, in U.S., Environmental Protection Agency (EPA)'s MOtor Vehicle Emission Simulator (MOVES) is a state-of-the-science emission modeling system that estimates emissions for mobile sources at the national, county, and project level for criteria air pollutants, greenhouse gases, and air toxics. The emission calculation is based on predefined speed bins and vehicle specific power (VSP) distribution, which is suitable from microscopic to macroscopic. Abou-Senna et al. (2013) integrated MOVES and VISSIM microscopic simulation to predict emissions from vehicles on limited-access highways. Liu et al. (2013) input vehicle operational parameters extracted from GPS trajectories into MOVES for estimating vehicle emission factors in Shanghai. International Vehicle Emission (IVE) model is also a VSP-based algorithm, which generated 60 VSP bins to calculate emissions. The model was evaluated by utilizing an available dataset of remote sensing measurements on a large number of vehicles in Hangzhou, China, and found that it performed well on estimating HC and CO emissions (Guo et al., 2007). Nagpure et al. (2011) applied the fundamental equations of IVE model to estimate the emission rates from the light-duty commercial vehicles, and revealed the increases of CO and VOCs and a decrease of NO_x emission rate along with altitude. Computer Program to Calculate Emissions from Road Transport (COPERT) is another widely used emission model developed by European Environment Agency (EEA) using speed as the main independent variable (Broderick & O'Donoghue, 2007; Lozhkina & Vladimir, 2015). Since COPERT complies with the European emission standard which is similar to the one adopted in China, the package is capable for emission estimation in Chinese cities (Lang et al., 2016). Bellasio et al.

(2007) implemented COPERT to calculate the emission inventory of Sardinia, Italy, and proved its reliability by comparison with field experiments. However, all above referred models require sufficient data when applied to a citywide level, which is unfortunately difficult to be obtained in practical.

On the country, vehicle-based GPS data as a source of driving trajectory record linking to the estimation and influencing mechanism of vehicular emissions, have gained increased attentions (Liu et al., 2009), which covers city wide road network continuously and consequently is believed more accurate. However, studies based on data from the new emerging car-hailing mode are scarce, while most similar studies were focused on taxi vehicle trajectories. Wang et al. (2008) utilized GPS to collect the driving pattern data including 140 taxis, and combined them with IVE to estimate the total emissions of CO, VOC, NO_x, and PM from vehicles in Shanghai. However, this study was just of coarsely resolved data collection with limited vehicle number and land scope. Nyhan et al. (2016) obtained GPS trajectory data from over 15,000 taxis in Singapore, which enabled to quantify the instantaneous driving cycle parameters in high spatio-temporal resolution with the input of real-time emission model in calculating CO_2, NO_x, VOCs, and PM. Luo et al. (2016) extracted speed information from taxi GPS data of Shanghai to quantify spatial and temporal hotspot emissions using COPERT model, disclosing a distribution of dual-core cyclic structure. However, their study failed to distinguish vehicles of different type and emission standard, and the final the output contains solely plotting-based figures.

The existing literatures about on-road emissions mainly were associated with three gaps as follows. First, the majority of studies requires huge amount of survey data as the input for the simulation stage. Since some research are in a rather large scale, the temporal difference was neglected, by just analyzing day, month or year as a whole. Second, most studies just estimate emissions from the original GPS sources without further reckoning on various vehicle types and emission standards. Finally, the results are mainly qualitative plotting or simple statistics without explanation with additional information such as land use and built environment. In fact, from the perspective of sustainable human settlements, it's necessary to explore the correlation between the road emissions and the air pollution concentrations, which should be the ultimate objective in analyzing on-road emissions.

The purpose of this study is to investigate the 24-h line source emission pattern of the road network so as to analyze the problematic segments with continuous or unconventional high emissions. Formulas from COPERT were selected to estimate NO_x emission of different vehicle types and emission standards. The geographical methods including *geographical detector*, *MORAN's Index* and spatial regression were applied to analyze the spatial–temporal characteristics, as well as impacts from the built environment factors in the vicinity. In comparison, the previous study of the same authors (Zhang et al., 2017) has focused on discussing the spatiotemporal distribution and the cause of road congestion based on taxi GPS trajectories, revealing how land use and built environment affect road congestions. The main objective is to propose a modeling frame to combine Fuzzy C-Means (FCM) clustering and spatial regression, while this chapter mainly focuses on road emissions, and utilizes COPERT model to calculate emission factor of road traffic, which is relatively innovative for the transportation big data analytics. One of the key innovation is to approximate the traffic flow composition for all roads by incorporating the surveyed results from 40 selected segments (both primary and secondary arterials), which extends the applicability of the sampled Didi on-line car-hailing data. Then, the 24-h trend pattern of spatial correlation between road emissions and air pollution concentration was further analyzed to emphasize the spatial and temporal relationship of NO_x traffic emission to the overall atmospheric air quality. More specifically, the two sub-objectives of this study are listed as follows:

1. To use Didi massive car-hailing data to extract line source emission pattern of road segments, and then classify segment-based 24-h emission pattern into various clusters through FCM clustering method.
2. To assess the impact of selected built environment and land use factors on segment-based emission rate using the spatial autoregressive moving average (SARMA) regression model, and determine the factors that may significantly contribute to the traffic emissions.

2 Materials and Data

2.1 Didi Service GPS Data

The massive Didi hailing-car GPS data from August 8 to 12, 2016 (Monday to Friday), sunny, were provided by Didi Inc., including both trajectories and orders from the registered taxis and the regular Didi hailing cars. The raw trajectory datasets include *Order ID*, *Signal Received Time*, *Longitude*, *Latitude*, *Speed*, etc. (Sun et al., 2014), with the time interval as 3 s. The overall number of records from August 8 to 12, 2016, are 9,973,471 (2,855,013 taxi records), 10,755,260 (3,184,219 taxi records), 9,544,299 (2,800,285 taxi records), 9,211,886 (2,911,388 taxi records), and 10,296,053 (2,931,417 taxi records) repectively, all in a similar volume level. The trajectories from both taxis and regular cars were first aggregated, so that the hourly average speeds for each day were obtained, as presented in Fig. 1. It was found that the 24-h average speed demonstrates a high similarity (with the largest difference at 4.97%), indicating that the 5-day data can represent multi-day patterns from the perspective of time-of-day variation.

Then, the space projection of GPS records for both AM and PM peak hours were obtained and are shown in Fig. 2, which also demonstrate rather similarities for both morning and evening peaks. Thus, the temporal and spatial similarity of multi-day data have been validated, and the 5-day datasets were chosen as the research subjects for the entire trajectories.

By focusing on certain amount of traffic, only segments, defined as the link between two intersections, from the primary and secondary roads within the Central City (Outer Ring) of Shanghai were considered. A criterion was set as the segment

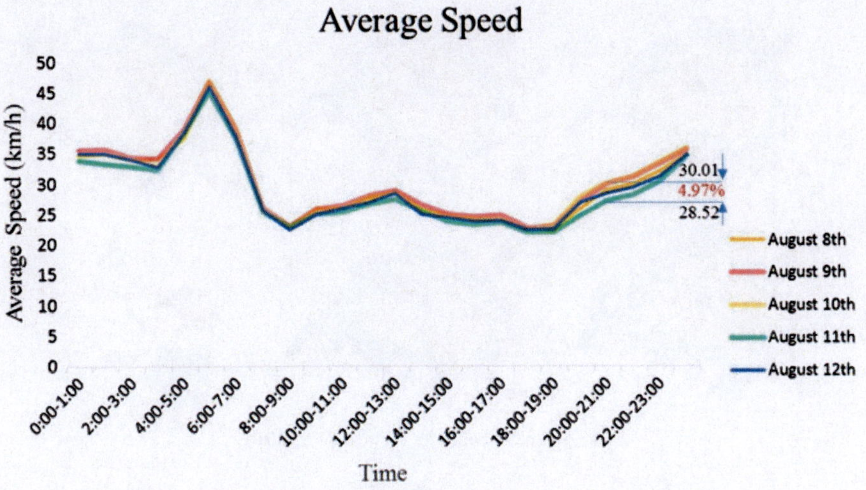

Fig. 1 24-h average speeds of Didi operational cars

length has to be longer than 300 m to avoid the excessive influence of intersections(Sun et al., 2020; Sun et al., 2021a). To this end, a total of 853 road segments were screened out by this stage.

2.2 *Data Processing*

As the raw data contain *Order ID*, *Signal Received Time*, *Longitude*, *Latitude*, *Speed*, etc., the flawed records are mainly caused by location deviation and data missing. Consequently, the trajectory points that do not match to the nearest roads within 15 m, stay within same location for over five minutes, or are with a speed over 120 km/h or with a sudden distance deviation over 100 m were eliminated. Moreover, ten vehicles per hour were set as the threshold for each segment to avoid error propagation, and finally an overall 536 segments were kept with these criteria. Then, the **hourly average speed** of each segment was calculated by averaging the speeds of all Didi-hailed cars passed through, and the **sampled traffic flow** for each segment was also extracted. More details for data processing procedure can be found from the previous study (Zhang et al., 2017).

To calculate the line source emissions, the Didi hailed cars alone are not enough. Supplementary field surveys were carried out to approximate the traffic composition, e.g. passenger cars, autobuses and trucks. As the urban expressway system in Shanghai has three rings: inner ring, middle ring and outer ring, which typically distinguish the city center from the suburb, locations for the proposed traffic composition survey were scheduled to be scattered at each zone along the expressways. Finally, 40 road segments were selected to investigate the vehicle composition, as shown in Fig. 3, among which 10 roads were covered for each area zone, with five

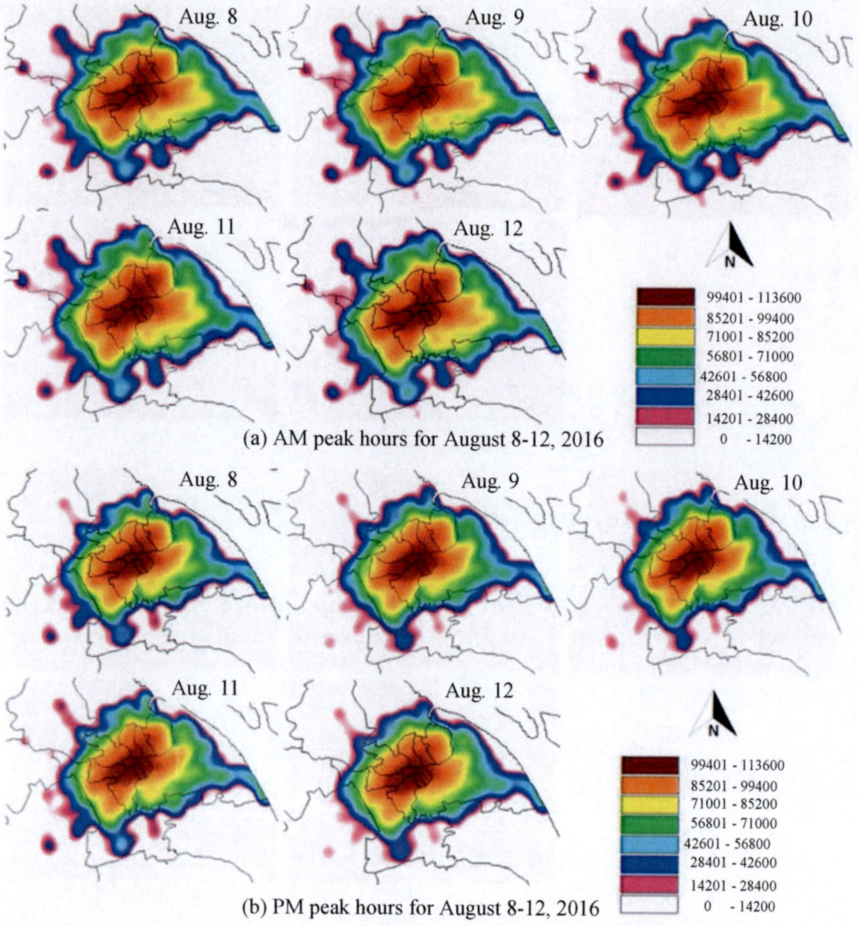

Fig. 2 Geospatial distribution of trajectory points for AM and PM peak hours

primary and five secondary, respectively. The results of surveying traffic composition for primary and secondary arterial segments within each zoning area are provided in Table 1. It can be figured out that passenger cars generally consist of more than 85% traffic volume, except for the primary roads in Outside Outer Ring, Outer-middle Rings, and Middle-inner Rings, in which the percentage of passenger cars are between 80 and 85%. Also, the percentages of trucks within the outer ring, particularly within the middle ring, are smaller (within 5%), as trucks are forbidden to enter into many central urban areas in Shanghai.

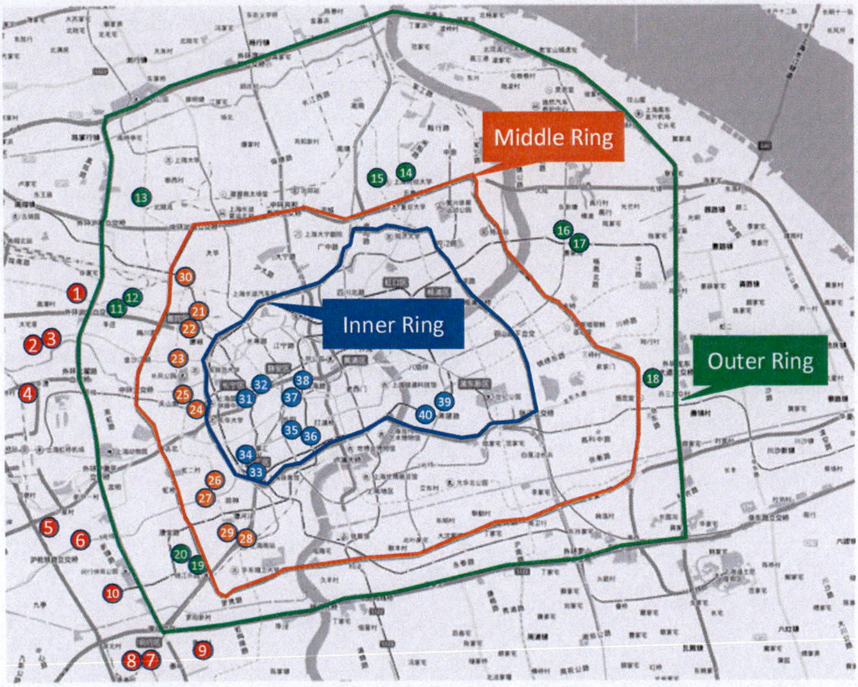

Fig. 3 Traffic volume and composition investigations on 40 selected road segments

Table 1 Surveying traffic composition for each zoning area per vehicle type (%)

Zoning area		Passenger car		Autobus	Truck	Overall
		Taxi	Ordinary			
Outside outer ring	Primary	14.3	68.1	8.2	9.4	100
	Secondary	13.6	74.5	6.3	5.6	100
Outer-middle rings	Primary	16.3	67.2	10.4	6.1	100
	Secondary	14.1	71.7	9.6	4.6	100
Middle-inner rings	Primary	6.9	73.9	16.3	2.9	100
	Secondary	8.4	76.9	12.3	2.4	100
Within inner ring	Primary	7.7	81	9.4	1.9	100
	Secondary	8.1	80.6	8.6	2.7	100

2.3 *Traffic Volume Estimation*

With the surveys, the total number of vehicles as well as the traffic compositions in each category, namely **passenger car, autobus** and **truck** on the 536 filtered roads were obtained. By combining with the Didi-hailed taxis data, which are all passenger cars, the percentage of the Didi hailing cars to the total passenger cars on each road segment were calculated, respectively. The ratio value was estimated as 12.9%, with no large differences across the zoning area and road type. Then, the number of on-road passenger cars, autobuses and trucks for the rest 496 segments can be estimated as follows:

1. For each segment k, calculate the number of overall passenger cars based on the estimated ratio, 12.9%:

$$Q1_k = q1_k/0.129 \tag{1}$$

where $Q1_k$ is the number of passenger cars (including both Didi and the regular passenger cars) on road segment k; and $q1_k$ is the number of Didi hailing vehicles (including taxis and ordinary on-hailing cars) on road segment k.

2. For each segment k, calculate the number of autobuses and trucks according to the surveyed traffic composition provided in Table 1:

$$Q2_k = Q1_k \times R2_k/R1_k \tag{2}$$

$$Q3_k = Q1_k \times R3_k/R1_k \tag{3}$$

where $Q2_k$ and $Q3_k$ are the number of autobuses and trucks on road segment k, and R_{1k}, R_{2k}, and R_{3k} are the proportions of passenger cars (including both Didi and the regular passenger cars), autobuses and trucks on the road segment k, which should be located on the corresponding zoning area and the road type.

2.4 Calculation of NO_x Line Source Emission

As the emission standard in China is similar to that in Europe, for example, Euro IV is equal to China Stage IV, this study mainly focused on the standards from China Stage III to V, issued in 2008, 2009 and 2014, respectively within Shanghai. Each newly-purchased vehicle should comply with the emission standard of the year, from which the standard distribution could be estimated. According to *Shanghai Municipal Comprehensive Transportation Operation Annual Report* (SRIUTD, 2015, 2016), the annual retired rate of vehicles in Shanghai is around 6%, including passenger cars, autobuses, and trucks. This study first obtained the number of newly-purchased vehicles of each model, and then incorporated the annual retiring rate to estimate the number of vehicles of China Stage III to V. Since China Stage I standard had been obsoleted in 2015 (Wang, 2015), the number of vehicles in Stage II equals to the total vehicle amount minus the number of vehicles from Stages III to V, with the results summarized in Table 2.

In this study, the emission rates of vehicles were calculated by type and standard, as follows.

Table 2 Emission standard distribution per vehicle type (%)

China stage	II	III	IV	V	Overall
Passenger car	17.34	12.22	55.63	14.81	100
Autobus	11.79	16.74	49.82	21.65	100
Truck	45.71	12.02	37.85	4.42	100

$$TE_k = \sum_{i=1}^{3}\sum_{j=2}^{5} EF_{i,j,k} \times q_{i,j,k} \times L_k \tag{4}$$

$$EF_k = TE_k / q_k \tag{5}$$

where, TE_k represents the total NO_x emission on road segment k (unit: g). $EF_{i,j,k}$ is the emission factor (unit: g/m^3) of vehicle type i and emission standard j on segement k. $q_{i,k}$ is the number of type i vehicles on segment k. L_k is the length of segment k (unit: m). EF_k is the average emission factor (unit: g/m^3) of a single vehicle on segment k, which delegates the pollution pattern (Chen et al., 2020; Sun et al., 2021b).

NO_x emission factor, $EF_{i,j,k}$, of vehicles could be estimated from hourly average speed according to COPERT (Nikoleris et al., 2011), in which the passenger car, autobus, and truck are corresponding to passenger vehicle, bus and light commercial trucks (articulated vehicles under traffic control), respectively. Road grade is not considered here, as the surface roads for analysis in the study are relatively flat, and COPERT is mainly a speed-dependent meso-level model, which is based on empirical formulas, concerning little about acceleration and detailed vehicle types represented by engine capacities. Moreover, the loadings of buses and trucks were assumed as 50% in this study, and the parameters used to estimate NO_x emission are provided in Table 3, with the calculation formulas as in Eqs. (6) and (7).

$$EF = (\mathrm{a} + \mathrm{c} \times V + \mathrm{e} \times V^2 + \mathrm{f}/V)/(\mathrm{g} + b \times V + d \times V^2) \tag{6}$$

$$EF = \mathrm{a} \times V^b + \mathrm{c} \times V^d \tag{7}$$

where, EF is the average emission factor, V is the hourly average speed of a certain road segment, and a, b, c, d, e, f, g are coefficient parameters.

The hourly NO_x emission factor EF_k for the 536 road segments were calculated, and then aggregated to the average emission factor as a vector with 24 dimensions, $\{EF_1, EF_{2,} \ldots EF_{24}\}$. Figure 4 presents the vehicle average emission factor by each hour, in which the emission pattern demonstrates obvious higher values during peak hours, indicating the severest pollution, although the emission pattern is related to multiple factors, including speed, traffic composition, and traffic control (Pan et al., 2017; Sun & Zhang, 2018).

Table 3 Parameters for estimating NO_x emissions in COPERT

Vehicle type	China stage	Eq.	a	b	c	d	e	f	g
Passenger car	II	(6)	2.84E−1	−2.34E−2	−8.69E−3	4.43E−4	−1.68E−4	0	1
	III		9.29E−2	−1.22E−2	−1.49E−3	3.97E−5	0	0	1
	IV		1.06E−1	0	−1.58E−3	0	0	0	1
	V		1.89E−1	1.57	8.15E−2	2.73E−2	−2.49E−4	−2.68E−1	1
Bus	II	(7)	1.09E01	−2.18E−1	1.11E02	−1.13	–	–	–
	III		3.26E02	−1.77	4.01E01	−5.77E−1	–	–	–
	IV		2.47E01	−5.30E−1	2.19E03	−3.54	–	–	–
	V	(6)	−4.81E01	1.44E01	2.26E−1	6.62E01	1	−6.33E−1	2.18E−1
Truck	II	(7)	2.03	0	−3.18E−1	0	2.41E−4	0	0
	III		Decrease by 16% compared with China Stage II						
	IV		Decrease by 32% compared with China Stage III						
	V		5.14E−1	−1.41E−2	−9E−2	–	–	3.43	–

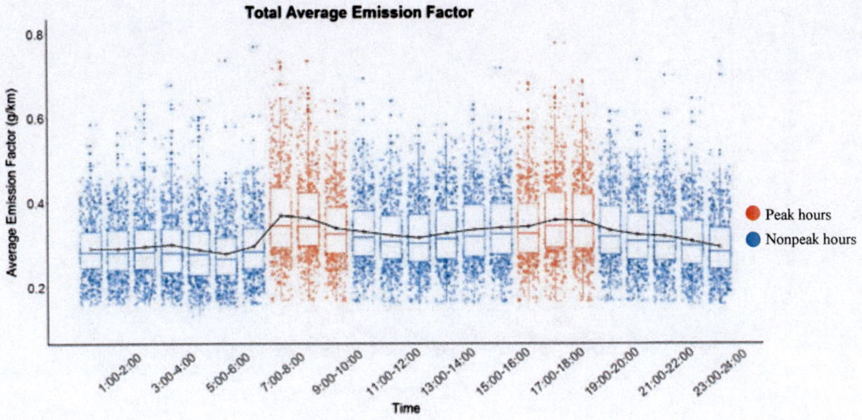

Fig. 4 Average emission factor for 24-h daily

3 Result Analysis and Discussion

3.1 Clustering Analysis Based on Road Segment Emission Factors

Clustering analysis, an unsupervised machine learning approach, was used to classify road segments into different categories based on their emission patterns. First, the 24-h emission pattern of each segment *i* was expressed by a 24-dimension vector: $X_i = \{EF_1, EF_2, \ldots, EF_{24}\}$. FCM Clustering method (Bezdek et al., 1984) was introduced to classify each segment-based 24-dimension vector into different groups by probabilities, which may be further integrated with mathematical regression, so as to assess the connection between the 24-h emission and the built environment and land use attributes.

As the cluster number has to be predetermined and requires to be evaluated by validity indexes to reach a meaningful and explicable result, the value was tested from 2 to 7. Four cluster validity indexes were introduced and calculated as the selection criteria to assess the cluster results, taking both similarity within a cluster and difference across clusters into account: Fuzzified PBM (PBMF) (Pakhira et al., 2005), Minimum centroids' distance (MCD) (Zhu & Nandi, 2014), Partition coefficient (PC) (Bezdek, 1973) and Fukuyama-Sugeno index (FSI) (Pal & Bezdek, 1995). Finally, four clusters was found that can balance cluster dispersion and in-cluster compactness, and consequently the best clustering number was chosen as four. Based on the maximum probability, the number of objects fallen within each cluster are 130, 202, 103 and 101, respectively.

Figure 5 presents the 24-h average emission factor of the four clusters, in which the membership is expressed by the color and size of the bubble. Figure 5a has the lowest emission rate per vehicle with a mean of 0.238 g/km. Cluster 2 has the

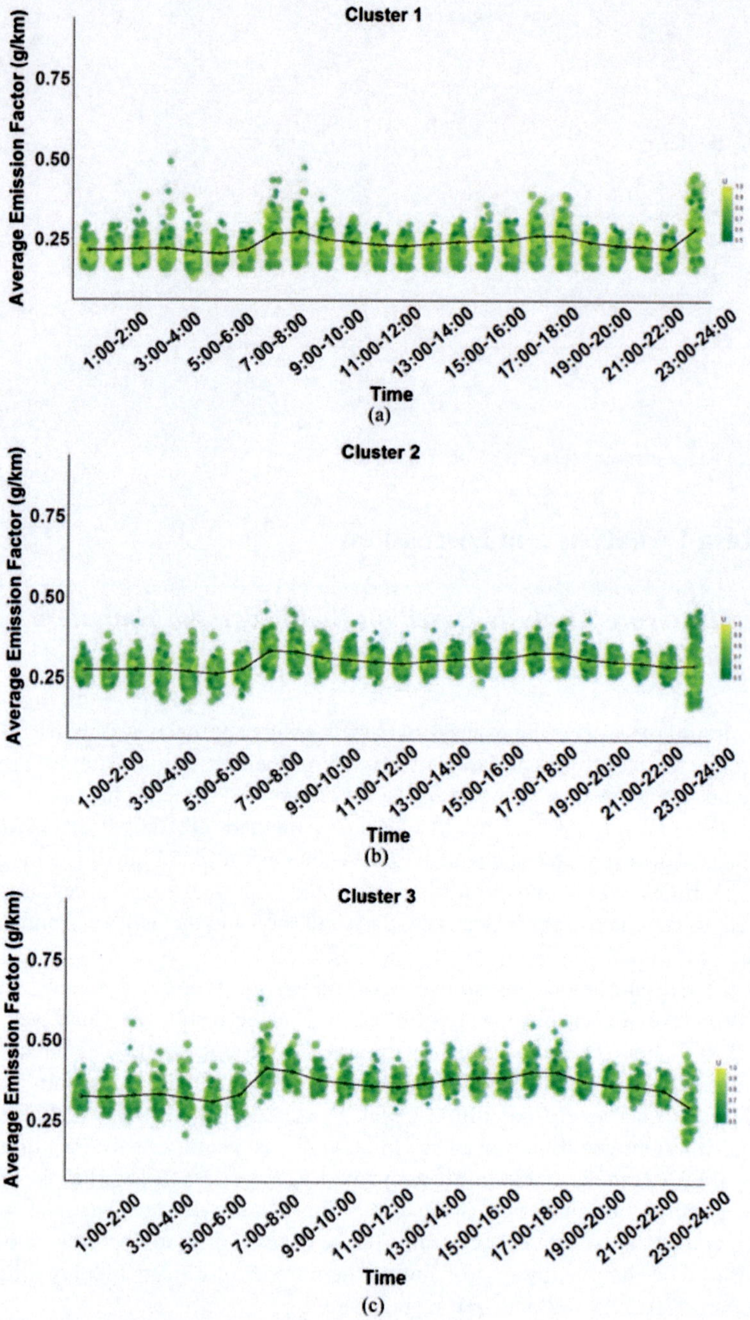

Fig. 5 Average NO_X emission by each cluster. **a** Cluster 1 (# of points = 130). **b** Cluster 2 (# of points = 202). **c** Cluster 3 (# of points = 103). **d** Cluster 4 (# of points = 101)

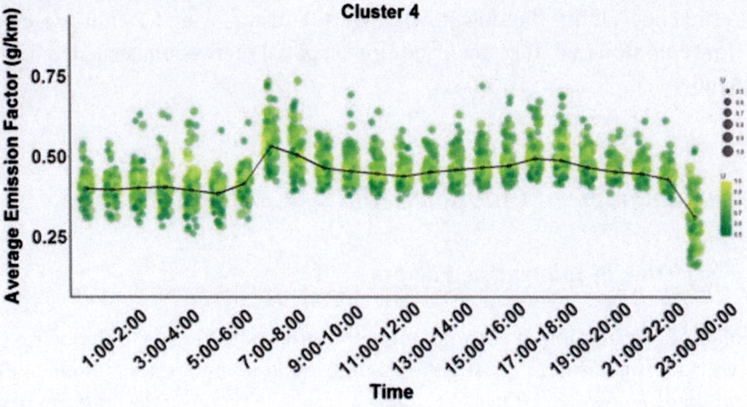

(d)

Fig. 5 (continued)

mean value of 0.286 g/km with a stable dispersion. Lower rate and stable trend mean perfect balance between traffic volume and operational speed. Cluster 3 with a mean of 0.358 g/km, which represents a notable increase of average emission. Cluster 4 has the highest value, 0.439 g/km, and the largest deviation, which has obvious extreme peak-hour phenomenon.

The spatial distribution of clustering result for all segments are illustrated in Fig. 6, in which each segment was allocated to the affiliated cluster based on highest membership value. It is shown that segments with high emission rates (Cluster 4) mainly concentrate between inner and middle rings, carrying majority of commuting vehicles. Additionally, this region also has the highest proportion of heavy-duty vehicles according to the field observation. Cluster 3 with the second highest emission rate is mainly located on outskirts and city center, which may be due to the low

Fig. 6 Clustering result on segment emission factors by primary membership

travel efficiency within the downtown CBDs. Clusters 1 and 2 indicate comparatively low emission rate, for which additional spatial regression is needed for further explanation.

3.2 Spatiotemporal Characteristics of Emission Pattern

3.2.1 Selection of Influencing Factors

To explain the formation of extreme high emission rate, various influencing factors were chosen for further investigations with spatial analysis methods including *Geographical Detector*, *MORAN's I* and *SARMA*. The possible built environment factors were selected as follows:

F1: Road type (primary or secondary road). During the numeric analysis, 1 indicates the primary road and 2 for secondary road. Road type may largely impact the travel speed of vehicles passed by, which is the key factor for emissions.

F2: Road segment length. Here, the segment was defined as the road component between two intersections. The starting and ending intersections were intentionally left out, as those are generally considered as the severest polluted nodes (He et al., 2015), and should be analyzed separately.

F3: Distance to the nearest ramp (defined by Euclidean distance). Traffic volume closed to ramp area is always with strong oscillation, which may cause serious congestion and emission problems (He et al., 2016). Particularly, low speed and braking nearby would induce extremely high emissions and deserve further investigations.

F4: Number of bus stations along the road segment per 100 m. Roads with large number of bus stations are always related to the commuting corridors or even transportation hubs in the vicinity, indicating more vehicles and possible aggravated traffic congestion and emissions (Long et al., 2010).

F5: Number of schools within 500 m per 100 m in segment length. In general, elementary schools or even middle schools may produce traffic pressure on the surrounding roads during daily attending or leaving periods by the picking-up behavior of parents or school buses (Yu & Liu, 2011), which may impact on-road emissions, and is particularly significant for large cities.

F6: Distance to the nearest metro station. Metro stations always carry large amount of passenger flow and transship of buses or passenger cars, which may obstruct the roads around (Huang et al., 2017).

F7: Distance to the nearest hospital. Hospital as one of the most important built environment factors and public service facility would attract large amount of traffic, which has been assessed as a key factor for on road emissions (Zhong & Bushell, 2017).

F8: The relative location to the urban expressway rings. The urban expressway system in Shanghai has three rings, inner ring, middle ring and outer ring, which typically distinguish the center with the suburb, and are considered as the spatial

Table 4 Variables for analysis

Variable	Abbreviation	Min	Mean	Max
F1: Road type	Rd_type	1	–	2
F2: Road segment length (m)	Rd_len	300	1314	4510
F3: Distance to the nearest ramp (m)	Dist_ramp	8.33	990.90	4304
F4: Number of bus stations along the road segment per 100 m (stations/100 m)	Num_bus	0	0.24	2.47
F5: Number of schools within 500 m per 100 m (schools/100 m, r = 500 m)	Num_scho	0	0.46	2.92
F6: Distance to the nearest metro station (m)	Dist_metro	8.05	814.9	3213
F7: Distance to the nearest hospital (m)	Dist_hosp	13.47	979.8	6207
F8: Relative location to the expressway rings	Loca_ring	0	–	3
F9: Number of public parking lots within 500 m	Parking lot	0	3.15	39
F10: Commercial area proportion (%)	Com_pro	0	5.25	59.12
F11: Residential area proportion (%)	Res_pro	0	29.29	99.41
F12: Transportation area proportion (%)	Trans_pro	0	15.63	100

layout of the city. The regions divided by the three expressway rings are denoted by 0–3 for the zones from within the inner ring to outside of the outer ring.

F9: Number of public parking lots within 500 m. The impact of public parking lots on congestion and emissions is similar to that of hospitals, as both are important traffic generation and attraction nodes.

F10–12: Land use factors. Emissions are sensitive to trip speed and vehicular characteristics, which may be largely affected by land use status (Cao et al., 2006; Hong & Goodchild, 2014). Here, land use mainly considers commercial area (**F10**), residential area (**F11**) and transportation area (**F12**), as other attributes, such as education and hospital have been considered within the built environment factors before (**F5** and **F7**). Transportation land cover mainly includes railway station, airport, transportation hubs, etc. The three factors are quantitatively expressed by the area proportion within a radius of 500 m.

The 12 variables for analysis are summarized in Table 4.

3.2.2 Geographical Detector

To assess the impact of the built environmental and land use factors to certain clusters, Geographical detector (Wang et al., 2010) was introduced to preliminarily explain the extent that each factor may be responsible for clustering, as shown in Eq. (8):

$$PD_R = 1 - \frac{1}{n\sigma_R^2}\sum_{i=1}^{4} n_{Di}\sigma_{Di,R}^2 \tag{8}$$

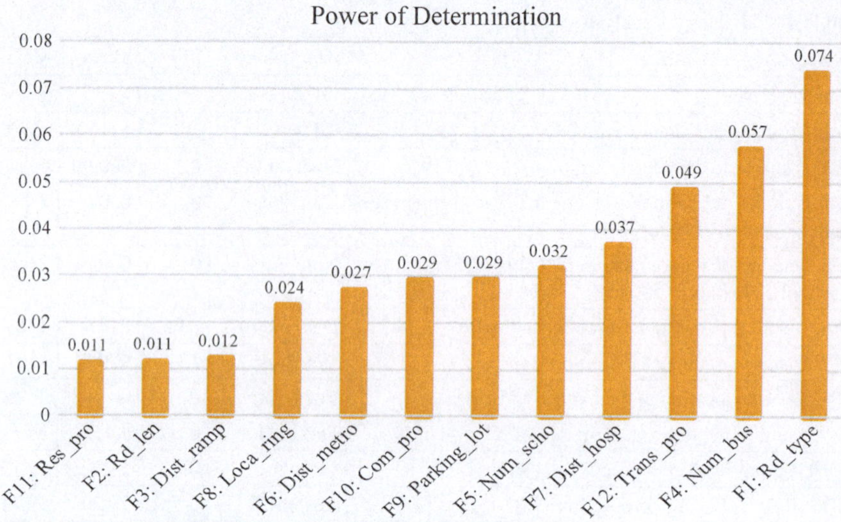

Fig. 7 Explanatory power of each determinant in ascending sequence

where PD_R is the power of determinant on clustering result for factor *R*, σ_R^2 is the global variance of factor *R* within the overall spatial region, $\sigma_{Di,R}^2$ is the divisional variation in cluster *Di*. Equation (8) interprets the ratio of n_{Di} weighted divisional variation over the global variance. Larger PD_R indicates factor *R* differs largely across clusters, and the determinant power of *R* is stronger. If $P_{D,R}$ equals to 1, factor *R* alone could perfectly differentiate objects.

Figure 7 presents the explanatory power of determination for each factor ascendingly. **F1: Road type** (0.074) has the highest *PD*, which may greatly impact vehicles' emission rate. **F4: Number of bus stations** alongside got the second (0.057), as oversize autobuses not only reduce the operation speed of road but also bring higher emission rate compared with other vehicle types. The third factor is the proportion of transportation land (F12: 0.049), which mainly reflects the occurrence of transportation hubs, as huge amount of mixed traffic flow pick-up passengers or cargos may occur along with hubs. **F7: Distance to the nearest hospital** (0.037), **F5: Number of schools around** (0.032), **F9: Number of public parking lots** (0.029), **F10: Proportion of commerical land** (0.029), and **F6: Distance to the nearest metro station** (0.027) have similar explanatory power. These are all hotspots of traffic attraction, which may cause traffic congestion or brake along with high emissions. As shown in Fig. 7, **F8: Relative location to the expressway rings** (0.024) also has a considerable impact on emissions, while the rest factors have little power of determinant. Moreover, factors might also cause individual impact on single cluster, which is unable to be explained by *PD*, and consequently further regression is required for detailed investigation.

Table 5 *MORAN's I* for clusters

Index	Cluster 1	Cluster 2	Cluster 3	Cluster 4
MORAN's I	0.232	0.324	0.199	0.536
P value	7.008e−09	1.449e−15	5.486e−07	2.2e−16

3.2.3 MORAN's I Test

GLOBAL MORAN's I (Moran, 1950) tests whether adjacent objects have spatial similarities with each other, ranging [−1, 1]. The definition of *MORAN's I* is given in Eq. (9) as follows:

$$I = \frac{N\sum_{i=1}^{n}\sum_{j=1}^{n}(X_i - \overline{X})(X_j - \overline{X})}{(\sum_{i=1}^{n}\sum_{j=1}^{n}w_{ij})\sum_{i=1}^{n}(X_i - \overline{X})} \quad (i \neq j) \tag{9}$$

where X_i and X_j are observed values, indicting the membership of clusters here; w_{ij} is the spatial weight matrix describing the spatial relationship between objects. $I = 0$ means total independence, $I > 0$ reflects positive correlation, and vice versa. As road network has spreading characteristics, one segment may influence another object far away. Consequently, w_{ij} should be based on distance decay as below (Gibbons & Overman, 2012):

$$w_{ij} = \begin{cases} \exp\left[-0.5\left(\frac{d_{ij}}{b}\right)^2\right], & d_{ij} < b \\ 0, & otherwise \end{cases} \tag{10}$$

where d_{ij} is the distance between the two adjacent objects, and b is chosen as 1000 m in this study which was set as the upper limit of buffer size when analyzing impact of road emissions, as suggested by Liu et al. (2016). *MORAN's I* mainly tests whether the membership of objects to a cluster has aggregation effect spatially.

Table 5 provides the *MORAN's I* value as well as the corresponding statistical *p*-value calculated for each cluster. It was found that all clusters have a positive *I* value and passed the 5% level of significance test, which means the segments belonging to the same cluster would spatially gather together. This further indicates the effectiveness of FCM clustering. The values of all indexes are relatively low, as the global *MORAN's I* was used here for each cluster, which was calculated based on the membership of all 536 road segments to certain cluster.

3.2.4 SARMA Regression of Road Segments

The advantage of using FCM is to conduct regression based on continuous membership (dependent variable). In addition to the referred 12 factors, neighboring segments may also influence traffic demand on a road as well as emissions, and the lagging influence and spatial dependency may play important roles. To this end, a spatial lag model called mixed spatial autoregressive moving average model (SARMA) (Anselin et al., 1996) is introduced to consider both dependence and errors with closer objects having a greater impact. The structure of SARMA is shown in Eq. (11):

$$
\begin{aligned}
y_i &= (1-\rho W)^{-1}X\beta + (1-\rho W)^{-1}u \\
u &= (1+\lambda W)\varepsilon,\ \varepsilon \sim N(0,\sigma^2 I)
\end{aligned}
\tag{11}
$$

where y_i is the segment's membership belonging to Cluster i, and X is the vector of 12 environment characteristics, both were standardized to $[-1, 1]$ before regression. ρ is the spatial autoregressive parameter measuring neighborhood effects, $\rho > 0$ means positive correlation and vice versa. λ is the spatial error coefficient, disclosing and quantifying the inherent similarity or dissimilarity, ε is the random error term, W is the spatial weight matrix as presented in Eq. (10), β is the coefficient vector, and σ^2I is the variance of the random error. The result of SARMA regression is shown in Table 6.

For **Cluster 1** (Lowest emission rate), almost all significant factors (i.e. F1: Road_type, F2: Rd_len, F3: Dist_ramp, F6: Dist_metro, F8: Loca_ring, F9: Parking_lot, F12: Trans_pro) cause less impact on operational speed. Although primary roads (F1) carry more traffic, they may associate with higher speed, and consequently cause less pollution. Longer segments (F2) are less affected by signal control, where vehicles suffer little braking and lining up, thus emitting less NO_x. Ramps (F3) are the merging points of traffic flow, and vehicles on segments far away from ramps have low emission rate. The role of distance to metro stations (F6) is similar to that of ramps. Relative location to expressway rings (F8) is positive, indicating low emission rate segments intersperse on outskirts. More parking lots around (F9) would impede the traffic, and reduce the membership to Cluster 1. The proportion of transportation land (F12) is a special factor, since hubs such as airports or rail stations, may have better traffic organization, resulting in better operation and less single-vehicle emission.

For **Cluster 2** (Moderate emission rate), emission rate per vehicle rises slightly. Road length and number of parking lots around remain key factors as Cluster 2. A new significant built environment is the number of schools within 500 m, which representes the amount of picking ups during attending and leaving periods, especially in large cities. This would largely disrupt the vehicle speed passing by, as well as aggravate emission rate. The proportion of residential land is significantly positive. High density of residential land may cause tidal phenomena, however the roads around residential land may have extreme lower traffic during non-peak hours, driving down the average emission rate.

Emission rate in **Cluster 3** (Abnormal emission rate) have higher deviation within 24 h, and obvious rush hour effect appears. The significant factors in Cluster 3 all somewhat demonstrate characteristics of commuting pressure. Longer roads composite commuting corridors, even though some may not be arterials. The traffic problems near ramps and metro stations are intensified during peak hours, as the

Table 6 Results of *SARMA* regression for cluster membership prediction

Factor	Cluster 1 Lowest			Cluster 2 Moderate			Cluster 3 Abnormal			Cluster 4 Highest		
	Coefficient	Standard error	Significance	Coefficient	Standard error	Significance	Coefficient	Standard error	Significance	Coefficient	Standard error	Significance
(Intercept)	0.7	0.0178	***	0.5758	0.0183	***	0.5736	0.0165	***	0.6088	0.0258	***
F1: Road_type	−0.0455	0.0119	***	0.0015	0.0091		0.0070	0.0117		0.0026	0.0125	
F2: Rd_len	0.0165	0.0128	***	0.0202	0.0104	*	0.0199	0.0131	**	−0.0417	−0.0144	***
F3: Dist_ramp	0.011	0.0122	*	0.0109	0.0089		−0.0073	0.012	***	−0.0203	0.0135	
F4: Num_bus	0.0025	0.0115		0.0082	0.0095		−0.0057	0.0125		0.0351	0.0126	***
F5: Num_scho	0.012	0.0134		−0.004	0.0112	**	0.0057	0.0167		−0.004	0.0165	
F6: Dist_metro	0.0269	0.0132	**	−0.0038	0.0113		−0.0083	0.0137	***	−0.0153	0.0144	
F7: Dist_hosp	−0.0113	0.0142		−0.0028	0.0114		−0.0027	0.0149		0.007	0.0138	
F8: Loca_ring	0.0226	0.0159	**	−0.0125	0.0127		−0.0039	0.0186		−0.0314	0.0148	**
F9:Parking_lot	−0.0186	0.0144	*	−0.018	0.0114	*	−0.0152	0.0165		0.0277	0.0165	*
F10: Com_pro	0.004	0.0129		0.0038	0.0097		0.0027	0.0149		0.0012	0.0156	*
F11: Res_pro	0.0063	0.0139		0.0039	0.0103	**	0.0207	0.0125	*	−0.0162	0.0138	
F12: Trans_pro	0.0221	0.0125	**	0.0021	0.0099		−0.0037	−0.0122	*	−0.014	0.0131	
Rho	0.0164	0.0369	*	0.1424	0.0351	**	0.004	0.0427	*	0.1389	0.0156	***
Lambda	0.1173	0.1032	***	−0.9419	0.0972		−0.1209	−0.1236	*	−0.0017	0.1186	**
LR test	1.833		*	15.171		***	0.808		*	11.58		***
Log likelihood	87.99			132.82			77.45			69.79		
AIC	−143.98			−233.65			−122.9			−107.57		

Note * indicates $p < 0.1$, ** indicates $p < 0.05$, *** indicates $p < 0.01$

closer to these areas means higher emission rate. Residential land here largely emphasizes the time dependency.

Cluster 4 (Highest emission rate) is the key to explain high NO_x emission. Short segments generally experience more signal control and are related to high emission rates. Relative location to ring is negative, indicating vehicles within the middle and inner rings have higher emission rate, which is consistent with rational recognition of downtown congestion. However, the parking lot effect changes to positive, indicating that more parking lots may lead vehicles hunting for possible parking lots and parking positions, which in turn induces increased congestion and vehicle emissions. A new factor is the number of bus stations alongside. According to COPERT, buses generally have higher emission rate and frequently entering and exiting station also influence the traffic flow, and thus more stations mean higher emissions. A higher proportion of commercial land always attracts more traffic and tends to become the center of congestion accompanied by high NO_x emission.

3.3 Correlations Between Atmospheric Concentration (NO_2) and Line Source Emission

To investigate the potential correlation between on-road emissions and atmospheric pollutant concentration, the hourly NO_2 concentrations from the atmospheric environmental monitoring stations within the studying area were extracted from the website of Shanghai Metropolitan Environment Monitoring Center (www.semc.gov.cn/aqi/home/Index.aspx). Since NO within atmosphere could be easily oxidized into NO_2, it is reasonable to use NO_2 as a surrogate of NO_X emissions, with the geo-spatial concentration trend compared with that from traffic emissions. The station-based NO_2 concentrations were summarized in hourly, with the unit of mg/m^3, and the unit-based line source emission was approximated using Eq. (12), with all variables introduced in Sect. 3.2:

$$EF_k = TE_k/L_k \quad (12)$$

Variation trends from the NO_X line source emission and the NO_2 concentration were analyzed based on density analysis and Kriging interpolation, respectively. To guarantee high precision of pollution diffusion, only the area within 500 m of 536 roads was investigated. The interpolated maps based on the modeled line source emission (NO_x) and the station monitored NO_2 were obtained and are shown in Fig. 8a, b (9 am as an example). The two figures demonstrate obvious correlationship based on Pearson correlation coefficient, and the city center is always the hotspot region of the emission.

Further analysis was conducted to verify the visualized correlation, which was carried out based on a raster of high resolution, 360 m * 360 m. As shown in Fig. 9, road emission and NO_2 concentrations have a positive correlation within all 24 h

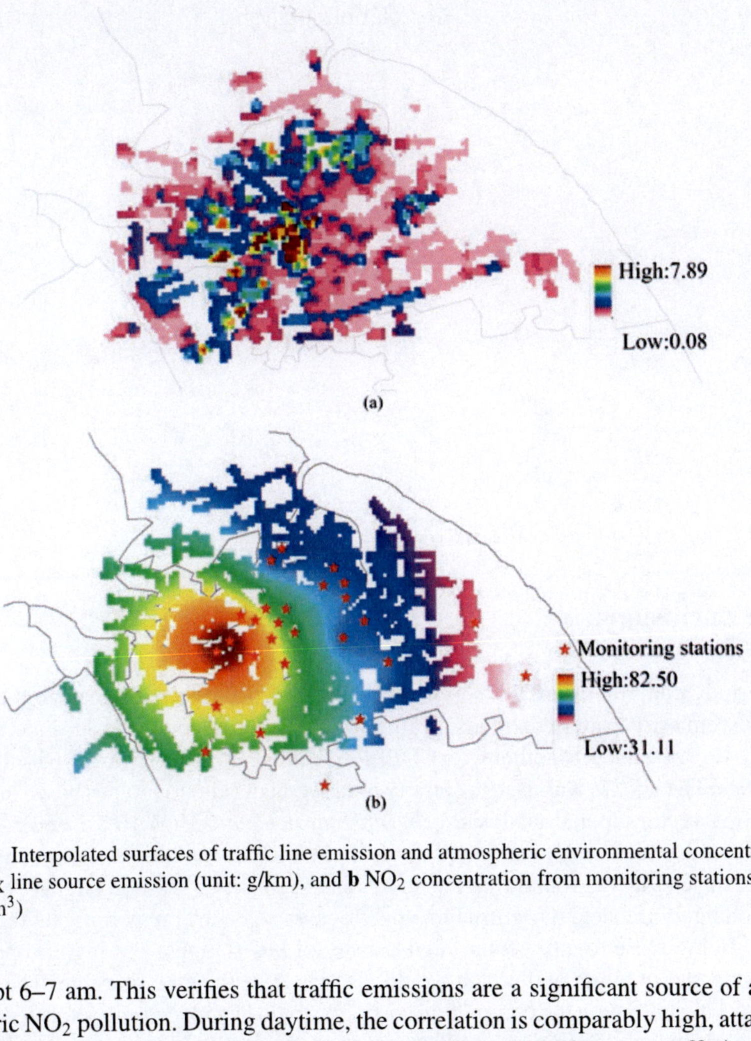

Fig. 8 Interpolated surfaces of traffic line emission and atmospheric environmental concentration, **a** NO_X line source emission (unit: g/km), and **b** NO_2 concentration from monitoring stations (unit: mg/km^3)

except 6–7 am. This verifies that traffic emissions are a significant source of atmospheric NO_2 pollution. During daytime, the correlation is comparably high, attaining to the largest value at afternoon (14–15 pm), even though the afternoon traffic is generally not as heavy as those of morning and evening peaks. Especially, all peak hours demonstrate high correlation as well. The sudden drop down at 0 am and 6 am was difficult to explain, which may due to the intraday meteorological conditions, such as wind direction, temperature or humidity. Moreover, relative low data accuracy may also be one of the potential reasons as well.

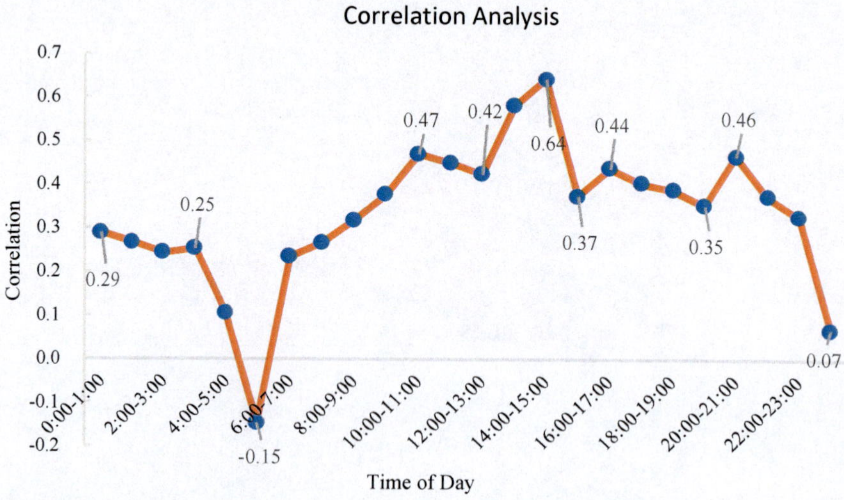

Fig. 9 Correlation of line emission and concentration

4 Conclusions

This study combines the Didi service data with traffic surveys to analyze the 24-h average emission pattern on roads. Traffic volume, vehicle composition, and average speed were extracted, together with COPERT to calculation the segment-based emission rate. Then, FCM was used to classify road segments into four types based on 24-h emission vector. Spatial analysis, *geographical detector*, *MORAN's I* and *SARMA* were introduced, so as to verify and disclose the spatial similarity and built environment impacts. It was found that the short road segments with more signal controls would largely impact NO_x emission, and green wave control may mitigate the situation. Being close to city center, high density of bus stations, and high proportion of residential or commercial land would also increase the emission rate. The high-density bus stations on the commuting corridor should be reduced and relocated to the neighboring roads. To mitigate traffic emissions, bus stations is not recommended to be located within 500 m to the parking lot area. Since ramps are crucial to high emission rate, the land use in the vicinity should avoid to having high trip generations. Moreover, as high proportion of transportation land use tend to reduce the traffic emissions, which may emphasize the importance of the planning and construction of integrated transport hub within city instead of suburban or urban fringe areas. However, as the emission factors of autobuses and trucks are generally much higher than the regular passenger cars (Sun et al., 2017), it is essential to have them electrified for reducing daily traffic emissions. The proposed framework could be used for any spatial problems investigation, particularly to estimate the traffic emission rate from Didi service GPS data and COPERT model by incorporating different vehicle types, which is an important innovation. The result could guide urban planning and assist to identify road segments with serious traffic emission problems.

This study also compared the hourly emission trend with the Kriging interpolated atmosphere NO_2 concentration, which found positive correlation and revealed the temporal variation of transportation emission contribution. While the results are promising, further studies need to be conducted to improve the performance of the model. First, this study just chose Shanghai as a case study, the conclusions should be generalized after analyzing more cities and regions. Second, the correlation of urban traffic line source emissions and atmospheric pollution concentration was just a preliminary investigation. The five days trajectories chosen as the research samples are mainly representatives for typical day situations instead of comprehensive circumstances. Additional data sources, including both vehicle related data and atmospheric factors, as well as dispersion models may be incorporated for further research and investigations. Finally, as many driving behaviors may be extracted from raw trajectory and order data, it's also necessary to analyze the spatiotemporal relationship between congestion and emissions, which could better explain formation of abnormal emissions from the perspective of traffic operation.

References

Abou-Senna, H., Radwan, E., Westerlund, K., & Cooper, C. D. (2013). Using a traffic simulation model (VISSIM) with an emissions model (MOVES) to predict emissions from vehicles on a limited-access highway. *Journal of the Air and Waste Management Association, 63*(7), 819–831. https://doi.org/10.1080/10962247.2013.795918

Anselin, L., Bera, A. K., Florax, R., & Yoon, M. J. (1996). Simple diagnostic tests for spatial dependence. *Regional Science and Urban Economics, 26*(1), 77–104. https://doi.org/10.1016/0166-0462(95)02111-6

Bellasio, R., Bianconi, R., Corda, G., & Cucca, P. (2007). Emission inventory for the road transport sector in Sardinia (Italy). *Atmospheric Environment, 41*(4), 677–691. https://doi.org/10.1016/j.cmpb.2013.01.002

Bezdek, J. C. (1973). Cluster validity with fuzzy sets. *Journal of Cybernetics, 3*(3), 58–73. https://doi.org/10.1080/01969727308546047

Bezdek, J. C., Ehrlich, R., & Full, W. (1984). FCM: The fuzzy c-means clustering algorithm. *Computers and Geosciences, 10*(2–3), 191–203. https://doi.org/10.1016/0098-3004(84)90020-7

Broderick, B. M., & O'Donoghue, R. T. (2007). Spatial variation of roadside C2–C6 hydrocarbon concentrations during low wind speeds: Validation of CALINE4 and COPERT III modelling. *Transportation Research Part D: Transport and Environment, 12*(8), 537–547. https://doi.org/10.1016/j.trd.2007.07.010

Cao, X., Mokhtarian, P. L., & Handy, S. L. (2006). Neighborhood design and vehicle type choice: Evidence from Northern California. *Transportation Research Part D: Transport and Environment, 11*(2), 133–145. https://doi.org/10.1016/j.trd.2005.10.001

Chen, F., Yin, Z., Ye, Y., & Sun, D. J. (2020). Taxi hailing choice behavior and economic benefit analysis of emission reduction based on multi-mode travel big data. *Transport Policy*, 97, 73–84

Defries, T. H., Sabisch, M., Kishan, S., Posada, F., German, J., & Bandivadekar, A. (2014). In-use fuel economy and CO_2 emissions measurement using OBD data on US light-duty vehicles. *SAE International Journal of Engines, 7*(3), 1382–1396. https://doi.org/10.4271/2014-01-1623

Gallus, J., Kirchner, U., Vogt, R., Börensen, C., & Benter, T. (2016). On-road particle number measurements using a portable emission measurement system (PEMS). *Atmospheric Environment, 124*, 37–45. https://doi.org/10.1016/j.atmosenv.2015.11.012

Gibbons, S., & Overman, H. G. (2012). Mostly pointless spatial econometrics? *Journal of Regional Science, 52*(2), 172–191. https://doi.org/10.1111/j.1467-9787.2012.00760.x

Guo, H., Zhang, Q. Y., Shi, Y., & Wang, D. H. (2007). Evaluation of the international vehicle emission (IVE) model with on-road remote sensing measurements. *Journal of Environmental Sciences, 19*(7), 818–826 (2007). http://doi.org/10.1016/S1001-0742(07)60137-5

He, H. D., Wang, J. L., Wei, H. R., Ye, C., & Ding, Y. (2016). Fractal behavior of traffic volume on urban expressway through adaptive fractal analysis. *Physica A: Statistical Mechanics and Its Applications, 443*, 518–525. https://doi.org/10.1016/j.physa.2015.10.004

He, H. D., Zhang, C., Wang, W., Hao, Y., & Ding, Y. (2015). Feedback control scheme for traffic jam and energy consumption based on two-lane traffic flow model. *Transportation Research Part D: Transport and Environment*. https://doi.org/10.1016/j.trd.2015.11.005

Hong, J., & Goodchild, A. (2014). Land use policies and transport emissions: Modeling the impact of trip speed, vehicle characteristics and residential location. *Transportation Research Part D: Transport and Environment, 26*(1), 47–51. https://doi.org/10.1016/j.trd.2013.10.011

Huang, X., Cao, X., Yin, J., & Cao, X. (2017). Effects of metro transit on the ownership of mobility instruments in Xi'an, China. *Transportation Research Part D: Transport and Environment, 52*(B), 495–505. http://doi.org/10.1016/j.trd.2016.09.014

International Joint Commission (IJC). (2012). *Canada-United States air quality agreement progress report*. Bilateral Air Quality Committee, Canada.

Kousoulidou, M., & Sarmaras, Z. (2008). Road-transport emission projections to 2020 in European urban environments. *Atmospheric Environment, 42*(32), 7465–7475. https://doi.org/10.1016/j.atmosenv.2008.06.002

Lang, L., Zhou, Y., & Cheng, S. (2016). Unregulated pollutant emissions from on-road vehicles in China, 1999–2014. *Science of the Total Environment, 573*, 974–984. https://doi.org/10.1016/j.scitotenv.2016.08.171

Liu, C., Henderson, B. H., Wang, D., Yang, X., & Peng, Z. R. (2016). A land use regression application into assessing spatial variation of intra-urban fine particulate matter ($PM_{2.5}$) and nitrogen dioxide (NO_2) concentrations in City of Shanghai, China. *Science of the Total Environment, 565*, 607–615. https://doi.org/10.1016/j.scitotenv.2016.03.189

Liu, H., Chen, X., Wang, Y., & Han, S. (2013). Vehicle emission and near-road air quality modeling for Shanghai, China. *Transportation Research Record: Journal of the Transportation Research Board, 2340*, 38–48. https://doi.org/10.3141/2340-05

Liu, K., Yamamoto, T., & Morikawa, T. (2009). Feasibility of using taxi dispatch system as probes for collecting traffic information. *Journal of Intelligent Transportation Systems, 13*(1), 16–27. https://doi.org/10.1080/15472450802644447

Long, L., Wang, X., Feng, B., Zhang, Y., & Yang, H. (2010). Exposure levels of $PM_{2.5}$ and CO at bus stations in urban Guangzhou. *Environmental Science & Technology, 33*(9), 140–145. http://doi.org/10.3969/j.issn.1003-6504.2010.09.029

Lozhkina, O. V., & Vladimir, N. (2015). Estimation of road transport related air pollution in Saint Petersburg using European and Russian calculation models. *Transportation Research Part D: Transport and Environment, 36*, 178–189. https://doi.org/10.1016/j.trd.2015.02.013

Luo, X., Dong, L., Dou, Y., Zhang, N., Ren, J., & Li, Y. (2016). Analysis on spatial-temporal features of taxis' emissions from big data informed travel patterns: A case of Shanghai, China. *Journal of Cleaner Production, 142*, 926–935. https://doi.org/10.1016/j.jclepro.2016.05.161

Mausami, D., & Weitz, M. (2017). *Inventory of U.S. greenhouse gas emissions and sinks: 1990–2015*. United States Environmental Protection Agency.

Moran, P. A. (1950). Notes on continuous stochastic phenomena. *Biometrika, 37*(1–2), 17–23. https://doi.org/10.2307/2332142

Nagpure, A. S., Gurjar, B. R., & Kumar, P. (2011). Impact of altitude on emission rates of ozone precursors from gasoline-driven light-duty commercial vehicles. *Atmospheric Environment, 45*(7), 1413–1417. https://doi.org/10.1016/j.atmosenv.2010.12.026

Nikoleris, T., Gupta, G., & Kistler, M. (2011). Detailed estimation of fuel consumption and emissions during aircraft taxi operations at Dallas/Fort Worth international airport. *Transportation Research Part D: Transport and Environment, 16*(4), 302–308. https://doi.org/10.1016/j.trd.2011.01.007

Ning, Z., & Chan., T. L. (2007). On-road remote sensing of liquefied petroleum gas (LPG) vehicle emissions measurement and emission factors estimation. *Atmospheric Environment, 41*(39), 9099–9110. http://doi.org/10.1016/j.atmosenv.2007.08.006

Nyhan, M., Sobolevsky, S., Kang, C., Robinson, P., & Corti, A. (2016). Predicting vehicular emissions in high spatial resolution using pervasively measured transportation data and microscopic emissions model. *Atmospheric Environment, 140*, 352–363. https://doi.org/10.1016/j.atmosenv.2016.06.018

Pakhira, M. K., Bandyopadhyay, S., & Maulik, U. (2005). A study of some fuzzy cluster validity indices, genetic clustering and application to pixel classification. *Fuzzy Sets and Systems, 155*(2), 191–214. https://doi.org/10.1016/j.fss.2005.04.009

Pal, N. R., & Bezdek, J. C. (1995). On cluster validity for the fuzzy c-means model. *IEEE Transactions on Fuzzy Systems, 3*(3), 370–379. https://doi.org/10.1109/91.413225

Pan, W., Xue, Y., He, H. D., & Lu, W. Z. (2017). Traffic control oriented impact on the persistence of urban air pollutants: A causeway bay revelation during emergency period. *Transportation Research Part D: Transport and Environment, 51*, 304–313. https://doi.org/10.1016/j.trd.2017.02.001

Shanghai Research Institute of Urban & Rural Construction and Transportation Development (SRIUTD). (2015). *Shanghai municipal comprehensive transportation operation annual report*. Shanghai, CN.

Shanghai Research Institute of Urban & Rural Construction and Transportation Development (SRIUTD). (2016). *Shanghai municipal comprehensive transportation operation annual report*. Shanghai, CN.

Silva, L. C., Dedini, F. G., Correa, F. C., Eckert, J. J., & Becker, M. (2015). Measurement of wheelchair contact force with a low cost bench test. *Medical Engineering and Physics, 32*(2), 163–170. https://doi.org/10.1016/j.medengphy.2015.11.014

Sun, D. J., Zhang, C., Zhang, L., Chen, F., & Peng, Z. R. (2014). Urban travel behavior analyses and route prediction based on floating car data. *Transportation Letters, 6*(3), 118–125. https://doi.org/10.1179/1942787514Y.0000000017

Sun, D. J., & Zhang, Y. (2018). Influence of avenue trees on traffic pollutant dispersion in asymmetric street canyons: Numerical modeling with empirical analysis. *Transportation Research Part D: Transport and Environment*. *65*, 784–795

Sun, D. J., Zhang, Y., Xue, R., & Zhang, Y. (2017). Modeling carbon emissions from urban traffic system using mobile monitoring. *Science of the Total Environment, 599–600*, 944–951. https://doi.org/10.1016/j.scitotenv.2017.04.186

Sun, D. J., Yin, Z., & Cao, P. (2020). An improved CAL3QHC model and the application in vehicle emission mitigation schemes for urban signalized intersections. Building and Environment, 183, 107213.

Sun, D. J., Zheng, Y., & Duan, R. (2021a). Energy consumption simulation and economic benefit analysis for urban electric commercial-vehicles. Transportation Research Part D: *Transport and Environment, 101*, 103083.

Sun, D. J., Wu, S., Shen, S., & Xu, T. (2021b). Simulation and assessment of traffic pollutant dispersion at an urban signalized intersection using multiple platforms. Atmospheric Pollution Research, 12(7), 101087.

Wang, H., Chen, C., Huang, C., & Fu, L. (2008). On-road vehicle emission inventory and its uncertainty analysis for Shanghai, China. *Science of the Total Environment, 398*, 60–67. https://doi.org/10.1016/j.scitotenv.2008.01.038

Wang, J. F., Li, X. H., Christakos, G., Liao, Y. L., & Zhang, T. (2010). Geographical detectors-based health risk assessment and its application in the neural tube defects study of the Heshun region, China. *International Journal of Geographical Information Science, 24*, 107–127. https://doi.org/10.1080/13658810802443457

Wang, X. Y. (2017). "Yellow label" vehicles banned since October, and the State Standard II vehicles with 10 years and above forbidden to enter into the outer ring area. http://money.163.com/15/0421/17/ANO7M1QB00253B0H.html. April 21, 2015. Accessed on October 25, 2017 (in Chinese)

Yu, L., & Liu, Y. (2011). Traffic characteristics analysis and suggestions on school bus operation for primary school students in Beijing. *Journal of Transportation Systems Engineering and Information Technology, 11*(5), 193–200. https://doi.org/10.3969/j.issn.1009-6744.2011.05.029

Zhang, K. S., Sun, D. J., Shen, S., & Zhu, Y. (2017). Analyzing spatiotemporal congestion pattern on urban roads based on taxi GPS data. *Journal of Transport and Land Use, 10*(1), 675–694. https://doi.org/10.5198/jtlu.2017.954

Zhang, Y., Wang, X., Li, G., Yang, W., Huang, Z., & Zhang, Z. (2015). Emission factors of fine particles, carbonaceous aerosols and traces gases from road vehicles: Recent tests in an urban tunnel in the Pearl River Delta, China. *Atmospheric Environment, 122*(2), 876–884. https://doi.org/10.1016/j.atmosenv.2015.08.024

Zhong, S., & Bushell, M. (2017). Impact of the built environment on the vehicle emission effects of road pricing policies: A simulation case study. *Transportation Research Part A: Policy and Practice, 103*, 235–249. https://doi.org/10.1016/j.tra.2017.06.007

Zhu, Z., & Nandi, A. K. (2014). Blind digital modulation classification using minimum distance centroid estimator and non-parametric likelihood function. *IEEE Transactions on Wireless Communications, 13*(8), 4483–4494. https://doi.org/10.1109/TWC.2014.2320724